METHODS IN MOLECULAR BIOLOGY

Series Editor
John M. Walker
School of Life and Medical Sciences
University of Hertfordshire
Hatfield, Hertfordshire, AL10 9AB, UK

For further volumes:
http://www.springer.com/series/7651

Gastrointestinal Physiology and Diseases

Methods and Protocols

Edited by

Andrei I. Ivanov

Department of Human and Molecular Genetics, VCU Institute of Molecular Medicine,
Virginia Commonwealth University, Richmond, VA, USA

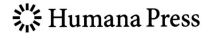 Humana Press

Editor
Andrei I. Ivanov
Department of Human and Molecular Genetics
VCU Institute of Molecular Medicine
Virginia Commonwealth University
Richmond, VA, USA

ISSN 1064-3745   ISSN 1940-6029   (electronic)
Methods in Molecular Biology
ISBN 978-1-4939-3601-4   ISBN 978-1-4939-3603-8   (eBook)
DOI 10.1007/978-1-4939-3603-8

Library of Congress Control Number: 2016939592

Printed on acid-free paper

This Humana Press imprint is published by Springer Nature
The registered company is Springer Science+Business Media LLC New York

Preface

The gastrointestinal tract is a unique, multifunctional organ in the human body. It is responsible for intake, digestion, and absorption of food, and excretion of bodily waste. It also houses a myriad of commensal and potentially pathogenic microorganisms that have profound effects on host development and homeostasis. Since these microorganisms have to be confined within the gut lumen, the gastrointestinal tract serves as the major impediment protecting internal tissues from invasion by harmful luminal microbes and exposure to their toxins. The gastrointestinal tract is also the largest organ of the immune system and is populated by specialized cells trained for border surveillance and recognition of external dangers.

Normal function of the gastrointestinal tract is frequently compromised by genetic factors, infections, stress, life habits, etc., that give rise to various diseases. Remarkably, impaired functions of this organ not only result in specific gastrointestinal disorders such as gastric ulcer, inflammatory bowel disease, or gastrointestinal tumors but also contribute to the development of other human pathologies including certain neurological and cardiovascular diseases, as well as diabetes. Collectively, these factors establish the study of the normal functions and disorders of the gastrointestinal system as one of the most important and exciting topics of modern biology and medicine.

The aim of *Gastrointestinal Physiology and Diseases: Methods and Protocols* is to provide an expert, step-by-step guide to a variety of techniques for examining the activity and regulation of the gastrointestinal system and for modeling the most common digestive diseases. This book is intended to target a large cohort of physiologists, cell and developmental biologists, immunologists, and physician-scientists working in the field of gastroenterology and beyond. This volume contains comprehensive and easy to follow protocols that are designed to be helpful to both seasoned researchers and newcomers to the field.

The protocols included in this volume are separated into five different parts. Part I (Chapters 1–9) describes in vitro and ex vivo techniques to study different aspects of the functions and differentiation of the gut mucosa, with particular emphasis on modern approaches to the growth, differentiation, and study of complex intestinal and gastric organoids. Part II (Chapters 10–15) outlines powerful in vivo imaging approaches to study biochemical alterations in epithelial cells, and to visualize leukocyte trafficking of in the gut during tissue inflammation and neoplasia. Part III (Chapters 16–20) presents protocols for the isolation, characterization, and therapeutic transfer of different types of intestinal immune cells. Part IV (Chapters 21–25) describes different animal models of gastrointestinal mucosal inflammation and injury. It describes classical models of chemically induced and infectious colitis in mice and also presents examples of the use of other model organisms in studying digestive disorders. Part V (Chapters 26–29) presents state-of-the-art animal models for studying tumor induction and development in the colon, stomach, and oral cavity.

I would like to thank all of the contributors for sharing their expertise and for carefully guiding readers through all the nuanced details of their respective techniques. I am very grateful to the series editor, Dr. John Walker, for his help during the editing process.

Richmond, VA, USA   *Andrei I. Ivanov*

Contents

Contributors

ANEESH ALEX • *Department of Electrical and Computer Engineering, Packard Laboratory, Lehigh University, Bethlehem, PA, USA; Center for Photonics and Nanoelectronics, Lehigh University, Bethlehem, PA, USA*

ERICA ALEXEEV • *Department of Medicine and the Mucosal Inflammation Program, The University of Colorado School of Medicine, Aurora, CO, USA*

DAVID ARTIS • *Jill Roberts Institute for Research in IBD, Joan and Sanford I. Weill Department of Medicine, Department of Microbiology and Immunology, Weill Cornell Medical College, Cornell University, New York, NY, USA*

CAITLYN W. BARRETT • *Department of Neurology, University of Pittsburgh, Pittsburgh, PA, USA*

TERRENCE A. BARRETT • *Division of Gastroenterology, Department of Medicine, University of Kentucky, Lexington, KY, USA*

ALEXANDER BELLER • *German Rheumatism Research Center (DRFZ), A Leibniz Institute, Berlin, Germany*

DAVID BENTREM • *Department of Surgery, Northwestern University Feinberg School of Medicine, Chicago, IL, USA; Robert H. Lurie Comprehensive Cancer Center, Northwestern University Feinberg School of Medicine, Chicago, IL, USA; Jesse Brown Veterans Affairs Medical*

CENTER, CHICAGO, IL, USA

CLAUDIA BEREK • *German Rheumatism Research Center (DRFZ), A Leibniz Institute, Berlin, Germany*

NINA BERTAUX-SKEIRIK • *Department of Molecular and Cellular Physiology, University of Cincinnati College of Medicine, Cincinnati, OH, USA*

KIRANDEEP BHULLAR • *Division of Gastroenterology, Department of Pediatrics, BC Children's Hospital, Vancouver, BC, Canada*

VALERIE BLANC • *Department of Medicine, Washington University School of Medicine, St. Louis, MO, USA*

ELSE S. BOSMAN • *Division of Gastroenterology, Departments of Pediatrics, BC Children's Hospital, Vancouver, BC, Canada*

JOSEPH BURCLAFF • *Division of Gastroenterology, Departments of Medicine, Washington University School of Medicine, St. Louis, MO, USA; Division of Gastroenterology, Departments of Developmental Biology, Washington University School of Medicine, St. Louis, MO, USA*

ERIC L. CAMPBELL • *Mucosal Inflammation Program, Department of Medicine, University of Colorado, Aurora, CO, USA*

JOMARIS CENTENO • *Department of Molecular and Cellular Physiology, University of Cincinnati College of Medicine, Cincinnati, OH, USA*

JUSTIN M. CHAN • *Division of Gastroenterology, Department of Pediatrics, BC Children's Hospital, Vancouver, BC, Canada*

TALAL A. CHATILA • *Division of Immunology, Children's Hospital Boston, Boston, MA, USA*

VAN TRUNG CHU • *German Rheumatism Research Center (DRFZ), A Leibniz Institute, Berlin, Germany; Immune Regulation and Cancer; Max-Delbrück-Center for MolecularMedicine, Berlin, Germany*

HANS CLEVERS • *Hubrecht Institute/KNAW, University Medical Center Utrecht, Utrecht, The Netherlands; Princess Maxima Center for Pediatric Oncology, University Medical Center Utrecht, Utrecht, The Netherlands*

NICHOLAS O. DAVIDSON • *Department of Medicine, Washington University School of Medicine, St. Louis, MO, USA*

TIMOTHY L. DENNING • *Institute for Biomedical Sciences, Georgia State University, Atlanta, GA, USA*

RUI FENG • *Department of Gastroenterology and Hepatology, The First Affiliated Hospital, Sun Yatsen University, Guangzhou, China*

YANG-XIN FU • *Department of Pathology and Committee on Immunology, The University of Chicago, Chicago, IL, USA*

DUKE GEEM • *Department of Pathology & Laboratory Medicine, Emory University, Atlanta, GA, USA*

JASON R. GOLDSMITH • *Department of Internal Medicine, University of Michigan, Ann Arbor, MI, USA*

ELIAS GOUNARIS • *Department of Surgery, Northwestern University Feinberg School of Medicine, Chicago, IL, USA; Robert H. Lurie, Comprehensive Cancer Center, Northwestern University Feinberg School of Medicine, Chicago, IL, USA*

XIAOHUAN GUO • *Tsinghua University School of Medicine, Beijing, China*

DIPICA HARIBHAI • *Section of Rheumatology, Department of Pediatrics, Medical College of Wisconsin, Milwaukee, WI, USA*

AKIHITO HARUSATO • *Institute for Biomedical Sciences, Georgia State University, Atlanta, GA, USA*

YOSHITAKA HIPPO • *Division of Molecular Carcinogenesis, Chiba Cancer Center Research Institute, Chiba, Japan; Division of Animal Studies, National Cancer Center Research Institute, Tokyo, Japan*

JAN HÜLSDÜNKER • *Department of Hematology, Oncology and Stem Cell Transplantation, Freiburg University Medical Center, Albert Ludwigs University Freiburg, Freiburg, Germany; Spemann Graduate School of Biology and Medicine (SGBM), University of Freiburg, Freiburg, Germany; Faculty of Biology, University of Freiburg, Freiburg, Germany*

YASUSHIGE ISHIHARA • *Molecular Diagnostic Technology Group, Advanced Core Technology Department, Research and Development Division, Olympus Tokyo, Tokyo, Japan*

CHRISTIAN JOBIN • *Division of Gastroenterology, Department of Medicine, College of Medicine, University of Florida, Gainesville, FL, USA*

RHEINALLT M. JONES • *Department of Pediatrics, Emory University School of Medicine, Atlanta, GA, USA*

LOTHAR JUST • *Institute of Clinical Anatomy and Cell Analysis, Eberhard Karls University Tübingen, Tübingen, Germany*

YUJI KAMIOKA • *Department of Pathology and Biology of Diseases, Graduate School of Medicine, Kyoto University, Kyoto, Japan; Innovative Techno-Hub for Integrated Medical Bio-Imaging, Kyoto University, Kyoto, Japan*

DANIEL J. KAO • *Department of Medicine and the Mucosal Inflammation Program, The University of Colorado School of Medicine, Aurora, CO, USA*

DOUGLAS J. KOMINSKY • *Department of Microbiology and Immunology, Montana State University, Bozeman, MT, USA*

JAGANATHAN KOWSHIK • *Department of Biochemistry and Biotechnology, Faculty of Science, Annamalai University, Annamalainagar, TN, India*

JOHN F. KUEMMERLE • *Division of Gastroenterology, Hepatology and Nutrition, Virginia Commonwealth University, Richmond, VA, USA*
CALVIN KUO • *Hematology Division, Department of Medicine, Stanford University School of Medicine, Stanford, CA, USA*
JORDI M. LANIS • *Department of Medicine and the Mucosal Inflammation Program, The University of Colorado School of Medicine, Aurora, CO, USA*
XINGNAN LI • *Hematology Division, Department of Medicine, Stanford University School of Medicine, Stanford, CA, USA*
LIPING LUO • *Department of Pediatrics, Emory University School of Medicine, Atlanta, GA, USA*
GORDON MACGREGOR • *Department of Cellular and Molecular Physiology, Yale University School of Medicine, New Haven, CT, USA*
YOSHIAKI MARU • *Division of Molecular Carcinogenesis, Chiba Cancer Center Research Institute, Chiba, Japan*
MICHIYUKI MATSUDA • *Department of Pathology and Biology of Diseases, Graduate School of Medicine, Kyoto University, Kyoto, Japan; Laboratory of Bioimaging and Cell Signaling, Graduate School of Biostudies, Kyoto University, Kyoto, Japan*
JASON C. MILLS • *Division of Gastroenterology, Departments of Medicine, Washington University School of Medicine, St. Louis, MO, USA; Division of Gastroenterology, Departments of Developmental Biology, Washington University School of Medicine, St. Louis, MO, USA; Division of Gastroenterology, Departments of Pathology & Immunology, Washington University School of Medicine, St. Louis, MO, USA*
REI MIZUNO • *Department of Pathology and Biology of Diseases, Graduate School of Medicine, Kyoto University, Kyoto, Japan; Department of Surgery, Graduate School of Medicine, Kyoto University, Kyoto, Japan*
KEVIN MUITE • *Department of Pathology and Committee on Immunology, The University of Chicago, Chicago, IL, USA*
SIDDAVARAM NAGINI • *Department of Biochemistry and Biotechnology, Faculty of Science, Annamalai University, Annamalainagar, TN, India*
ILKE NALBANTOGLU • *Department of Pathology and Immunology, Washington University School of Medicine, St. Louis, MO, USA*
PETER H. NECKEL • *Institute of Clinical Anatomy and Cell Analysis, Eberhard Karls University Tübingen, Tübingen, Germany*
MARKUS F. NEURATH • *Department of Medicine 1, University of Erlangen-Nuremberg, Erlangen, Germany*
HANG THI THU NGUYEN • *UMR 1071 Inserm, University of Auvergne, Clermont-Ferrand, France*
JENNIFER M. NOTO • *Division of Gastroenterology, Department of Medicine, Vanderbilt University Medical Center, Nashville, TN, USA*
AKIFUMI OOTANI • *Hematology Division, Department of Medicine, Stanford University School of Medicine, Stanford, CA, USA*
KAORU ORIHASHI • *Division of Molecular Carcinogenesis, Chiba Cancer Center Research Institute, Chiba, Japan*
BOBAK PARANG • *Division of Gastroenterology, Department of Medicine, Vanderbilt University, Nashville, TN, USA; Department of Cancer Biology, Vanderbilt University, Nashville, TN, USA*
RICHARD M. PEEK • *Division of Gastroenterology, Department of Medicine, Vanderbilt University Medical Center, Nashville, TN, USA*

M. BLANCA PIAZUELO • *Division of Gastroenterology, Department of Medicine, Vanderbilt University Medical Center, Nashville, TN, USA*

APRIL R. REEDY • *Department of Pathology, Emory University School of Medicine, Atlanta, GA, USA*

JUDITH ROMERO-GALLO • *Division of Gastroenterology, Department of Medicine, Vanderbilt University Medical Center, Nashville, TN, USA*

JOSE B. SAENZ • *Division of Gastroenterology, Departments of Medicine, Washington University School of Medicine, St. Louis, MO, USA*

YOSHIHARU SAKAI • *Department of Surgery, Graduate School of Medicine, Kyoto University, Kyoto, Japan*

MICHAEL A. SCHUMACHER • *Division of Gastroenterology, Children's Hospital Los Angeles, Los Angeles, CA, USA*

GERALD SCHWANK • *Hubrecht Institute/KNAW, University Medical Center Utrecht, Utrecht, The Netherlands; Institute of Molecular Health Sciences, ETH Zurich, Zürich, Switzerland*

RAMESH A. SHIVDASANI • *Department of Medical Oncology, Dana-Farber Cancer Institute, Boston, MA, USA; Department of Medicine, Harvard Medical School, Boston, MA, USA*

MANISHA SHRIVASTRAV • *Department of Surgery, Northwestern University Feinberg School of Medicine, Chicago, IL, USA*

ELIA D. TAIT WOJNO • *Baker Institute for Animal Health, Department of Microbiology and Immunology, College of Veterinary Medicine, Cornell University, Ithaca, NY, USA*

NAOKI TAKEMURA • *Department of Mucosal Immunology, School of Medicine, Chiba University, Chiba, Japan; Division of Innate Immune Regulation, International Research and Development Center for Mucosal Vaccines, Institute of Medical Science, The University of Tokyo, Tokyo, Japan*

SARAH TOMKOVICH • *Department of Microbiology and Immunology, University of North Carolina at Chapel Hill, Chapel Hill, NC, USA; Division of Gastroenterology, Department of Medicine, College of Medicine, University of Florida, Gainesville, FL, USA*

SATOSHI UEMATSU • *Department of Mucosal Immunology, School of Medicine, Chiba University, Chiba, Japan; Division of Innate Immune Regulation, International Research and Development Center for Mucosal Vaccines, Institute of Medical Science, The University of Tokyo, Tokyo, Japan*

BRUCE A. VALLANCE • *Division of Gastroenterology, Department of Pediatrics, BC Children Hospital, Vancouver, BC, Canada*

SADASIVAN VIDYASAGAR • *Department of Radiation Oncology, University of Florida Shands Cancer Center, Gainesville, FL, USA*

BENNO WEIGMANN • *Department of Medicine 1, University of Erlangen-Nuremberg, Erlangen, Germany*

ALYSSA K. WHITNEY • *Mucosal Inflammation Program, Department of Medicine, University of Colorado, Aurora, CO, USA*

CALVIN B. WILLIAMS • *Section of Rheumatology, Department of Pediatrics, Medical College of Wisconsin, Milwaukee, WI, USA*

CHRISTOPHER S. WILLIAMS • *Division of Gastroenterology, Department of Medicine, Vanderbilt University, Nashville, TN, USA; Department of Cancer Biology, Vanderbilt University, Nashville, TN, USA; Vanderbilt Ingram Cancer Center, Nashville, TN, USA; Veterans Affairs Tennessee Valley Health Care System, Nashville, TN, USA*

JOANNA WROBLEWSKA • *Department of Pathology and Committee on Immunology, The University of Chicago, Chicago, IL, USA*

YANA ZAVROS • *Department of Molecular and Cellular Physiology, University of Cincinnati College of Medicine, Cincinnati, OH, USA*

ROBERT ZEISER • *Department of Hematology, Oncology and Stem Cell Transplantation, Freiburg University Medical Center, Albert Ludwigs University Freiburg, Freiburg, Germany; Klinik für Innere Medizin I, Schwerpunkt Hämatologie, Onkologie und Stammzelltransplantation, Universitätsklinikum Freiburg, Freiburg, Germany*

CHAO ZHOU • *Department of Electrical and Computer Engineering, Packard Laboratory, Lehigh University, Bethlehem, PA, USA; Center for Photonics and Nanoelectronics, Lehigh University, Bethlehem, PA, USA; Bioengineering Program, Lehigh University, Bethlehem, PA, USA*

Part I

In Vitro and Ex Vivo Systems to Study Gastrointestinal Functions and Diseases

Chapter 1

CRISPR/Cas9-Mediated Genome Editing of Mouse Small Intestinal Organoids

Gerald Schwank and Hans Clevers

Abstract

The CRISPR/Cas9 system is an RNA-guided genome-editing tool that has been recently developed based on the bacterial CRISPR-Cas immune defense system. Due to its versatility and simplicity, it rapidly became the method of choice for genome editing in various biological systems, including mammalian cells. Here we describe a protocol for CRISPR/Cas9-mediated genome editing in murine small intestinal organoids, a culture system in which somatic stem cells are maintained by self-renewal, while giving rise to all major cell types of the intestinal epithelium. This protocol allows the study of gene function in intestinal epithelial homeostasis and pathophysiology and can be extended to epithelial organoids derived from other internal mouse and human organs.

Key words Small intestinal organoids, Intestinal stem cells, CRISPR/Cas9, Genome editing

1 Introduction

1.1 The CRISPR/Cas9 Genome-Editing Tool

Clustered regularly interspaced short palindromic repeats (CRISPRs) are classes of repeated DNA sequences found in bacteria and archaea. Together with CRISPR-associated (Cas) genes they are part of an adaptive bacterial immune defense system, which confers resistance to foreign genetic elements such as phages [1]. The CRISPR/Cas immune defense process involves three steps. First, upon infection foreign DNA sequences are inserted as new spacers into the CRISPR locus. Second, the locus is transcribed into a single noncoding precursor CRISPR RNA (pre-crRNA) and is processed into short stretches of mature crRNA. Third, the mature crRNA forms a ribonucleoprotein complex with Cas proteins, which specifically recognizes and destroys the invading foreign DNA [2].

So far, three types of CRISPR systems have been discovered. In contrast to the type I and type III CRISPR/Cas systems, the type II system relies on a single Cas protein for DNA interference, but—in addition to the crRNA—requires a tracrRNA bound to Cas9 [3]. The specificity of the CRISPR/Cas9 ribonucleoprotein

Andrei I. Ivanov (ed.), *Gastrointestinal Physiology and Diseases: Methods and Protocols*, Methods in Molecular Biology, vol. 1422, DOI 10.1007/978-1-4939-3603-8_1, © Springer Science+Business Media New York 2016

complex to the invading DNA is mediated through Watson-Crick base pairing of a 20-nucleotide long stretch that is complementary between the crRNA and the invading DNA. The HNH and RuvC-like nuclease domains of the Cas9 protein then eliminate the foreign DNA by generating a double-strand break (DSB). In principle, any DNA sequence that is followed by a protospacer-associated motif (PAM), a conserved sequence of 2–5 nucleotides, can be recognized and cut by the CRISPR/Cas9 complex [3].

In 2012, the labs of Emanuelle Charpentier and Jennifer Doudna together adapted the type II CRISPR system for genome editing [4]. By combining the tracrRNA with the scRNA, a synthetic single guide RNA (sgRNA) was generated, which effectively targets Cas9 to a DNA sequence of interest and leads to the site-specific generation of a DSB [4]. Like in previously developed genome-editing tools, DSBs generated by CRISPR/Cas9 can modify the targeted DNA locus in two ways. First, in the absence of a homologous DNA template the DSBs can generate small insertions or deletions, as they are repaired by the error-prone non-homologous end-joining (NHEJ) pathway [5]. Second, in the presence of an exogenous homologous DNA template, the DSBs can be repaired by the homology directed repair pathway (HDRP), which allows to introduce specific DNA sequences and thus to precisely modify the genomic region [5]. In 2013, several research groups have demonstrated successful CRISPR/Cas9-mediated genome editing in a number of different organisms, ranging from plants to human cells [6, 7]. The easy design, high targeting efficiency, and low off-target mutation frequency of the CRISPR/Cas9 system rapidly made it the most commonly used genome-editing tool [3].

1.2 Mouse Small Intestinal Organoids

Mouse small intestinal organoids are in vitro-grown three-dimensional epithelial structures that closely resemble the in vivo gut epithelium. They can be established from single Lgr5+ stem cells, which are embedded in Matrigel and supplied with a cocktail of tissue-specific growth factors [8]. Like the in vivo gut epithelium, intestinal organoids contain a crypt-like compartment with self-renewing Lgr5+ stem cells, and a villus-like compartment with differentiated enterocytes, paneth cells, and enteroendocrine cells [8]. Minor changes in growth factor composition allow the growth of organoids from a range of human epithelial tissues [9–12]. Epithelial organoids are genetically and phenotypically stable [13] and can be genetically modified by CRISPR/Cas9-based genome editing [14–16].

In this protocol, we describe step by step how to edit the genomes of mouse small intestinal organoids using CRISPR/Cas9 in combination with DNA templates for homologous recombination (HR). In addition, we provide information on how to adapt the protocol for genome editing in organoids derived from different tissues.

2 Materials

2.1 Small Intestinal Organoid Culture Components

1. ECM matrix (*see* **Note 1**).

   (a) Corning® Matrigel® Growth Factor Reduced (GFR) Basement

   Membrane Matrix, cat no. 356231, or

   (b) Cultrex® BME2 RGF organoid matrix, cat no. 3533-005-02.

2. Intestinal organoid medium (*see* **Note 2**).

   (a) Advanced DMEM/F12 (Life Technologies).

   (b) GlutaMax (Life Technologies).

   (c) HEPES.

   (d) Penicillin-Streptomycin.

   (e) N2 supplement (Life Technologies, cat no. 17502–044).

   (f) B27 supplement (Life Technologies, cat no. 17504–044).

   (g) *N*–Acetylcysteine.

   (h) Murine recombinant EGF (Life Technologies, cat no. PMG8044).

   (i) Murine recombinant Noggin (PeproTech, cat no. 250–38).

   (j) Human recombinant R-spondin1 (PeproTech, cat no. 120–38).

   Preparation of the intestinal organoid medium: First supplement 500 ml of Advanced DMEM/F12 with 5 ml 100× Glutamax, 5 ml 1 M HEPES, and 5 ml 100× Penicillin-Streptomycin. This Advanced DMEM/F12+++ medium is stable at 4 C° for at least one month. To prepare 20 ml of intestinal organoid medium, supplement the Advanced DMEM/F12+++ medium with 400 μl of 50× B27, 200 μl of 100× N2, 50 μl of 500 μg/ml *N*-acetylcysteine, 2 μl of 500 μg/ml mouse EGF, 20 μl of 100 μg/ml mouse recombinant Noggin, and 20 μl of 1 mg/ml human recombinant R-spondin1. Intestinal organoid medium is stable for at least two weeks at 4 C°.

3. Recovery™ Cell Culture Freezing Medium (Life Technologies).

4. Cryopreservation cell freezing containers.

5. 24-well and 48-well cell culture plates.

2.2 Transfection and Clonal Selection of Small Intestinal Organoids

1. Lipofectamine 2000 (Life Technologies).

2. Opti-MEM (Life Technologies).

3. Y-27632 dihydrochloride (Sigma-Aldrich).

4. CRISPR/Cas9 plasmids: pSpCas9(BB)-2A-GFP(PX458), pSpCas9(BB)-2A-Puro (PX459) V2.0 (available via Addgene).

5. Trypsin replacement solution (TrypLE) (Life Technologies).

6. Murine recombinant Wnt-3a (Millipore, cat no. GF154).

7. 4-Hydroxytamoxifen (Sigma-Aldrich).

8. Purelink Genomic DNA Extraction kit.

9. Puromycin.

10. Nicotinamide.

11. Parafilm.

12. Refrigerated centrifuge with microtiter plate carrier.

13. Thermal cycler.

3   Methods

3.1   Design and Generation of CRISPR/Cas9 Genome-Editing Vectors

For the design and the cloning of CRISPR/Cas9 vectors, we advise to follow the Nature Protocol from the Zhang lab [5]. We generally use the pSpCas9(BB)-2A-GFP CRISPR/Cas9 plasmid, and insert the specific target sequence as described in their protocol.

The design of the template DNA for homologous recombination depends on the application. We generally synthesize plasmids with a 500 bp homology region up- and downstream of the desired nucleotide change. Close to the modified nucleotide (<50 bp), we insert a puromycin resistance cassette flanked with loxP sites. This setup allows efficient screening for homologous recombination events using antibiotics and subsequent cassette excision (*see* **Note 3**).

3.2   Establishment of Small Intestinal Crypt Cultures

The establishment of organoid cultures from freshly isolated murine small intestinal crypts is described in Sato and Clevers, 2013 [17]. This protocol explains all experimental steps in detail and lists all required reagents. Please follow this protocol to establish and passage murine small intestinal organoids.

3.3   Cryopreservation of Organoids

After establishing a new organoid line, we recommend cryopreserving the line, and always starting from an early passage when gene-editing experiments are repeated.

1. After establishing a new organoid line, passage the culture once or twice prior to cryopreservation (*see* **Note 4**).

2. Approximately 7 days after seeding, replace the mouse small intestinal organoid medium with 1–2 ml of cold basal culture medium, and disrupt Matrigel by gently pipetting with a p1000 pipette.

3. Transfer organoids from one well into a 15 ml falcon tube, and disrupt them by gently pipetting 10–15 times with a fire-polished Pasteur pipette.

4. Centrifuge organoids at $150 \times g$ for 5 min at 4 C°, remove the supernatant, and resuspend the cell pellet in 0.5 ml of ice-cold Recovery™ Cell Culture Freezing Medium.

5. Transfer the cell suspension into 1.5 ml cryogenic storage tube and put the tubes in a CoolCell® cryopreservation container. Immediately transfer the container to the −80 C° freezer, and keep it at −80 C° for 24 h. Afterwards tubes can be moved to the liquid nitrogen container for long-term storage.

3.4 Preparation of Organoids for Lipofection

1. Start with organoids from an early passage. If cryopreserved organoids are used, thaw a vial in a 37 C° water bath and immediately suspend in 10 ml of basal culture medium containing 10 % FBS. Centrifuge at $150 \times g$ for 5 min at 4 C°, remove supernatant, repeat the washing step with basal culture medium, and resuspend the pellet in 150 μl of ice-cold Matrigel.

2. Divide the Matrigel cell suspension as hemispheric droplets into three wells of a 24-well plate, and incubate in a 37 C° incubator for 10 min for Matrigel polymerization (*see* **Note 5**).

3. Add mouse intestinal organoid medium supplemented with Wnt-3a (100 ng/ml) and nicotinamide (10 mM) to the wells, and change medium every 2–3 days (*see* **Notes 6** and **7**). Approximately once a week organoids can be passaged.

4. After two passages organoids should be cystic, and are ready for transfection.

5. Replace the intestinal organoid medium with 1 ml of cold basal culture medium, and disrupt Matrigel by gently pipetting with a p1000 pipette. Transfer organoids from four wells (~100 organoids per well) into one 15 ml falcon tube, and break them by gently pipetting with a fire-polished Pasteur pipette.

6. Centrifuge at $150 \times g$ for 5 min, and wash the pellet with 5 ml of ice-cold basal culture medium to fully remove Matrigel (*see* **Note 8**).

7. Resuspend the pellet in 4 ml of pre-warmed TrypLE and incubate at 37 C° for 5 min in a water bath (*see* **Note 9**).

8. Centrifuge at $150 \times g$ for 5 min at 4 C°, and resuspend the pellet in 450 μl mouse small intestinal organoid medium supplemented with Wnt-3a (100 ng/ml), nicotinamide (10 mM), and the Rho kinase inhibitor Y-27632 (10 nM).

9. Transfer the cell suspension into one well of a 48-well plate, let the cells sink to the bottom, and analyze cell density under the microscope. Cells should be 70–90 % confluent.

<table>
<tr><td>

3.5 Lipofection of Organoids with CRISPR/Cas9 Plasmids

</td><td>

We recommend also reading the Lipofectamine® 2000 reagent protocol on the Life Technologies webpage: http://tools.lifetechnologies.com/content/sfs/manuals/Lipofectamine_2000_Reag_protocol.pdf

1. Dilute 2 µl of Lipofectamine reagent in 25 µl of Opti-MEM® medium.

2. Dilute plasmids (0.5 µg CRISPR/Cas9 vector and 0.5 µg HDR plasmid) in 25 µl of Opti-MEM® medium.

3. Mix diluted DNA and diluted Lipofectamine reagent, and incubate for 5 min.

4. Add the 50 µl Lipofectamine-DNA complex gently to one well of dissociated organoids.

5. Seal the plate with parafilm, and centrifuge 60 min at $600 \times g$ at 32 °C.

6. Discard the parafilm, and incubate the plate for another 2–4 h in a tissue culture incubator.

7. Collect the transfected cells in a 15 ml falcon tube, spin at $150 \times g$ for 5 min, and resuspend the pellet in 100 µl of ice-cold Matrigel.

8. Divide the Matrigel cell suspension into two wells of a 24-well plate, and incubate in the 37 C° incubator for 10 min for Matrigel polymerization.

9. Add mouse small intestinal organoid medium supplemented with Wnt-3a (100 ng/ml), nicotinamide (10 mM), and Y-27632 (10 nM), and place the plate into the tissue culture incubator.

</td></tr>
<tr><td>

3.6 Selection of Genome-Modified Organoids

</td><td>

1. 3 days after transfection start with the antibiotics selection (500 ng/ml puromycin) (*see* **Note 10**).

2. When drug-sensitive organoids start to grow out, pick individual organoids from the Matrigel under a binocular microscope using a p200 pipette, transfer them individually into 1.5 ml tubes, and split them by pipetting with the p200 pipette.

3. Centrifuge for 5 min at $900 \times g$, resuspend in 100 µl ice-cold Matrigel, and plate cells in a 24-well plate. After Matrigel polymerization add normal small intestinal organoid medium (*see* **Note 11**).

4. After expansion of clonal organoids, use 1–2 wells for genomic DNA isolation. For the isolation you can use the PureLink® Genomic DNA extraction kit from Life Technologies or other standard genomic DNA isolation procedures.

5. Analyze the genome by PCR and Sanger sequencing to identify clones with correct HR events. Use primer pairs that bind within the puromycin resistance cassette and up- or downstream of the

</td></tr>
</table>

HR arms. Alternatively, instead of using primers that bind within the puromycin cassette allele-specific primers for the inserted nucleotide change can be used. In general a 2bp difference at the 3′-end of the primer is sufficient for specificity.

6. Remove the antibiotics resistance cassette by expressing the Cre recombinase. We usually use small intestinal organoids established from mice expressing tamoxifen-inducible cre:ERT2 (Villin-CreERT2), which allows cassette excision simply by adding 4-Hydroxytamoxifen (100 nm) for 12 h to the culture media. Alternatively organoids can be transduced with a ready-to-use Cre recombinase adenovirus (Vector Biolabs), lipofected with a Cre-expressing plasmid (e.g., Addgene plasmid pCAG-Cre:GFP), or transduced with a CRE protein as described in D'Astolfo et al. [18].

4 Notes

1. Corning® Matrigel® and Cultrex® BME2 are both protein matrices secreted by Engelbreth-Holm-Swarm (EHS) mouse sarcoma cells, which in our hands work equally well for culturing mouse small intestinal organoids. Both matrices can be bought in bottles, which should be thawed on ice and aliquoted into 1 ml vials. If possible don't switch between batches within one experiment.

2. Instead of preparing your own murine small intestinal organoid medium, you can also use IntestiCult™ organoid growth medium (Stemcell Technologies, cat no. 06005).

3. It is also possible to use the PiggyBac transposon system instead of the Cre-lox system. In that case use a puromycin cassette flanked with PiggyBac sites, and a PiggyBac transposase for cassette excision [19]. The main advantage of the PiggyBac system is that after excision only a four bp TTAA sequence remains within the host genome.

4. Intestinal organoids can be passaged 1:5 every 7 days. If this ratio cannot be achieved, try to dissociate organoids into smaller pieces by vigorous pipetting. After crypt isolation, two passages are usually enough to obtain a whole 24-well plate (with one 50 μl drop of Matrigel per well containing ~100 organoids).

5. It can be sometimes difficult to form hemispheric droplets with Matrigel. In that case organoids sink to the bottom and attach to the tissue culture plate. Overnight pre-incubation of the cell culture plate in a 37 °C tissue culture incubator helps to form hemispheric droplets.

6. Small intestinal organoid transfection media contains Wnt-3a and Nicotinamide, which drives organoids from the budding structure to spherical organoids that solely consist of Lgr5 positive stem cells. This step greatly increases the number of organoids grown from single cells after lipofection. It usually takes one to two weeks until organoids are fully spherical. Instead of recombinant Wnt-3a the medium can also be supplemented with 50 % Wnt-3a conditioned medium [20].

7. Lipofection and CRISPR/Cas9 plasmid delivery by lipofection has also been performed in organoids derived from other tissues, such as the human small intestine, colon, and liver [14–16]. In principle the protocol described here can also be used to edit the genomes of these types of organoids. Human small intestinal and human colon organoids however are already grown in medium containing Wnt-3a and nicotinamide. Therefore no additional Wnt-3a and nicotinamide need to be added prior to lipofection. Also liver organoids consist solely of bipotent progenitor cells during expansion, and thus no extra growth factors are required before the transfection.

8. If Matrigel is not yet fully removed, repeat the washing step once more. Large amounts of Matrigel can block the single-cell dissociation in the next step.

9. Check organoids after incubation under the microscope. If you have clumps of 1–10 cells you can proceed. If the clumps are still bigger, prolong the incubation for another 5 min. Be aware that overdigestion significantly decreases cell survival.

10. The concentration of puromycin needed for selection might vary from line to line. We therefore recommend determining the optimal antibiotics concentration for the line prior to the transfection.

11. Withdrawal of Wnt-3a and nicotinamide from the media will revert the phenotype of small intestinal organoids from cystic structures to budding structures, which contain differentiated cell types. Organoids need approximately 1–2 weeks for full conversion back to the budding phenotype.

Acknowledgments

This work was funded by grants from the European Research Council (EU/232814-StemCeLLMark), the KNAW/3V-fund, the SNF (31003A_160230), and the Human Frontiers in Science Program long-term fellowship LT000422/2012.

References

1. Bhaya D, Davison M, Barrangou R (2011) CRISPR-Cas systems in bacteria and archaea: versatile small RNAs for adaptive defense and regulation. Annu Rev Genet 45:273–297

2. Sorek R, Lawrence CM, Wiedenheft B (2013) CRISPR-mediated adaptive immune systems in bacteria and archaea. Annu Rev Biochem 82:237–266

3. Sander JD, Joung JK (2014) CRISPR-Cas systems for editing, regulating and targeting genomes. Nat Biotechnol 32:347–355

4. Jinek M, Chylinski K, Fonfara I, Hauer M, Doudna JA, Charpentier E (2012) A programmable dual-RNA-guided DNA endonuclease in adaptive bacterial immunity. Science 337:816–821

5. Ran FA, Hsu PD, Wright J, Agarwala V, Scott DA, Zhang F (2013) Genome engineering using the CRISPR-Cas9 system. Nat Protoc 8:2281–2308

6. Cong L, Ran FA, Cox D, Lin S et al (2013) Multiplex genome engineering using CRISPR/Cas systems. Science 339:819–823

7. Mali P, Yang L, Esvelt KM, Aach J, Guell M, DiCarlo JE, Norville JE, Church GM (2013) RNA-guided human genome engineering via Cas9. Science 339:823–826

8. Sato T, Vries RG, Snippert HJ, van de Wetering M et al (2009) Single Lgr5 stem cells build crypt-villus structures in vitro without a mesenchymal niche. Nature 459:262–265

9. Boj SF, Hwang CI, Baker LA, Chio II et al (2015) Organoid models of human and mouse ductal pancreatic cancer. Cell 160:324–338

10. Huch M, Bonfanti P, Boj SF, Sato T et al (2013) Unlimited in vitro expansion of adult bipotent pancreas progenitors through the Lgr5/R-spondin axis. EMBO J 32:2708–2721

11. Jung P, Sato T, Merlos-Suarez A, Barriga FM et al (2011) Isolation and in vitro expansion of human colonic stem cells. Nat Med 17:1225–1227

12. Sato T, Stange DE, Ferrante M, Vries RG et al (2011) Long-term expansion of epithelial organoids from human colon, adenoma, adenocarcinoma, and Barrett's epithelium. Gastroenterology 141:1762–1772

13. Huch M, Gehart H, van Boxtel R, Hamer K et al (2015) Long-term culture of genome-stable bipotent stem cells from adult human liver. Cell 160:299–312

14. Drost J, van Jaarsveld RH, Ponsioen B, Zimberlin C et al (2015) Sequential cancer mutations in cultured human intestinal stem cells. Nature 521:43–47

15. Matano M, Date S, Shimokawa M, Takano A, Fujii M, Ohta Y, Watanabe T, Kanai T, Sato T (2015) Modeling colorectal cancer using CRISPR-Cas9-mediated engineering of human intestinal organoids. Nat Med 21:256–262

16. Schwank G, Koo BK, Sasselli V, Dekkers JF et al (2013) Functional repair of CFTR by CRISPR/Cas9 in intestinal stem cell organoids of cystic fibrosis patients. Cell Stem Cell 13:653–658

17. Sato T, Clevers H (2013) Primary mouse small intestinal epithelial cell cultures. Methods Mol Biol 945:319–328

18. D'Astolfo DS, Pagliero RJ, Pras A, Karthaus WR et al (2015) Efficient intracellular delivery of native proteins. Cell 161:674–690

19. Yusa K, Rashid ST, Strick-Marchand H, Varela I et al (2011) Targeted gene correction of alpha1-antitrypsin deficiency in induced pluripotent stem cells. Nature 478:391–394

20. Willert K, Brown JD, Danenberg E, Duncan AW, Weissman IL, Reya T, Yates JR III, Nusse R (2003) Wnt proteins are lipid-modified and can act as stem cell growth factors. Nature 423:448–452

Chapter 2

Lentivirus-Based Stable Gene Delivery into Intestinal Organoids

Yoshiaki Maru, Kaoru Orihashi, and Yoshitaka Hippo

Abstract

Lentivirus-based gene delivery works efficiently for the majority of mammalian cells cultured under standard two-dimensional conditions. By contrast, intestinal epithelial organoids embedded into three-dimensional extracellular matrix appear to be resistant to lentiviral transduction. We observed that Matrigel, a matrix that reconstitutes a basement membrane and is indispensable for cell survival and proliferation, prevents lentiviruses from binding to intestinal cells. In this chapter, we describe a simple method of a highly efficient gene transduction into intestinal organoids. This method involves organoid dispersion into single intestinal epithelial cells, mixing these individual cells with lentiviral particles, plating on Matrigel, and subsequent re-embedding into Matrigel. Under these conditions, the majority of the cells are exposed to the virus in the absence of the matrix barrier while remaining attached to the matrix. Using a GFP-labeled lentivirus, we demonstrate that this method allows for highly efficient infection of intestinal organoids after overnight incubation of Matrigel-attached cells with lentiviral particles.

Key words Matrigel, 3D culture, Organoids, Tumorigenesis, Stem cells, Viral transduction

1 Introduction

Intestinal stem cells and organoids can be propagated over a long period of time in a Matrigel supplemented with defined factors [1]. This finding implies that functional analysis of specific genes in a physiological setting might become feasible, even without generating genetically engineered mice, if only stem cells could be transduced with vectors encoding shRNA or cDNA. Lentiviruses can infect not only dividing cells but also nondividing cells, including dormant stem cells, resulting in very high infection efficiency [2]. Moreover, genes delivered are readily integrated into the host genome. Assuming that these two features would also lead to stable gene transduction in 3D culture, we initially simply overlaid culture media containing lentiviral particles encoding GFP on Matrigel, in which organoids were embedded. To our surprise, virtually no cells in organoids turned green by this approach.

Andrei I. Ivanov (ed.), *Gastrointestinal Physiology and Diseases: Methods and Protocols*, Methods in Molecular Biology, vol. 1422, DOI 10.1007/978-1-4939-3603-8_2, © Springer Science+Business Media New York 2016

Furthermore, either dissociation of organoids into single cells or mixing the viral particles with Matrigel did not improve transduction efficiency, thereby indicating the inhibitory effect of Matrigel on lentiviral infection. However, in the absence of Matrigel, both intestinal organoids and single cells died before being infected, confirming the requirement of Matrigel for primary cells survival [3]. Finally, we observed that mixing viral particles with single intestinal cells, but not with intact organoids, followed by overnight incubation on Matrigel, resulted in extremely high infection efficiency [4]. Based on this observation, we developed a simple experimental protocol to achieve a robust lentiviral-mediated gene delivery into intestinal organoids.

2   Materials

2.1 Production and Concentration of Lentiviruses

1. 293FT cell line (Life Technologies).
2. 293FT culture medium: DMEM high glucose, supplemented with 10 % fetal bovine serum (FBS) without antibiotics.
3. Opti-MEM® medium (Life Technologies).
4. Lipofectamine® 2000 (Life Technologies).
5. Virapower™ Lentiviral Packaging Mix (Life Technologies).
6. Lentiviral Vector (*see* **Note 1**).
7. 10 cm culture dish (uncoated and collagen coated).
8. 10 ml syringe.
9. 0.45 μm syringe filter.
10. PEG-it™ (Systems Biosciences).
11. 50 ml tube.
12. Cryotube.

2.2 Primary Organoid Culture and Preparation of Single Intestinal Cells

1. Matrigel® Matrix Basement Membrane (Corning, #354234). Store at 4 °C.
2. 1× Organoid culture media: Advanced DMEM/F12 supplemented with Penicillin, Streptomycin, Fungizone, L-glutamine, 50 ng/ml EGF (Peprotech, #315-09), 250 ng/ml R-Spondin1 (R&D Systems, #3474-RS), 100 ng/ml Noggin (Peprotech, #250-38), 1 μM Jagged-1 (ANASPEC, #61298), 10 μM Y27632. Store at 4 °C.
3. 12-well culture plate.
4. Accumax™ (Innova Cell Technologies). Store at 4 °C (*see* **Note 2**).
5. Phosphate buffered saline for cell culture, pH 7.4. Store at 4 °C.

6. 2× Organoid culture medium: This medium has the same composition as the 1× Organoid medium, but contains two-fold higher concentrations of all growth factors and inhibitors. Store at 4 °C.

7. Cell scraper.

8. 5 ml Eppendorf tube.

2.3 Lentiviral Infection of Intestinal Cells

1. 1.5 ml Eppendorf tube.

2. TransDux™ (Systems Biosciences). Store at –20 °C (*see* **Note 3**).

3. CHIR-99021 (FOCUS Biomolecules), 20 mM in DMSO. Store at –20 °C.

3 Methods

Lentiviruses must be handled in an area with the biosafety level 2 (BSL2). All nonviral cell culture experiments must be conducted on a clean bench.

3.1 Production and Concentration of Lentiviral Particles

1. Propagate 293FT cells in the 293FT culture media.

2. Plate 293FT cells on 10 cm collagen-coated dishes one day prior to transfection (*see* **Note 4**).

3. Transfect 293FT cells at 90 % confluence with a lentiviral vector and Virapower™ Lentiviral Packaging Mix. Use Opti-MEM® and Lipofectamine 2000 for the transfection (*see* **Note 5**).

4. Replace the transfection mixtures with 293FT culture media after 8 h of transfection.

5. Collect the supernatant with 10 ml syringe, 48 h after the medium change.

6. Pass the supernatant through a 0.45 μm syringe filter to recover only viral particles into a 50 ml tube.

7. Add 1/4 volume of PEG-it™ to the filtered supernatant. Invert several times and store at 4 °C for 24 h.

8. Centrifuge the tube for 30 min at $2300 \times g$ in 4 °C to spin down viral particles. Discard the supernatant.

9. Resuspend the pellet with 1 ml of Opti-MEM, divide the suspension into four 250 μl aliquots, and place into cryo tubes. Store at –80 °C until use (*see* **Note 6**).

3.2 Primary Organoid Culture and Preparation of Single Cells

1. Isolate intestinal crypts from 3- to 5-week-old mice using 2 mM EDTA, as previously described [1, 4].

2. Dissociate crypts into single cells by adding 1 ml of Accumax to the pellet (*see* **Note 7**). Vortex for 3 s and incubate for 5 min at 37 °C in a water bath.

3. Add 5 ml of cold PBS to wash out the enzyme. Centrifuge for 5 min at $400 \times g$ to precipitate single cells and discard the supernatant.

4. Resuspend single cells with 1× organoid culture media. Carefully disperse the cells by pipetting for 5–10 times with P1000 pipetman (*see* **Note 8**).

5. Add 80 μl of ice-cold liquid Matrigel into each well of a 12-well plate and incubate at 37 °C for 20 min to completely solidify Matrigel (*see* **Note 9**).

6. Plate 800 μl of single-cell suspensions into each well of the Matrigel-coated 12-well plate (*see* **Note 10**).

7. Next day (~16 h later), remove the medium with dead cells by aspiration. Overlay 60 μl of fresh cold Matrigel on the attached cells and incubate at 37 °C for 20 min to solidify Matrigel (*see* **Note 11**).

8. Overlay 800 μl of 1× organoid culture media to start 3D culture. Medium is changed every 3–4 days.

9. When organoids become subconfluent, collect the entire content of the well (organoids, Matrigel, and medium) into a 5 ml tube using a cell scraper (*see* **Note 12**).

10. Centrifuge for 5 min at $400 \times g$. Carefully aspirate the medium and Matrigel (*see* **Note 13**).

11. Dissociate organoids into single cells by adding 1 ml of Accumax to the pellet. Vortex for 3 s and incubate for 5 min at 37 °C in a water bath.

12. Add 5 ml of cold PBS to wash out the enzyme. Spin down the single cells and discard the supernatant (*see* **Note 14**).

13. Resuspend single cells with 2× organoid culture media supplemented with 2.5 μM CHIR-99021. Thoroughly disperse the cells by pipetting 10–15 times with a P1000 Pipetman (*see* **Note 15**).

14. Prepare a single-cell suspension at the concentration of $4 \sim 6 \times 10^5$ cell/ml (*see* **Note 16**).

3.3 Lentiviral Infection (See a Flowchart in Fig. 1)

1. Add 80 μl of cold liquid Matrigel into each well, and incubate at 37 °C for 20 min to solidify Matrigel (*see* **Note 17**).

2. Combine 250 μl of a single-cell suspension in 2× organoid culture and 250 μl of viral particles in a 1.5 ml Eppendorf tube. Add 2.5 μl of TransDux™ and mix well by tapping (*see* **Note 18**).

3. Plate 500 μl of mixture of viral particles and single cells on solidified Matrigel. Incubate overnight at 37 °C (*see* **Note 19**).

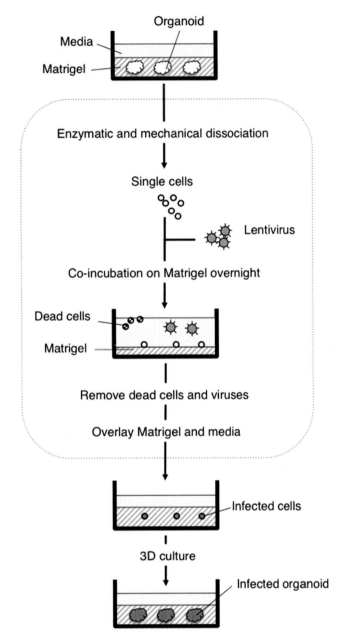

Fig. 1 Lentiviral infection of primary intestinal cells at glance. The figure depicts flowchart of key procedures that are required for efficient infection of intestinal organoids with lentiviruses

4. Next day (~16 h later), remove the medium with dead cells by aspiration. Overlay 60 μl of ice-cold Matrigel to cover the cells attached to Matrigel at the bottom. Incubate at 37 °C for 20 min to solidify Matrigel (*see* **Note 20**).

5. Resume 3D culture by overlaying 800 μl of 1× organoid culture media.

Bright field   **Fluorescence**

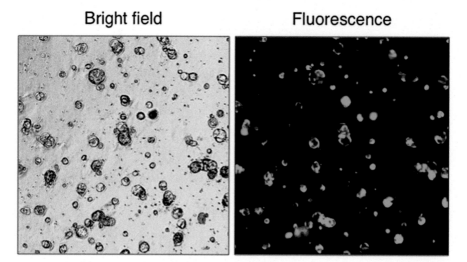

Fig. 2 Efficient gene transduction of primary intestinal cells. Intestinal cells obtained from dispersed mouse organoids were incubated with a lentivirus bearing a GFP-encoding vector, pCDH-CMV-MCS-EF1-copGFP. The cells were re-embedded into Matrigel and images were acquired 36 h later. Note that nearly 100 % of newly formed intestinal organoids are GFP-positive

6. If a GFP expressing vector is introduced, more than 90 % of the cells should turn green, as soon as 36 h after beginning of the infection (Fig. 2).

7. Start selection of transduced cells a few days after infection, if an introduced vector harbors a certain drug-resistant gene (*see* **Note 21**).

4   Notes

1. We always include a GFP-encoding vector such as pCDH-CMV-MCS-EF1-copGFP, so that infection efficiency in each experiment can be monitored alongside by fluorescence microscopy.

2. Accumax has protease, collagenolytic and DNase activity, and gently disperses cells while inflicting a little damage. Normal cells do not tolerate dispersion by trypsin-EDTA.

3. TransDux facilitates viral infection of mammalian cells similarly to polybrene, but it works more efficiently for primary cells.

4. If plated on uncoated dishes, 293FT cells could easily detach after transfection, leading to a low yield of viral particles.

5. For Lipofectamine 2000-mediated transfection, we follow the manufacturer's instructions.

6. We usually make frozen stocks of lentiviruses, and rarely use freshly produced viral particles, because it is difficult to synchro-

nize viral production with the organoid harvesting. This described virus production procedure results in tenfold concentration of viral supernatants, which compensates for approximately tenfold decrease of virus titer during the frozen stock thawing.

7. The aim of this step is to separate an undifferentiated population from differentiated or more committed populations that will never contribute to development of organoids.

8. At this step, there is no need for thorough dispersion, which could reduce colony-forming capacity. As the magnitude of dispersion appears to significantly vary among researchers, even with the same number of times of pipetting, the best number of times of pipetting for each researcher must be empirically determined.

9. We initially rinse the surface of each well with cold PBS to more evenly spread liquid Matrigel before its polymerization.

10. We typically divide a cell suspension obtained from the small intestine of a single mouse into six wells of the 24-well plate.

11. Relatively undifferentiated cells easily attach to Matrigel and remain viable, while differentiated cells do not attach and die shortly. Accordingly, stem or progenitor cells are likely to be enriched at the beginning of the 3D culture.

12. We observed that the timing of organoid harvest is critical for the efficient infection. The best results can be obtained by using actively proliferating organoids with extensive budding of crypt-like structure and little debris or dead cells in the lumen. Typically, the best timing is around days 4–7 of primary 3D culture and this time window lasts only 1–2 days.

13. The organoids should be visible as a white pellet at the tube bottom. Softened Matrigel will form a middle layer, and the culture medium will form a top layer. At this step, medium should be completely removed, but some Matrigel may be still present in the pellet.

14. Matrigel is no longer solid or visible as a pink pellet.

15. To maximize the cell contact area with viral particles, it is imperative to thoroughly dissociate organoids into single cells and keep cell density at a low level because cell aggregation could hamper efficient infection. However, we noted that overly long enzymatic treatment and excessive pipetting could eventually lead to a very low colony-forming capacity. Addition of a GSK3β inhibitor, CHIR-99021, which has been recently shown to significantly improve the colony-forming capacity of single intestinal cells [5], will help to achieve both high infection efficiency and formation of a large number of organoids (Fig. 3).

16. Accurate counting of viable cell numbers, with a hemocytometer and Trypan blue staining, may become difficult because organ-

CHIR-9902

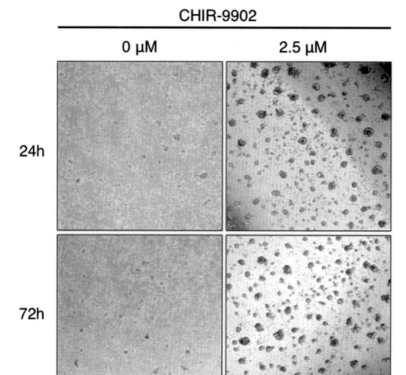

Fig. 3 A GSK3β inhibitor, CHIR-99021, enhanced the development of intestinal organoid. Addition of 2.5 μM CHIR-99021 to a single-cell suspension significantly increased colony-forming capacity after re-embedding the cells into Matrigel. Images were taken at 24 h and 72 h after the re-embedding

oids contain many dead cells, and single cells keep dying during counting. Therefore counting must be performed very promptly.

17. This step is identical to those described in Subheading 3.1, **step 8**.

18. Pre-mixing of cells and viruses before plating is critical because we observed lower infection efficiency when these two components were separately placed on polymerized Matrigel.

19. This step is essentially the same as described in Subheading 3.1, **step 9**. By attaching to Matrigel, individual cells can survive overnight incubation with lentiviruses, which is sufficient for high efficiency viral transduction.

20. We observe that the majority of GFP-positive cells usually remain green, even at 3 weeks after infection, strongly suggesting a stable transduction of stem cells [4].

21. For a longer culture involving subculture, follow first Subheading 3.2, **steps 9–14**, and then Subheading 3.2, **steps 5–8**.

Acknowledgment

This work was supported by Grants-in-Aid for Scientific Research (B) 26290044 from the Japan Society for the Promotion of Science (JSPS).

References

1. Sato T, Vries RG, Snippert HJ et al (2009) Single Lgr5 stem cells build crypt-villus structures in vitro without a mesenchymal niche. Nature 459:262–265

2. Sakuma T, Barry MA, Ikeda Y (2012) Lentiviral vectors: basic to translational. Biochem J 443:603–618

3. Lee GY, Kenny PA, Lee EH, Bissell MJ (2007) Three-dimensional culture models of normal and malignant breast epithelial cells. Nat Methods 4:359–365

4. Onuma K, Ochiai M, Orihashi K et al (2013) Genetic reconstitution of tumorigenesis in primary intestinal cells. Proc Natl Acad Sci U S A 110:11127–11132

5. Wang F, Scoville D, He XC et al (2013) Isolation and characterization of intestinal stem cells based on surface marker combinations and colony-formation assay. Gastroenterology 145:383–395

Chapter 3

Co-culture of Gastric Organoids and Immortalized Stomach Mesenchymal Cells

Nina Bertaux-Skeirik, Jomaris Centeno, Rui Feng, Michael A. Schumacher, Ramesh A. Shivdasani, and Yana Zavros

Abstract

Three-dimensional primary epithelial-derived gastric organoids have recently been established as an important tool to study gastric development, physiology, and disease. Specifically, mouse-derived fundic gastric organoids (mFGOs) co-cultured with Immortalized Stomach Mesenchymal Cells (ISMCs) reflect expression patterns of mature fundic cell types seen in vivo, thus allowing for long-term in vitro studies of gastric epithelial cell physiology, regeneration, and bacterial-host interactions. Here, we describe the development and culture of mFGOs, co-cultured with ISMCs.

Key words Mouse fundic gastric organoids, Co-culture, Immortalized stomach mesenchymal cells

1 Introduction

In vitro three-dimensional organoid culture systems have been developed to investigate the development, physiology, and disease of multiple organs including liver [1, 2], pancreas [3], Barrett's esophagus [4], stomach [5–9], small intestine [5, 10], and colon [4, 11]. In an effort to enhance our understanding of the development of gastric disease, our lab has established a protocol to generate primary cultured whole fundic gland-derived gastric organoids [5, 7, 8, 12, 13]. Cultures derived from whole dissected glands are distinct from similar corpus-derived cultures generated from single-cell preparations of Troy+ve populations [14]. Troy-positive cells are expressed at the corpus gland base in a subset of differentiated chief cells and generate long-lived gastric organoids that are differentiated toward the mucus-producing cell lineages of the neck and pit regions. Here we report a culture method of fundic-derived gastric organoids devised for the maintenance of mature cell lineages observed throughout the fundic epithelium in vivo. Maintained in matrigel and gastric organoid growth media, organoids are co-cultured with immortalized stomach mesenchymal cells (ISMCs) [15].

Andrei I. Ivanov (ed.), *Gastrointestinal Physiology and Diseases: Methods and Protocols*, Methods in Molecular Biology, vol. 1422, DOI 10.1007/978-1-4939-3603-8_3, © Springer Science+Business Media New York 2016

Gastric organoids recapitulate the differentiated cell types normally found in the stomach and have a polarized epithelium with a defined lumen. These gastric organoids have been used as a model system for the study of *Helicobacter pylori*-induced disease [8, 12], signaling pathways involved in proliferation of the epithelium [13], and gastric physiology [7].

2 Materials

2.1 Immortalized Stomach Mesenchymal Cells (ISMCs)

1. Growth media: DMEM High Glucose, 10 % Fetal Calf Serum, and 1 % Penicillin/Streptomycin.
2. Freezing medium: Recovery™ Cell Culture Freezing Medium (Life Technologies).
3. Washing Solution: Dulbecco's phosphate buffered saline (DPBS) without calcium or magnesium.
4. Cell Releasing Solution: Trypsin-EDTA 0.25 % phenol red.
5. Corning® 75 cm² tissue culture treated flasks.

2.2 Mouse-Derived Fundic Gastric Organoids (mFGOs)

1. Storage and Washing buffer for gastric glands and organoids: ice-cold Dulbecco's phosphate buffered saline (DPBS) without calcium or magnesium.
2. Conical tubes, 50 mL, 15 mL, and 5 mL volume.
3. Chelation buffer for isolation of mouse gastric glands: ice-cold Dulbecco's phosphate buffered saline (DPBS) without calcium or magnesium with 5 mM EDTA, pH 7.4.
4. Dissociation buffer for isolation of mouse gastric glands: ice-cold Dulbecco's phosphate buffered saline (DPBS) without calcium or magnesium with 43.4 mM Sucrose and 54.9 mM D-Sorbitol.
5. Basement membrane for growth of organoids: Growth Factor Reduced, Phenol Red-free Matrigel™ (Fisher).
6. Plates for growth of organoids: Costar™ Transwell™ Clear Polyester Membrane Inserts for 12-Well Plates.
7. Growth and culture medium: Advanced Dulbecco's modified Eagle medium/F12 medium (Life Technologies, 12634–010) supplemented with 2 mM Glutamax, 100 U/ml Penicillin/Streptomycin, 10 mM HEPES Buffer, 1 mM nAcetylcysteine (Sigma), 1× N2 (Life Technologies), 1× B27 (Life Technologies), 50 % Wnt-conditioned medium, 10 % R-spondin–conditioned medium supplemented with gastric growth factors including 100 ng/ml bone morphogenetic protein inhibitor Noggin (PeproTech), 10 nM [Leu-15]-gastrin I (Sigma), 50 ng/ml epidermal grow factor (EGF) (PeproTech), and 100 ng/ml fibroblast growth factor 10 (FGF10) (PeproTech).

8. For staining: 4 % paraformaldehyde made up in regular 1×
   PBS, EdU proliferation assay kit (Life Technologies).

9. Hoechst (Life Technologies, H3570), 1:1000 solution in
   DPBS.

3 Methods

3.1 Isolation of Mouse Gastric Glands

All mouse studies were approved by the University of Cincinnati
Institutional Animal Care and Use Committee (IACUC) that
maintains an American Association of Assessment and Accreditation
of Laboratory Animal Care (AAALAC) facility.

1. Euthanize mouse of at least 6 weeks of age. Using a pair of
   sterile surgical scissors, with the mouse stomach facing up,
   make an incision into the lower right abdominal cavity. Grasp
   the stomach (underneath the liver) with forceps, and use a
   small pair of scissors to cut the stomach from the esophagus
   and the duodenum. Open the stomach along the greater curvature, all the way through both the opening to the esophagus
   and the duodenum, so that the stomach can be laid flat. Wash
   the stomach twice in a weigh-boat filled with ice-cold DPBS
   without calcium or magnesium (w/o Ca/Mg), and then place
   into a 50 mL conical tube with 30 mL of ice-cold DPBS w/o
   Ca/Mg for transport to the micro-dissecting table.

2. Remove the stomach carefully and pin the stomach (basolateral side upwards, and luminal side down) open on a silicon
   filled dish so that the forestomach is pointing toward the top
   of the dish (*see* **Note 1**). Working quickly but carefully, use fine
   point curved forceps to lift the serosal muscle layer in segments
   and use micro-dissecting curved scissors to cut and remove the
   muscle from the epithelium (*see* **Note 2**). *See* Fig. 1a for reference. For fundic-specific gastric organoids, cut and separate
   the fundic tissue with a wide margin from the forestomach and
   antrum to prevent gland contamination, and for antral specific
   organoids, isolate the antrum from any remaining pyloric
   sphincter tissue (*see* **Note 3**). Using scissors cut the fundus or
   antrum into small (less than 5 mm²) pieces, and scoop into a
   sterile 15 mL conical tube prefilled with 5 mL of chelation buffer. Place tube on gentle rocker at 4 °C for 2 h.

3. Thaw desired volume of Matrigel™ on ice or at 4 °C during
   this 2 h EDTA incubation step (*see* **Note 4**).

4. After incubation with EDTA chelation buffer, remove EDTA
   buffer after allowing tissue pieces to settle to the bottom of the
   tube. Replace this solution with 5 mL of dissociation buffer.
   Shake the tube by hand, with the bottom of the tube pointed
   toward the ground for 2 min (*see* **Note 5**). *See* Fig. 1b for pic-
   ture of the gland isolation.

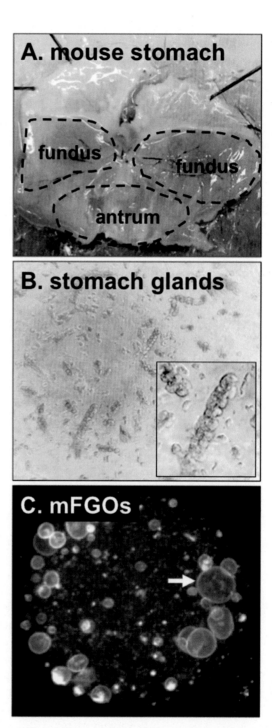

Fig. 1 Development of Mouse Fundic Gastric Organoids. Mouse stomach is harvested, serosal muscle layer is removed (**a**) and sections of the desired portion of the stomach are collected. The stomach glands (fundic depicted) are harvested (**b**), embedded into Matrigel and provided with growth medium. Mouse organoids formation can be visualized by day six of culture (**c**)

5. Set the tube back on ice briefly to let larger fragments of tissue settle to the bottom of the tube, but not long enough for gastric glands to settle. Using a 1000 μL pipet, remove the 5 mL of volume without disturbing the large tissue fragments at the bottom of the tube. Transfer this volume into a sterile 5 mL tube and centrifuge at 150 g for 5 min at 4 °C.

6. Remove supernatant carefully, so as not to disturb the pellet of glands (*see* **Note 6**). Keeping the tube on ice, transfer desired volume of Matrigel™ to the glands (*see* **Note 7**) and mix gently to avoid air bubbles.

7. Using a 200 μL pipet, pipet 40 μL of the glands in Matrigel™ directly onto the Transwell™ membrane (*see* **Note 8**). Incubate the plate at 37 °C for 15 min to allow Matrigel™ to solidify. After this incubation add 0.5 mL of growth media to the top of the transwell, and 1 mL of growth media to the bottom of the transwell. Allow organoids to grow for 4 days (*see* **Note 9**). *See* Subheading 3.3 for co-culture technique with ISMCs.

3.2 Growth of Immortalized Stomach Mesenchymal Cells

1. Prepare culture growth media with desired volume of DMEM Hi Glucose, 10 % Fetal Calf Serum, and 1 % Penicillin/Streptomycin (PS).

2. Thaw the Immortalized Stomach Mesenchymal Cells (ISMCs) stock at 37 °C.

3. In a 75 cm² cell culture flask, add 29 mL of the growth media and pipette 1 mL of thawed ISMC stock solution for a total volume of 30 mL (*see* **Note 10**).

4. After cell addition, slide the flask in forward and backward motion to mix the cells and media. Incubate cells at 37 °C.

5. Cell adhesion should be visible after 24 h under light microscope (*see* **Note 11**).

6. After cells reach 95–100 % confluence, to passage to 12-well plates for co-culture, wash cells with 5 mL of warm DPBS w/o Ca/Mg, remove by vacuum, and treat with 5 mL of 0.25 % trypsin at 30 °C to release cells from flask. Add 10 mL of culture medium to inactive trypsin. Transfer to 50 mL conical tube, and spin at 1000 rpm for 5 min. Remove supernatant, and resuspend in 10 mL growth medium.

7. Count cells using a hemocytometer, and determine appropriate dilution factor to suspend ISMCs to seed 12,000 cells per well in the 12-well plate (*see* **Note 12**). Let cells adhere overnight (or 12–24 h) before culturing with organoids.

8. Passage the rest of the ISMCs onto the T75 cm² flasks for further growth to continue providing fresh ISMCs to organoids.

3.3 Co-culture of ISMCs with Mouse Gastric Organoids

1. At day 4 of growth, transfer Transwell™ membrane inserts to the 12-well plate seeded with ISMCs (*see* **Note 13**). Provide fresh organoid growth medium to both the top and bottom compartment of the Transwell™, taking care to remove old media using a 1000 µL pipet, without disturbing the organoids.

2. Every other day (every 48 h) the organoid culture must be supplied with a fresh 12-well plate of ISMCs; otherwise the organoids will not have enough growth factors to continue growth (*see* **Note 14**).

3. Allow organoids to continue growing up to 7–10 days (post gland embedding in Matrigel™), and harvest for experimentation, staining, or passaging to maintain growth.

3.4 Immuno-fluorescence and EdU Proliferation Assay

1. To stain for proliferating cells, you must complete an EdU uptake step prior to fixation. Otherwise, proceed directly to fixation and staining (**step 3**). Thaw EdU component at 37 °C, and prepare solution in DMEM basic medium with a ratio of 1:500.

2. For EdU uptake, remove half of the growth media volume in the Transwell™, 250 mkl from upper compartment and 500 mkl from bottom compartment and replace with same volume of EdU-DMEM solution, incubate at 37 °C for 1 h.

3. (Begin here if EdU uptake is not desired) For fixation, remove all the liquid from the Transwell™, add prewarmed 4 % paraformaldehyde 500 µL to upper and 1 mL to bottom compartments and incubate for 15 min at room temperature (*see* **Note 15**).

4. Remove the 4 % paraformaldehyde from upper and bottom compartments and replace with the same volume of DPBS. Wrap with Parafilm™ and store at 4 °C until staining.

5. For EdU staining, follow manufacture's directions.

6. For permeabilization, incubate in 0.5 % Triton X-100 for 20 min at room temperature, adding 1 mL liquid to the bottom portion of the Transwell™, and 500 µL liquid to the insert with organoids (*see* **Note 16**).

7. Wash 2 times with DPBS, adding 1 mL liquid to the bottom portion of the Transwell™, and 500 µL liquid to the insert with organoids, and resuspend in primary antibody solution (made up in DPBS) and incubate at 4 °C overnight, again adding 1 mL of liquid to the bottom portion of the Transwell™, and 500 µL of liquid to the insert with organoids.

8. Wash 2 times with 1 mL DPBS, and resuspend cells in secondary antibody at 4 °C overnight, using same volumes as in **step 7**.

9. Wash two times DPBS, and Hoechst solution if desired, and incubate for 30 min in the dark at room temperature, using the same volumes as in **step 7**.

10. Wash two times with DPBS; leave membrane submerged in 1 mL of DPBS to prevent cells from drying on the top insert, and in 2 mL of DPBS in the bottom portion of the Transwell™.

11. For imaging, place the insert (with the organoids still suspended in DPBS) on a glass coverslip. Store wrapped in Parafilm™ in DPBS (1 mL in the insert, 2 mL in the bottom compartment) at 4 °C.

4   Notes

1. To prevent the tissue from drying out, you can pour 3–5 mL of ice-cold DPBS w/o Ca/Mg over the pinned down stomach.

2. Be careful not to let the curved micro-dissecting scissors cut into the epithelium. To prevent this from happening, hold the scissors in parallel to the stomach as you are cutting away the muscle layer.

3. The antrum will appear more translucent and thinner than the fundus, and thus this distinction will make it easier to delineate where to cut if making region-specific organoids. If making whole-stomach organoids, combine fundus and antrum tissue.

4. If thawing a full bottle of Matrigel™, place in an ice bucket at 4° overnight. Mix gently with a 1000 μL pipet, and aliquot into desired volumes, keeping all materials on ice to prevent Matrigel™ from solidifying.

5. Shake up and down twice per one second for best results. If, after shaking, you do not see gastric glands under the microscope, you can shake again for 1–2 min to increase the yield. This can be done by removing 50 μL from freshly shaken tube to check the density of the gastric glands, underneath a 10× objective on a light microscope. Extended shaking (>5–10 min) will result in low organoid yield, as glands will fragment into single cells.

6. It is easiest to remove the supernatant by vacuum, and then pull off the remaining liquid with a 1000 μL pipet.

7. It is important to avoid air bubbles by first pipetting the desired volume, and then decreasing the volume on the pipet, making sure to completely dissolve the pellet of glands in the Matrigel evenly, while keeping the tube on ice.

8. Do not pipet more than 40 µL on to the membrane because the Matrigel™ will spread instead of forming a domed bubble shape, and organoids will not form.

9. Sphere formation should occur within 24 h of growth.

10. Must pipette slowly to prevent cell breakage.

11. Cells should be passaged after cells have adhered and reached 95–100 % confluence.

12. To passage ISMCs onto the 12-well plate, add 1 mL of cell solution to each well. For example, to seed one plate of cells, create a suspension of 12 mL of volume, for 1 mL per well.

13. To transfer, first remove the DMEM growth medium from the 12-well plate with the ISMCs, then carefully transfer the transwell membrane to the 12-well plate containing the ISMCs, making sure to keep everything sterile in the hood. Make sure to provide fresh organoid growth medium on the top and bottom of the transwell.

14. This means that the organoid growth media will be fully replaced every other day, when the organoids on the Transwell™ inserts are switched to a fresh 12-well plate of ISMCs.

15. The paraformaldehyde must be prewarmed; otherwise the Matrigel™ will dissolve, and the organoids will be lost during the washes for staining.

16. Do not remove liquid by vacuum, instead take great care and use a 1000 µL pipet, always placing new volume down the side of the insert, not directly on top of the organoids.

Acknowledgments

This work was supported by NIH 1R01DK083402 grant (Zavros), NIH 5T32GM105526 grant (Bertaux-Skeirik), and the Albert J. Ryan Fellowship (Bertaux-Skeirik).

References

1. Huch M, Dorrell C, Boj SF et al (2013) In vitro expansion of single Lgr5+ liver stem cells induced by Wnt-driven regeneration. Nature 494:247–250

2. Skardal A, Devarasetty M, Rodman C, Atala A, Soker S et al (2015) Liver-tumor hybrid organoids for modeling tumor growth and drug response in vitro. Ann Biomed Eng 43:2361–2373

3. Huch M, Bonfanti P, Boj SF et al (2013) Unlimited in vitro expansion of adult bi-potent pancreas progenitors through the Lgr5/R-spondin axis. EMBO J 32:2708–2721

4. Sato T, Stange DE, Ferrante M et al (2011) Long-term expansion of epithelial organoids from human colon, adenoma, adenocarcinoma, and Barrett's epithelium. Gastroenterology 141:1762–1772

5. Mahe MM, Aihara E, Schumacher MA et al (2013) Establishment of gastrointestinal epithelial organoids. Curr Protoc Mouse Biol 3:217–240

6. Barker N, Huch M, Kujala P et al (2010) Lgr5(+ve) stem cells drive self-renewal in the stomach and build long-lived gastric units in vitro. Cell Stem Cell 6:25–36

7. Schumacher MA, Aihara E, Feng R et al (2015) The use of murine-derived fundic organoids in studies of gastric physiology. J Physiol 593:1809–1827

8. Bertaux-Skeirik N, Feng R, Schumacher MA et al (2015) CD44 plays a functional role in helicobacter pylori-induced epithelial cell proliferation. PLoS Pathog 11:e1004663

9. Wroblewski LE, Piazuelo MB, Chaturvedi R et al (2015) Helicobacter pylori targets cancer-associated apical-junctional constituents in gastroids and gastric epithelial cells. Gut 64:720–730

10. Sato T, Vries RG, Snippert HJ et al (2009) Single Lgr5 stem cells build crypt-villus structures in vitro without a mesenchymal niche. Nature 459:262–265

11. Jung P, Sato T, Barriga FM et al (2011) Isolation and in vitro expansion of human colonic stem cells. Nat Med 17:1225–1227

12. Schumacher MA, Feng R, Aihara E et al (2015) Helicobacter pylori-induced Sonic Hedgehog expression is regulated by NFκB pathway activation: the use of a novel in vitro model to study epithelial response to infection. Helicobacter 20:19–28

13. Feng R, Aihara E, Kenny S et al (2014) Indian Hedgehog mediates gastrin-induced proliferation in stomach of adult mice. Gastroenterology 147:655–666

14. Stange DE, Koo BK, Huch M et al (2013) Differentiated Troy+chief cells act as reserve stem cells to generate all lineages of the stomach epithelium. Cell 155:357–368

15. Jayewickreme CD, Shivdasani RA (2015) Control of stomach smooth muscle development and intestinal rotation by transcription factor BARX1. Dev Biol 405:21–32

Chapter 4

An Air–Liquid Interface Culture System for 3D Organoid Culture of Diverse Primary Gastrointestinal Tissues

Xingnan Li, Akifumi Ootani, and Calvin Kuo

Abstract

Conventional in vitro analysis of gastrointestinal epithelium usually relies on two-dimensional (2D) culture of epithelial cell lines as monolayer on impermeable surfaces. However, the lack of context of differentiation and tissue architecture in 2D culture can hinder the faithful recapitulation of the phenotypic and morphological characteristics of native epithelium. Here, we describe a robust long-term three-dimensional (3D) culture methodology for gastrointestinal culture, which incorporates both epithelial and mesenchymal/stromal components into a collagen-based air–liquid interface 3D culture system. This system allows vigorously expansion of primary gastrointestinal epithelium for over 60 days as organoids with both proliferation and multilineage differentiation, indicating successful long-term intestinal culture within a microenvironment accurately recapitulating the stem cell niche.

Key words Gastrointestinal tissue culture, Organoid, Three dimensional culture, Air–liquid interface, Stem cell niche

1 Introduction

The intestine is one of the organs with the highest cell turnover, with the intestinal epithelial lining undergoing complete regeneration every 5–7 days. The rapid turnover of intestine is due to the presence of intestinal stem cell (ISC) populations at the crypt base and at other locations such as the crypt "+4" position [1, 2]. Similar to intestine, homeostatic cell renewal of the stomach and pancreas is supported by stem cell populations located in epithelial compartments, the pyloric glands and the pancreatic ducts [3, 4]. The stem cell niche of gastrointestinal organs provides a microenvironment composed of both epithelial and mesenchymal/stromal cells, in which stem cells undergo self-renewal or differentiation.

Because of inherent limitations of 2D epithelial cell culture, there has been increasing interest in using 3D cell culture models to mimic multicellular morphological and functional features of the parental gastrointestinal epithelium. Embryonic organ culture has

Andrei I. Ivanov (ed.), *Gastrointestinal Physiology and Diseases: Methods and Protocols,* Methods in Molecular Biology, vol. 1422, DOI 10.1007/978-1-4939-3603-8_4, © Springer Science+Business Media New York 2016

demonstrated the 3D architecture and cellular differentiation pattern of intestine. However, this in vitro methodology has been restricted to embryonic tissue and has shown limited cellular viability for less than 14 days [5]. Recent studies have described methodology for 3D "organoid" culture of the epithelium from intestinal crypts, pancreatic ducts, stomach, liver, and/or stem cell populations isolated from primary gastrointestinal tissues. These submerged Matrigel cultures requires various growth factor supplementation to supply paracrine signaling [3, 4, 6–9], possibly because of their purely epithelial nature. We have been able to robustly culture intestinal explants containing both epithelial and mesenchymal cells into spheroid-like organoids using an air–liquid interface (ALI) methodology that does not require exogenous growth factor supplementation [10–12]. In this system, we have obtained gastrointestinal organoids from neonatal or adult mouse tissues exhibiting multilineage differentiation and sustained growth for >60 days [10–12]. Furthermore, this ALI method allows robust culture from genetically engineered mice and in vitro Cre-dependent oncogene activation, tumor suppressor inactivation, or viral transduction. This 3D ALI organoid system has tremendous potential for bridging the gap between 2D cell line-based research and animal model-based in vivo studies in gastrointestinal biology.

2   Material

Prepare all solutions using milli-Q water. Filter solutions using vacuum driven filter (pore size: 0.22 μm). Store all reagents at 4 °C.

2.1   Collagen Gel Matrix Components

1. Solution A: Cellmatrix type I-A (Porcine type I collagen, Nitta Gelatin Inc.) or (Cultrex® Rat Collagen I, Trevigen), ready to use. Store at 4 °C (see Note 1).

2. Solution B: 10× concentrated sterile culture medium (Ham's F-12). To make 100 ml solution B, dissolve one bag of Ham's F-12 Nutrient Mix powder (Life Technologies) in 100 ml of sterile milli-Q water and filter through a 0.22 μm Corning filter. Store at 4 °C.

3. Solution C: sterile reconstitution buffer. Make up to 100 ml with sterile milli-Q water, 0.05 N NaOH, 200 mM HEPES, and 2.2 g NaHCO₃. Filter through a 0.22 μm Corning filter. Store at 4 °C (see Note 1).

2.2   3D Culture Assembly Components

1. Inner dish: Millicell culture plate inserts, 0.4 μm, 30 mm (PICM03050, Millicell-CM, Millipore) (see Note 2).

2. Outer dish: Easy Grip tissue culture dish, 60×15 mm (see Note 3).

2.3 Culture Medium and Buffer

1. Culture medium: Ham's F12 supplemented with 20 % fetal bovine serum and 50 μg/ml gentamicin reagent solution (10 mg/ml in distilled water).

2. Freezing medium: 90 % fetal bovine serum and 10 % DMSO.

3. Fixation buffer: 4 % PFA or 10 % buffered formalin phosphate.

2.4 Dissection Apparatus

Iris scissors, forceps (World Precision Instruments) sterilized by autoclaving within instant sealing sterilization pouch.

2.5 Primary Tissues

Intestine, stomach, and pancreas removed from neonatal mice using aseptic procedure (*see* **Note 4**).

3 Methods

3.1 Preparation of the Bottom Layer of Collagen Gel Matrix in the Inner Dish

1. Add ice-cold solution B to A at a volume ratio 1:8 (for total of 10 ml collagen gel: 8 ml solution A, 1 ml solution B) and mix well by pipetting up and down in a 50 ml conical tube on ice.

2. Add 1 volume of ice-cold solution C to the mixture of A and B (for total of 10 ml collagen gel: 1 ml solution C). Mix well on ice until the color of the mixture turns pink.

3. Avoid air bubbles through the procedure. Pipet bubbles out if there are any.

4. Keep the reconstituted collagen solution on ice (4 °C) until use.

5. Pour 1 ml of reconstituted collagen solution into each 30 mm diameter Millicell insert in a tissue culture hood (*see* **Note 5**).

6. Leave inserts in the hood at room temperature for 20–30 min until collagen solidifies completely. To expedite solidification, inserts with collagen can be placed in a 37 °C incubator.

3.2 Preparation of Primary Gastrointestinal Tissues

1. Following aseptic procedure, remove murine gastrointestinal tissues (intestine, stomach, or pancreas) and immediately immerse in ice-cold culture medium such as Ham's F12 medium (without serum) or PBS (*see* **Note 6**).

2. For intestine and stomach, inflate the removed tissue by injecting ice-cold medium into the lumen or the body of stomach. Open the tissue lengthwise and wash in ice-cold culture medium or PBS to remove all luminal contents.

3. Transfer the tissue to the hood. Rinse the tissue with ice-cold medium or PBS 2–3 times.

4. Mince the thoroughly washed tissue with iris scissors on a sterile area such as a tissue culture plate lid. The tissue should be finely minced until it shows viscous and almost homogenous appearance. The final size suitable for culture is represented by tissue fragments approximately 0.3 mm³ or under. This mincing procedure should be finished on ice and should not exceed 5 min to avoid cell damage and drying the tissue.

3.3 Assembly of ALI 3D Primary Culture System

1. Prepare reconstituted collagen solution for tissue-containing top layer as described above (Subheading 3.1). We typically use 1 ml of collagen solution for each 30 mm diameter Millicell insert.

2. Pipette the minced tissue into collagen solution and mix well on ice. Pour 1.0 ml of the tissue-containing collagen gel onto the inner dish with bottom layer gel matrix prepared in Subheading 3.1. A 0.5 cm length of murine gastrointestinal tissue is enough to prepare 1 ml of tissue-collagen gel mixture.

3. Place the completed inner dish (total volume of collagen is 2.0 ml) in a new empty 60 mm outer dish. Transfer the covered outer dish to a 37 °C incubator and allow the gel of the inner dish to solidify for 20–30 min.

4. After solidifying of the top layer tissue-containing gel, add 1.5 ml culture media into the outer dish in the hood. At this point, the culture media should not reach above the top cellular gel layer which allows the formation of the ALI microenvironment (*see* **Note 7**) (Fig. 1).

5. Transfer the culture assembly to an incubator at 37 °C with a humidified atmosphere of 5 % CO_2.

6. Change culture medium every 7 days. If desirable, growth factors or variable test compounds can be added in the culture medium.

7. To check growth of primary gastrointestinal organoids, use a light microscope or a stereomicroscope. Typically, formation of cystic organoids from primary cells can be observed within 3–7 days (Fig. 2).

3.4 Disaggregation, Subculture, and Cryopreservation of Primary Organoids

This procedure should be finished in tissue culture hood to avoid contamination.

1. Recover primary organoids from collagen gel by incubating organoid-containing collagen gel in PBS with 300 U/ml collagenase IV at 37 °C for 20–30 min. Shake the sample every 10 min to expedite disaggregation. Centrifuge at 1000 rpm for 5 min, 4 °C.

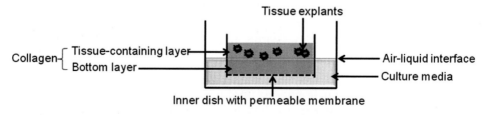

Fig. 1 Schematic representation of 3D culture assembly. The tissue explants are cultured in collagen gels under an air–liquid interface microenvironment

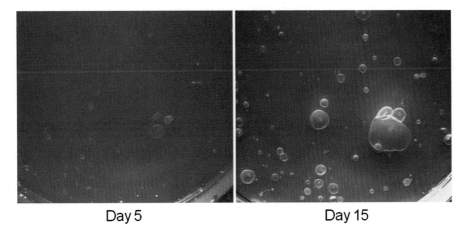

Day 5   Day 15

Fig. 2 Stereomicroscopy of mouse neonatal gastric organoid expansion at day 5 and day 15

2. If single cell suspension is desired, resuspend and incubate the cell pellet in 0.05 % trypsin/EDTA for 5 min at 37 °C. The disaggregation procedure can be followed by flow cytometry or viral transduction (*see* **Note 8**).

3. Add 5 ml of 100 % FBS to the cell suspension. Incubate the sample at room temperature for 5 min. Centrifuge at 1000 rpm for 5 min, 4 °C.

4. Repeat **step 3** two times.

5. If cryopreservation is desired, resuspend the final cell pellet in freezing medium. Keep it in a Corning freezing vial at −80 °C.

6. To subculture, resuspend the cell pellet in reconstituted collagen solution and replate on pre-made bottom layer collagen as described above (Subheading 3.3). We can usually expand one primary culture into 3–4 sub-cultures. Prepare 1 ml of collagen solution for each 30 mm diameter Millicell insert. Assemble the 3D culture system and transfer to a 37 °C incubator.

3.5 Embedding of Primary Organoids for Histological Analysis

1. Fix organoids by immersing the inner dishes in fixation buffer overnight.

2. Rinse inner dishes with PBS.

3. Cut the organoid-containing collagen gel along with membrane from inner dishes using a surgical blade. Place the gel/membrane on a pre-wet paper towel and cut into 4–5 stripes (about 1 cm or thinner). Peel the stripes from paper towel and lay them flat in a cassette. Keep the samples in PBS and proceed to paraffin embedding or generation of frozen section by OCT (*see* **Note 9**).

4. Cross sections can be applied for variable staining methods such as hematoxylin and eosin. The sections can also be used for immunohistochemistry for various antibodies (Fig. 3).

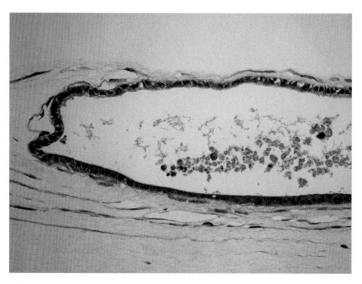

Fig. 3 Hematoxylin and eosin staining of mouse neonatal pancreatic organoids at day 14 of culture

4   Notes

1. Different Cellmatrix type I-A gels may require different pH of solution C for proper reconstitution. For collagen gel from Trevigen (Cultrex® Rat Collagen I, Lower Viscosity (3 mg/ml)), we still use the volume ratio 8:1:1 of Solution A, B, and C to prepare reconstituted collagen solution, but solution C needs to be adjusted to pH 11.0.

2. The inner dish should have a permeable or pored membrane bottom. Besides the Millicell culture plate inserts we usually use, other culture plate inserts with variable size can be used as inner dishes, such as Corning 6.5 mm diameter transwell or HTS 96-well transwell plates. The scale of materials and reagents should be adjusted in accordance with the intended use.

3. The 60 mm outer dish can be replaced by 6-well culture plates if desired. Scale down the culture medium to 1.0 ml.

4. In general, growth from neonatal intestinal tissue (harvested at postnatal day 0-day 7) is most optimal. For wild-type tissue, we have had success with samples from small and large intestine up to 26 weeks of age. We recommend adding R-spondin1 (500 ng/ml) or R-spondin conditional medium to improve initiation of organoids from wild-type adult tissue. Cre activation elicited vigorous growth of adult Apc$^{flox/flox}$ intestinal organoids with or without latent alleles of Kras or p53. Gender does not affect the efficacy of culture. We typically add ~10^7 pfu of adenovirus Cre-GFP to the inner dish for excision of floxed loci.

5. For HTS 96-well transwell plate, we find it difficult to follow a two collagen layer procedure. We usually mix reconstituted collagen with minced tissue or disaggregated organoids and apply 50–60 μl of the mixture directly to each transwell. 120 μl of culture medium will be applied to each outer well.

6. A 1 cm length of murine intestinal tissue is enough to make up to two inner dishes for culture. We don not use any extra steps to enrich epithelial cell population or crypts. The entire piece of tissue (including epithelial layer, lamina propria, and muscles) will be minced for primary plating.

7. It is very important that the culture media in the outer dish does not cover the cultured cells in the inner dish. Compared to submerged culture, the ALI configuration provides greater oxygenation for primary gastrointestinal tissue (*see* Ref. 13).

8. Lentiviral or retroviral transduction are usually completed by spinoculation. Resuspend the cell pellets (Subheading 3.4) in 0.5–1.0 ml of culture media (with 5 μg/ml polybrene) and add virus (~10–20 MOI) directly into the cell suspension. Spin at 2000 rpm, room temperature for 1 h. Leave the conical tubes with loosed cap in a 37 °C incubator for 4–6 h. Spin and replate the cells to collagen after the incubation.

9. Since the sample will be sectioned from the bottom of the cryomold, the sample stripes should be placed in the cryomold accordingly. The samples must be placed on the bottom of the cryomold on the edge of the long side. This will allow generation of optimal cross sections.

Acknowledgments

This work was supported by NIH grants U19AI116484, U01DE025188, U01DK085527, U01CA176299, U01CA151920, P30CA124435 to CJK.

References

1. Barker N, van Es JH, Kuipers J et al (2007) Identification of stem cells in small intestine and colon by marker gene Lgr5. Nature 449:1003–1007

2. Sangiorgi E, Capecchi MR (2008) Bmi1 is expressed in vivo in intestinal stem cells. Nat Genet 40:915–920

3. Barker N, Huch M, Kujala P et al (2010) Lgr5(+ve) stem cells drive self-renewal in the stomach and build long-lived gastric units in vitro. Cell Stem Cell 6:25–36

4. Huch M, Bonfanti P, Boj SF et al (2013) Unlimited in vitro expansion of adult bipotent pancreas progenitors through the Lgr5/R-spondin axis. EMBO J 32:2708–2721

5. Abud HE, Watson N, Heath JK (2005) Growth of intestinal epithelium in organ culture is dependent on EGF signaling. Exp Cell Res 303:252–262

6. Sato T, Clevers H (2013) Growing self-organizing mini-guts from a single intestinal

stem cell: mechanism and applications. Science 340:1190–1194

7. Pylayeva-Gupta Y, Lee KE, Bar-Sagi D (2013) Microdissection and culture of murine pancreatic ductal epithelial cells. Methods Mol Biol 980:267–279

8. Jin L, Feng T, Shih HP et al (2013) Colony-forming cells in the adult mouse pancreas are expandable in Matrigel and form endocrine/acinar colonies in laminin hydrogel. Proc Natl Acad Sci U S A 110:3907–3912

9. Smukler SR, Arntfield ME, Razavi R et al (2011) The adult mouse and human pancreas contain rare multipotent stem cells that express insulin. Cell Stem Cell 8:281–293

10. Ootani A, Li X, Sangiorgi E et al (2009) Sustained in vitro intestinal epithelial culture within a Wnt-dependent stem cell niche. Nat Med 15:701–706

11. Katano T, Ootani A, Mizoshita T et al (2013) Establishment of a long-term three-dimensional primary culture of mouse glandular stomach epithelial cells within the stem cell niche. Biochem Biophys Res Commun 432:558–563

12. Li X, Nadauld L, Ootani A et al (2014) Oncogenic transformation of diverse gastrointestinal tissues in primary organoid culture. Nat Med 20:769–777

13. DiMarco RL, Su J, Yan KS et al (2014) Engineering of three-dimensional microenvironments to promote contractile behavior in primary intestinal organoids. Integr Biol 6:127–142

Chapter 5

Organotypical Tissue Cultures from Fetal and Neonatal Murine Colon

Peter H. Neckel and Lothar Just

Abstract

The complex functions of the gastrointestinal tract rely on the coordinated interplay of several cell and tissue types involving epithelium, connective tissue, smooth muscles as well as cells of the immune and nervous system. It is therefore obvious, that these functions can hardly be investigated sufficiently using cell lines or two-dimensional cell cultures.

Here, we describe an easy to produce three-dimensional organotypical explants culture from fetal and neonatal murine colon. This model is suitable for in vitro testing of intestinal function or the evaluation of developmental or pathological processes.

Key words Organotypical tissue culture, Intestine, Colon, Explants, In vitro model, Organ culture

1 Introduction

Two-dimensional culture systems of intestinal cells dominate the experimental design in biomedical research due to the advantages they offer in terms of handling, analysis, and interpretation of results. However, a major restriction of these techniques is the lack of three-dimensional interactions of cultured cells with the cellular partners they would encounter in vivo [1, 2]. Thus, the establishment of three-dimensional cultures recapitulating structural and functional complexity of specific organs is one of the main challenges in the development of new cell culture models, especially for complex organs composed of many different tissue and cell types like the intestine.

The gut is organized in different distinct tissue layers and is composed of various cell types like epithelial cells, fibroblasts, smooth muscle cells, neurons, or glial cells. In order to establish an organotypical intestinal cell culture model, the different cellular partners have to be arranged in a topographically specific manner resembling the in vivo situation. Therefore, three-dimensional intestinal tissue cultures of various gut regions have been described

Andrei I. Ivanov (ed.), *Gastrointestinal Physiology and Diseases: Methods and Protocols*, Methods in Molecular Biology, vol. 1422, DOI 10.1007/978-1-4939-3603-8_5, © Springer Science+Business Media New York 2016

that were maintained as free-floating cultures [3, 4], cultured on different extracellular matrixes [5–9], or on chick chorioallantoic membrane [10] to provide an in vivo-like environment; for review *see* [11, 12].

In this protocol, we describe an easy to produce three-dimensional culture model for fetal and neonatal murine colon. The intestinal tissues that are cultured on membrane inserts at the air–liquid interface maintain the three-dimensional organization of gut layers for at least 2 weeks without loss of autonomous contractility [7]. This in vitro model can be used to bridge the gap between two-dimensional or cell culture-based testing and animal experiments by offering the advantages of both experimental setups: easy handling and affordability of cell culture and a complex three-dimensional microenvironment that models in vivo complexity of the organ. This model can be used to investigate the mechanisms underlying developmental processes, host–microbe interactions, or their pathological consequences, as well as have several other physiological and toxicological applications.

2 Materials

2.1 Tissue Culture Media

1. Dissection medium: Hanks balanced salt solution (HBSS) without Ca^{2+} and Mg^{2+} supplemented with penicillin (100 units/ml) and streptomycin (0.1 mg/ml).

   We have tested and used two different culture media: A complex culture medium with lower serum concentration and defined growth factors, and a simplified and cheap culture medium that is supplemented with a higher concentration of horse serum (*see* **Note 1**).

2. Complex culture medium: DMEM/F-12 medium supplemented with 10 % (v/v) horse serum, insulin/transferrin/selenium mix (1:100, Life Technologies), 1 mg/ml albumax (Life Technologies), 1 µM hydrocortisone (Sigma-Aldrich), 14 nM glucagon (Sigma-Aldrich), 1 nM triiodthyronine (Sigma-Aldrich), 200 µM ascorbat-2 phosphate, 20 µM linolic acid (Sigma-Aldrich), 10 nM estradiol (Sigma-Aldrich), and 5 ng/ml keratinocyte growth factor (Sigma-Aldrich).

3. Simplified culture medium with higher serum concentration: mix 50 ml of DMEM/F-12, 24 ml of HBSS with Ca^{2+} and Mg^{2+}, 25 ml of horse serum and supplement with 2 mM l-glutamine and additional 0.9 mg/ml $NaHCO_3$.

4. To culture neonatal gut slices, add 100 units/ml penicillin, 0.1 mg/ml streptomycin and optionally, 5 µg/ml ciprofloxacin (Fresenius Kabi Deutschland GmbH) to the complex culture medium.

5. 70 % ethanol.

2.2   Equipment

1. Mcllwain tissue chopper (Mickle Laboratory Engineering CO, Guildford, UK).

2. 6-Well plastic cell culture plates.

3. Millicell cell culture inserts with hydrophilic PTFE membrane (Merck Chemicals GmbH, Schwalbach, Germany) (*see* **Note 2**).

4. Standard cell culture equipment (i.e., cell culture incubator, water bath, balances).

5. Stereo microscope with minimum 50× magnification.

6. Dissecting set (at least one pair of scissors, a set of fine forceps, and two spatulas).

3   Methods

3.1   Tissue Culture Plate Preparation

1. Add 750 µl of culture medium into each well of the plastic 6-well plate and put a Millicell insert into the well (*see* **Note 3**).

3.2   Preparation of Intestinal Slices

All procedures involving animal handling and experimentation must be performed in agreement to international animal welfare guidelines and according to a protocol approved by the Institutional Animal Care and Use Committee.

1. Euthanize fetal and neonatal mice.

2. Immobilize the bodies by pinning them to the substratum.

3. Disinfect the skin of neonatal mice using 70 % ethanol to avoid contamination of the tissue culture.

4. Cut open the skin and abdominal wall to get access to the visceral organs. Use two sets of forceps—one for opening the skin, the other to open the peritoneum and removal of the gut.

5. Remove the entire bowel, transfer it into the Petri dish filled with dissection medium, and carefully remove mesenteries, fat, and other attached tissues in order to unfold the gut.

6. Use the Mcllwain tissue chopper to cut the gut into 250 µm thick transversal slices (Figs. 1 and 2; *see* **Notes 4** and **5**). After slicing by the tissue chopper, the tissue tension of each intestinal slice will alter localization of different tissue layers (Fig. 1).

7. By, using two spatulas, carefully transfer the intestinal slices into a new Petri dish filled with the dissection medium. Avoid damaging the tissue with the spatula (*see* **Note 6**).

8. Use a stereo microscope to evaluate the quality and homogeneity of the intestinal slices. Once you choose the best slices, gently remove rests of mucus and chyme from the intestinal lumen using fine forceps.

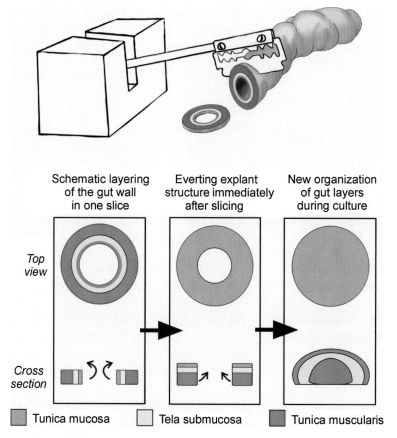

Fig. 1 Generation and remodeling of the intestinal tissue sections. The upper panel illustrates the process of slicing of gut segments using a tissue chopper. The lower left panel depicts the anatomical diagram of the alignment of different tissue layers in the gut slices. Such layer alignment changes during tissue remodeling in cultured intestinal explants. Due to tissue tension of the gut slices, the intestinal mucosa dislocates to the upper and outer surface of the explants immediately after slicing (lower middle panel). After one week in culture, the majority of smooth muscle cells is positioned in the central lower part of the culture, covered by the connective tissue cells and superficial mucosal epithelium (right lower panel)

9. Carefully transfer the intestinal slices to the PTFE membrane using two spatulas (Fig. 2). The slices can be easily moved on and off the spatula with a second scoop by utilizing the cohesion forces of medium surrounding the specimen. This prevents damage to tissue by rude contact with the used instruments. Depending on size and form of your specimen, you can transfer several slices onto one membrane. On our experience, up to five intestinal sections can be cultured on a single membrane (*see* **Notes 7** and **8**).

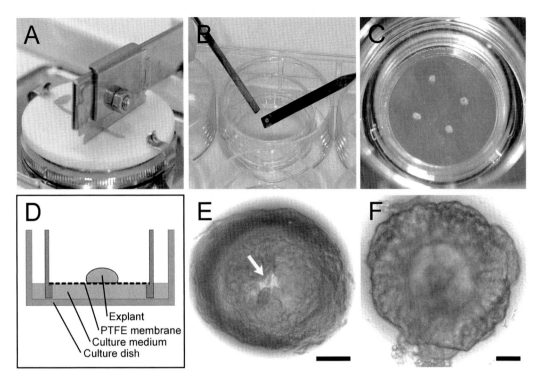

Fig. 2 Preparation of intestinal tissue slices. After cutting of gut segments with a tissue chopper (**a**), slices should be transferred on the membrane inserts using two spatulas (**b**). The cohesion forces of the medium surrounding the specimen can be used to gently move the sample onto the insert membrane, without risking any damage to the tissue. (**c**) Bright-field view of four freshly prepared intestinal slices cultured on a membrane insert. (**d**) Schematic cross-section view demonstrates the principle of tissue culturing at the air–liquid interface. (**e**) Bright-field images of colon explants directly after slicing. The arrow points to the tissue-free area in the center of the culture. (**f**) Intestinal explant after one week in culture. Scale bars: **e** = 200 μm; **f** = 100 μm

3.3 Culture of Intestinal Slices

1. Incubate the slices at 37 °C and 5 % CO_2 in a humidified incubator (*see* **Note 9**).

2. Change culture medium every second day (*see* **Note 10**). In order to do that, prewarm the culture medium in a water bath to 37 °C, lift the insert from the well using sterile forceps, add 750 μl of fresh medium into the well, and put back the insert trying to avoid trapping air bubbles under the PTFE membrane.

3. Carefully wash the surface of the intestinal explants every fourth day with 1 ml of culture medium to remove mucus and shed epithelial cells.

4. In order to study cell implantation into intestinal mucosa, add desired cells to the center of a freshly cut intestinal slice using a sterile micropipette (Fig. 3). Such cell implantation should be performed during the first 24 h after preparation of intestinal slices [7]. During tissue remodeling in isolated slices, grafted cells become embedded into the intestinal explant. A similar protocol can be used to deliver drugs or nanoparticles into the organotypical tissue culture.

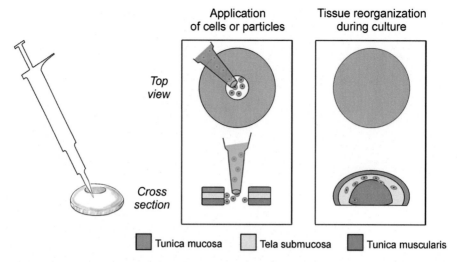

Fig. 3 Implantation of cells into the center of freshly prepared intestinal slice using a micropipette. After tissue remodeling within the isolated slice, grafted cells (*green*) became incorporated into the intestinal tissue

4   Notes

1. If your experimental setup interferes with high concentration of the serum, use the low serum complex culture medium.

2. Since the PTFE membrane is very thin and transparent, intestinal specimens cultured on this membrane can be examined and imaged using bright field and phase contrast microscopy.

3. It is important that the bottom surface of the PTFE membrane is uniformly wet and no air bubbles are trapped underneath the membrane insert.

4. Slice thickness is a liming factor for the air–liquid tissue cultures (on our experience, 250 μm is an optimal thickness; a good range is 200–400 μm). The insufficient supply of nutrients and oxygen will lead to cell necrosis in the center of the slices that are too thick.

5. Cutting extremely thin (<200 μm) slices has a disadvantage, since it produces asymmetric intestinal tissue sections that are difficult to examine and image.

6. It is important that the freshly cut slices do not dry out on the plate of the tissue chopper. Avoid damaging the intestinal tissue with the spatulas.

7. Do not load too many samples onto one membrane. Closely located specimens could be pulled together by cohesion forces.

8. After transfer of slices onto PTFE membrane, avoid any subsequent movement of the tissue.

9. The use of antibiotics in the culture medium is not necessary for fetal cultures. For neonatal specimens, we strongly suggest to add penicillin/streptomycin mixture and (optionally) ciprofloxacin to the tissue culture medium.

10. Change medium quickly to avoid drying intestinal tissue explants.

References

1. Barrila J, Radtke AL, Crabbe A, Sarker SF, Herbst-Kralovetz MM, Ott CM, Nickerson CA (2010) Organotypic 3D cell culture models: using the rotating wall vessel to study host-pathogen interactions. Nat Rev Microbiol 8:791–801

2. Freshney RI (2010) Three-dimensional culture. In: Culture of animal cells. John Wiley & Sons Inc., Hoboken, New Jersey, pp 481–495

3. Autrup H (1980) Explant culture of human colon. Methods Cell Biol 21:385–401

4. DeRitis G, Falchuk ZM, Trier JS (1975) Differentiation and maturation of cultured fetal rat jejunum. Dev Biol 45:304–317

5. Altmann GG, Quaroni A (1990) Behavior of fetal intestinal organ culture explanted onto a collagen substratum. Development 110:353–370

6. Bareiss PM, Metzger M, Sohn K et al (2008) Organotypical tissue cultures from adult murine colon as an in vitro model of intestinal mucosa. Histochem Cell Biol 129:795–804

7. Metzger M, Bareiss PM, Nikolov I, Skutella T, Just L (2007) Three-dimensional slice cultures from murine fetal gut for investigations of the enteric nervous system. Dev Dyn 236:128–133

8. Quinlan JM, Yu WY, Hornsey MA, Tosh D, Slack JM (2006) In vitro culture of embryonic mouse intestinal epithelium: cell differentiation and introduction of reporter genes. BMC Dev Biol 6:24

9. Schiff LJ (1975) Organ cultures of rat and hamster colon. In Vitro 11:46–49

10. Pomeranz HD, Gershon MD (1990) Colonization of the avian hindgut by cells derived from the sacral neural crest. Dev Biol 137:378–394

11. Bermudez-Brito M, Plaza-Diaz J, Fontana L, Munoz-Quezada S, Gil A (2013) In vitro cell and tissue models for studying host-microbe interactions: a review. Br J Nutr 109:S27–S34

12. Randall KJ, Turton J, Foster JR (2011) Explant culture of gastrointestinal tissue: a review of methods and applications. Cell Biol Toxicol 27:267–284

Chapter 6

Ussing Chamber Technique to Measure Intestinal Epithelial Permeability

Sadasivan Vidyasagar and Gordon MacGregor

Abstract

Epithelial cells are polarized and have tight junctions that contribute to barrier function. Assessment of barrier function typically involves measurement of electrophysiological parameters or movement of non-ionic particles across an epithelium. Here, we describe measurement of transepithelial electrical conductance or resistance, determination of dilution potential, and assessment of flux of nonionic particles such as dextran or mannitol, with particular emphasis on Ussing chamber techniques.

Key words Tight junction, Paracellular permeability, Transepithelial resistance, Diffusion potential

1 Introduction

Epithelia are sheets of cells that line body cavities and external surfaces in multicellular organisms. Epithelial cells have tight junctions that act as physical and chemical barriers to prevent seep of chemicals and maintain electrical gradients while allowing selective transport of solutes and water between internal and external compartments. The formation, quality, and permeability of tight junctions determines the resistance and integrity of the epithelial tissue [1]. Barrier function is largely studied using three methods: [1] measurement of transepithelial electrical conductance (G_t); [2] measurement of dilution potential, in which transepithelial voltage changes are recorded after exposure to solutions of different ionic strength; and [3] measurement of junctional permeability involving measurements of fluxes of non-ionic particles.

The Ussing method of measuring active ion transport and leaking resistance in epithelial tissues was first published in 1946. The method is nondestructive and label free and is an effective means of assessing barrier function. Epithelia can be classified as leaky or tight on the basis of electrical resistance measurements. For many years, the epithelium of choice for study of barrier function was

Andrei I. Ivanov (ed.), *Gastrointestinal Physiology and Diseases: Methods and Protocols*, Methods in Molecular Biology, vol. 1422, DOI 10.1007/978-1-4939-3603-8_6, © Springer Science+Business Media New York 2016

frog skin, which had negligible flux through the intercellular space. The transepithelial resistance (R_t) for frog skin is typically ~40,000 $\Omega \cdot cm^2$. Such barriers are relatively impermeable and are called tight epithelia. However, epithelia lining the small intestine have very low R (~4–80 $\Omega \cdot cm^2$) due to high intercellular permeation; these epithelia are considered leaky. Unlike the tight epithelia, leaky epithelia have a robust ability to transport water and electrolytes.

Transepithelial transport may occur via two routes: transcellular, in which molecules are transported through the cell, crossing the apical, and basolateral plasma membranes; and paracellular, which refers to transport between cells, via the cell junctions or the intercellular space [2] (Fig. 1). Transepithelial resistance is the sum of the transcellular resistance (R_{tc}) and the paracellular resistance (R_{pc}). Transcellular resistance (R_{tc}) is the sum of the resistances from the apical cell membrane (R_a) and the basolateral cell membrane (R_b). Paracellular (R_p) resistance is equal to the sum of the apical tight junction resistance (R_{tj}) and the lateral space resistance (R_{ls}).

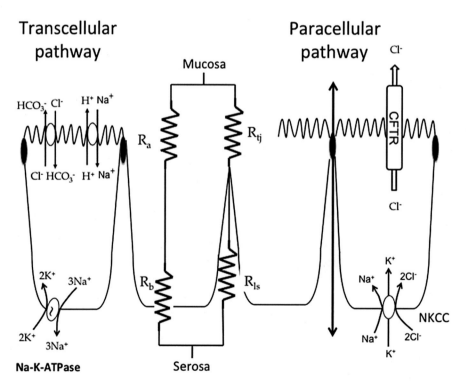

Fig. 1 A schematic diagram of the electrical circuit model of the series resistances across the trans- and paracellular pathways of an epithelial cell monolayer. Transcellular transport is controlled by transporters in the apical and basolateral surfaces and is largely responsible for the transcellular conductance (G_t). The resistance of the transcellular pathway ($R_a + R_b$) series elements is typically much higher than that of the parallel elements of the paracellular pathway in a leaky epithelium like the small intestine. Thus, the overall resistance of an epithelium is defined by R_{pc}, which is defined by the composition of the tight junction (TJ)

$$R_T = R_{pc} + R_{tc}$$

$$R_{pc} = R_{tj} + R_{ls}$$

$$R_{tc} = R_a + R_b$$

In parallel circuits, $\dfrac{1}{R_e} = \dfrac{1}{R_1} + \dfrac{1}{R_2}$

Therefore, $\dfrac{1}{R_T} = \dfrac{1}{R_{pc}} + \dfrac{1}{R_{tc}} = \dfrac{1}{R_{tj} + R_{ls}} + \dfrac{1}{R_a + R_b}$

and $R_T = \dfrac{(R_a + R_b)(R_{tj} + R_{ls})}{(R_{tj} + R_{ls}) + (R_a + R_b)} = \dfrac{1}{G_T}$, where G_T = transepithelial resistance

Therefore, $G_T = \dfrac{(R_{tj} + R_{ls}) + (R_a + R_b)}{(R_a + R_b)(R_{tj} + R_{ls})}$

Leaky epithelia are those with a transepithelial resistance (R_t) less than $1000\ \Omega \cdot cm^2$, those in which transcellular resistance (R_{tc}) is greater than paracellular resistance (R_{pc}), or those in which the paracellular conductance (G_{pc}) is greater than 50 % of the total tissue conductance (G_t) [3]. In "leaky" epithelia such as those found in the intestine, the ionic conductance through the paracellular pathway (as opposed to the transcellular pathway) accounts for >90 % of total G_t [4].

Electrophysiological measurements based on the Ussing system include transmembrane voltage (V_t), transepithelial membrane resistance (R_t), and short circuit current (I_{sc}). Transepithelial conductance (G_t) is calculated using Ohm's law, with $G_t = I_{sc}/V_t$. I_{sc} is the sum of the charges carried per unit time through the membrane by the actively transported ions (Fig. 2). This is equal to the current that must be applied to drop the membrane potential to zero, i.e., to equal concentrations of ions on either side of the membrane. Thus, the current is a direct measurement of the active transport ability for the membrane of interest. This measurement is called the short-circuit current, since it has exactly the same numerical value as the current which could be expected to flow from one membrane surface to the other through an ideal short-circuiting connection [5]. When the epithelium is bathed in identical solutions and is voltage clamped to zero, there is no net passive diffusion of ions, and there is no driving force for paracellular transport of ions or water. The G_t, or its reciprocal R_t, is a useful measure of the leakiness of a tissue. As such, G_t (or R_t) is a good measure of tissue viability or integrity of the intestinal preparation. G_t measurements also enable determination of net ion movement, based on the unidirectional fluxes of ions measured in individual tissues. G_t (or R_t) is also used to pair tissues based on similar conductance or resistance (~5 % difference is acceptable) during isotope flux

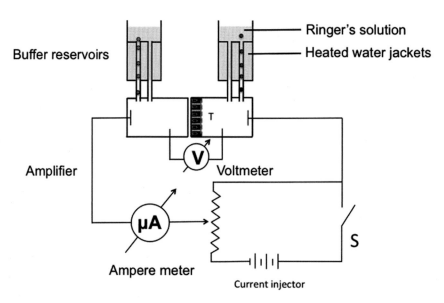

Fig. 2 Electrical circuit showing the transepithelial electrical measurements in the Ussing chamber setup. The buffer reservoirs are water jacketed to maintain uniform temperature. The buffer is gassed and uniformly circulated using a bubble lift system. The amplifier was used to perform the electrical measurements

studies to calculate net flux (J_{net}) from unidirectional mucosal-to-serosal flux (J_{ms}) and mucosal-to-serosal flux (J_{sm}), with $J_{net} = J_{ms} - J_{sm}$. G_t is further discussed below in relation to paracellular conductance (R_{pc}) across the intestinal epithelium.

The tight junction forms a semipermeable gate that restricts paracellular diffusion in a charge- and size-selective manner [6]. Measurements of paracellular flux of nonionic particles are often used to estimate intestinal permeability under a variety of conditions. The particles used for these assays are typically, fluorescent dyes conjugated to hydrophilic solutes (e.g., mannitol or dextran) that are not transcellularly transported. These particles can be used directly in Ussing chambers or can be administered to laboratory animals such as rats or mice via gastric gavage. Particles of differing sizes can be used to assess paracellular permeability. For example, fluorescein isothiocyanate (FITC)-dextran particles at sizes of 4 kDa, 10 kDa, and 40 kDa can be used to determine the level of damage to intestinal villi.

2   Materials

2.1   Harvesting and Mounting Intestinal Epithelial Tissue

1. Rat colon.

2. Ringer's solution: 140 mM Na^+, 119.8 mM Cl^-, 25 mM HCO_3^-, 2.4 mM HPO_4^-, 0.4 mM $H_2PO_4^-$, 1.2 mM Mg^{2+}, and 1.2 mM Ca^{2+}.

3. Razor blades.

4. Glass slides.

5. Ussing chambers.

2.2 Measuring G_t, R_t, and Dilution Potential of Intestinal Tissue

1. The VCC-MC8 Multichannel Voltage current clamp (Physiological Instruments, San Diego, CA) or the DVC-1000 Dual voltage current clamp (WPI, Sarasota, FL) are instruments that function as power supply, ammeter, voltmeter, rheostat, and switch.

2. Silver/silver chloride electrodes (*see* **Note 1**): Dissolve 4 % agar in Ringer's solution by boiling till the agar become golden brown. Pour the hot golden-brown agar into a silver/silver chloride pellet electrode for voltage sensing or into a silver chloride wire electrode, previously plated with chloride, for current sensing. To chloride plate, pass a current through the electrode and another silver wire, both immersed in a solution of 3 M potassium chloride, at a rate of 1 mA/cm^2. This causes chloride ions to be deposited on the electrode. Electroplate till the electrode becomes dark gray/black (*see* **Note 2**). Alternatively, electroplating can be done by immersing the electrode in household bleach (*see* **Note 3**).

3. Salt solutions: A simple NaCl-based solution can be used for measurements, containing: 150 mM NaCl, 2 mM CaCl$_2$, 1 mM MgCl$_2$, 10 mM glucose, 10 mM HEPES, pH 7.2, osmolarity 300 mOsm/L. The solution is warmed to 37 °C and gassed with 100 % O$_2$. In the case of the small intestine, the glucose can be replaced with glutamine to avoid transepithelial potential generated by Na-dependent glucose transport. For NaCl$_{0.5}$ solution, 75 mM of the NaCl is replaced with 150 mM mannitol or sucrose to maintain equal chamber osmolarity.

2.3 Assessing Epithelial Permeability to Nonionic Solutes

1. [^{3}H] mannitol.

2. FITC-dextran (molecular weight, 4 kDa, 10 kDa, or 40 kDa).

3 Methods

3.1 Harvesting and Mounting Intestinal Epithelial Tissue

1. We will describe using the distal part of the rat colon for studying paracellular permeability of the intestinal epithelium. The entire procedure should be carried out at temperatures below 4 °C by bathing the tissues in ice-cold Ringer's solution until the Lucite chambers are fixed onto the reservoirs of the Ussing chamber. For use of cultured epithelial cells *see* **Note 4**.

2. Create a midline incision and mobilize the distal colon of an adult rat from the surrounding tissues after cutting open the pelvic bones for easy approach.

3. Remove the colon and flush it with ice-cold Ringer's solution to remove the colonic contents.

4. Cut the colon along the mesenteric border with a small pair of scissors and fix it on a glass plate with a clip, with the serosal side facing up. Perianal skin can be used to fix the colon in order to maximize the length of the intact mucosa.

5. Spread the colon uniformly on the side using a finger wetted with Ringer's solution. Pull the proximal end of the colon with a glass slide to stretch the colon.

6. Make a longitudinal cut to the level of the submucosa using sharp razor blades. Carefully remove the muscular and serosal layers along the length of the colon.

7. Clip the remaining colon (partial strip; Fig. 3) to the glass plate with the mucosa facing up. Stretch the colon again lengthwise.

8. Make longitudinal cuts across half the thickness of the remaining colon and remove the submucosal tissue from the mucosa using a razor blade. With a finger wetted in Ringer's solution, slowly push the mucosa along its entire length, taking extreme care at Peyer's patches (aggregated lymphoid nodules) to achieve a continuous strip of the colon. At the distal end, cut the submucosa from the mucosa, leaving an intact mucosal sheet (Fig. 4).

9. Transfer the mucosa, mucosal side down, to a wax board covered with a clean plastic sheet. Stretch the mucosal layer in all directions and anchor it with pins. Push the Ussing chamber half containing its own pins into the stretched mucosa. Cut away the plastic sheet, leaving a wide margin around the perimeter of the Ussing chamber. Carefully remove the plastic sheet, taking care to leave the mucosa in place by holding the mucosal sheet against the Ussing chamber.

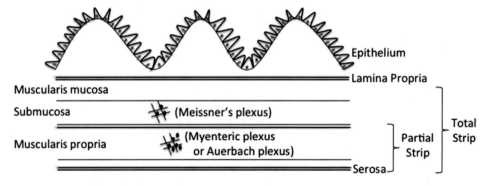

Fig. 3 Schematic drawing showing the layers of the colonic wall, indicating the layers that are removed in partial and total strips. Removal of the partial strip removes only the myenteric (Auerbach's) plexus, whereas removal of the total strip removes both Auerbach's and Meissner's (submucosal) nerve plexuses. Removal of the total strip yields a mucosal sheet with the epithelium, lamina propria, and a small portion of muscularis mucosa

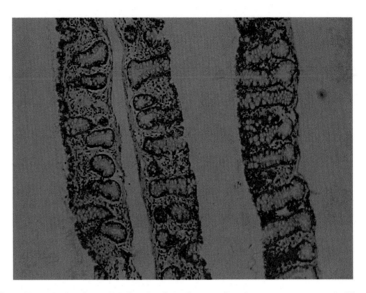

Fig. 4 Three strips from the distal colon of a rat, showing mucosa separated from underlying muscularis and serosa. Note the intact surface cells and crypts and lamina propria extending up to muscularis mucosa with no submucosa or underlying layers. These strips are mounted vertically between the two halves of the Ussing chamber

3.2 Measuring G_t and R_t of Intestinal Tissue

1. Mount epithelial tissue in an Ussing chamber so that both sides of the tissue are bathed with an equal volume of pre-warmed Ringer's solution (*see* **Note 5**). Maintain the tissue at 37 °C, 5 % CO_2, and 95 % O_2. Gassing allows adequate oxygenation and also for continuous mixing of the solution in the reservoir—a bubble-lift technique for mixing.

2. Connect a voltmeter in parallel and an ammeter in series with the tissue (Fig. 2).

3. Note the transmural potential difference (PD) on the voltmeter when the switch is open. This is the V_t (Fig. 2).

4. Close the rheostat and adjust the current till the voltmeter reads zero PD across the membrane.

5. Note the current output on the ammeter. The current required to bring the potential difference to zero is the I_{sc}.

6. Calculate G_t by dividing the I_{sc} by the V_t. $R_t = 1/G_t$. For considerations in interpreting results, *see* **Note 6**.

3.3 Measurements of the Dilution Potential

1. Configure the Ussing chamber so that V_2 is in contact with solution bathing the apical membrane of the intestine, and V_1 is in the basolateral chamber (Fig. 5). The reported open-circuit voltage will be $V_2 - V_1$ with respect to the blood side of the epithelium. In open-circuit mode, adjust the transepithelial PD to 0 mV.

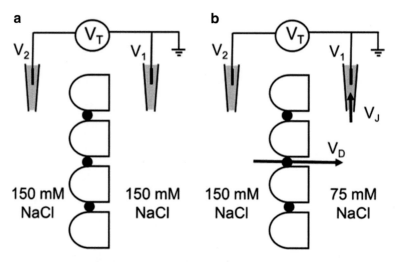

Fig. 5 Schematic diagram of the Ussing chamber, voltage electrodes, and epithelium. (**a**) The gastrointestinal epithelium is shown with the apical membrane on the *left* and the basolateral membrane on the *right*. The voltage electrode V_2 is placed in the apical compartment and V_1 in the basolateral compartment, and the difference, V_T, is calculated by $V_2 - V_1$. In the case of symmetrical 150 mM NaCl solutions, all electrode and junction potentials are nulled and $V_T = 0$ mV. (**b**) Replacing the basolateral Ussing chamber compartment with 75 mM NaCl (NaCl$_{0.5}$), will produce a measurable transepithelial potential (V_T). This is measured as the difference between the potential difference produced by the inequalities in the permeability of sodium and chloride through the paracellular pathway, known as the dilution potential (V_D), and the potential drop that occurs at the 3 M KCL electrode due to the different solution (V_J)

2. Replace the basolateral solution with one that contains twofold lower concentration of NaCl (NaCl$_{0.5}$); five partial subtotal replacements of 5 ml is sufficient for a complete solution exchange. Wait for about 30 s or until voltage measurement has stabilized.

3. Note transepithelial potential V_t, and immediately change the basolateral solution back to full NaCl solution (*see* **Note 7**). This transepithelial potential (V_t), is the sum of the potential difference generated by the transepithelial Na and Cl permeability, called the diffusion potential (V_D), and the liquid junction potential (V_L) generated at the 3 M KCl agar silver/silver chloride electrode (V_2) in contact with the basolateral NaCl$_{0.5}$ bath solution. This liquid junction potential of the V_2 electrode can be measured using the open circuit mode of the voltage clamp amplifier, using a 3 M KCl agar bridge to connect the sealed off hemi chambers of the Ussing apparatus (*see* **Note 8**).

4. The ion permeability ratio (β) for the tissue can be calculated using the Goldman–Hodgkin–Katz equation [7], where β is the ratio of the permeability of the tight junctions to chloride over the permeability to sodium ($\beta = P_{Cl}/P_{Na}$).

5. Calculate the dilution potential (V_D) using the following equation:

$$V_D = -\frac{RT}{F}\ln\left[\frac{(\alpha + \beta)}{1 + \alpha\beta}\right]$$

with $\beta = \dfrac{\alpha - x}{\alpha x - 1}$ and, $x = e^{-\left(\frac{VF}{RT}\right)}$

with $\beta = \dfrac{\alpha - x}{\alpha x - 1}$ and, $x = e^{-\left(\frac{VF}{RT}\right)}$ The activity coefficient for 150 mM NaCl is calculated at 0.752 and for 75 mM it is calculated as 0.797; therefore, $\alpha = 1.89$ and $RT/F = 26.71$ mV at 37 °C (*see* **Note 9**).

6. Further analyses and the absolute tissue permeability of sodium (P_{Na}) and chloride (P_{Cl}) can be calculated from the conductance measurement and a simplified Kimizuka–Koketsu equation [8].

3.4 Measuring Permeability to Nonionic Solutes Using Ussing Chambers

1. Add either fluorescent tracers (e.g., FITC dextran) or a small amount of radiolabeled substance (e.g., [³H] mannitol) to the Ringer's solution for both the serosal and the luminal baths, in equal concentrations in separate sets of tissues. Optimal concentrations of the tracers should be determined experimentally based on the sensitivity of detection equipment.

2. Equilibrate for 30–45 min, so that the labeled solute reaches a steady-state rate of flux into the opposite bath. Collect samples from the chamber opposite that to which the particles were added.

3. Measure [³H] mannitol in samples using a liquid scintillation counter (Beckman LS6500, USA). For FITC-dextran, determine concentrations by spectrophotofluorometry, with an excitation wavelength of 485 nm (20 nm bandwidth) and an emission wavelength of 528 nm (20 nm bandwidth).

4 Notes

1. Silver/silver chloride electrodes are universally used in physiologic techniques, as they have very stable potentials at a variety of currents and are therefore considered "nonpolarizable." These electrodes are also reversible, meaning they can pass current bidirectionally [9]. Only the silver chloride component may come into contact with the experimental solutions.

2. A drift in junction potential and/or a change in the color of the silver/silver chloride electrode suggests degradation due to use or storage. Degraded electrodes should be replated. When electroplating a previously chloride-plated silver wire, care must be taken to ensure that the chloride is evenly distributed. To achieve uniform plating, complete removal of residual silver

chloride is necessary. This can be achieved by an electrical method or a mechanical method. With the electrical method, the polarity of the electrode is reversed prior to making the silver wire positive in the chloride solution. This polarity reversal is maintained until the silver wire is stripped of all brown coating (~10 s). Electroplating then proceeds as described above. Mechanically, all previously coated chloride is scraped from the electrode with a blunt razor blade, taking care that the entire surface is again shiny.

3. Electroplating can also be done by immersing the electrode in household bleach (Clorox) for 15–30 min until the dark brown color is obtained. The electrodes are then rinsed in Ringer's solution prior to being filled with agar. In our hands, drifting junction potentials are more frequently observed with electrodes made using bleach; we therefore prefer the current-plating technique.

4. Chopstick electrodes (STX2 from World Precision Instruments, USA) in combination with an epithelial voltohmmeter (e.g., EVOM2 from World Precision Instruments, USA) can be used for measurements of transepithelial voltage and resistance in tissue culture wells. The lengths of these electrodes are unequal, so that the longer (external) electrode touches the bottom of the dish containing the external culture media while the shorter (internal electrode) ends above the membrane in the tissue culture cup or insert. The longer electrode ensures proper positioning between the electrode and the cell layer in the cup during the transmembrane measurement, critical if reproducible measurements are to be made. Electrolytes are also added to a blank cup or insert to control for the resistance across the membrane of the tissue culture insert or cup. As the resistance is inversely proportional to the area of the tissue, the resistance is reported as the product of resistance and area ($\Omega \cdot cm^2$). Endohm chambers (World Precision Instruments) in combination with an epithelial voltohmmeter (EVOM2, World Precision Instruments) can be used in place of the chopstick electrodes. The chamber and the cap of an Endohm chamber each contain a pair of concentric electrodes: a voltage-sensing silver/silver chloride pellet in the center plus an annular current electrode. The height of the top electrode can be adjusted to fit cell culture cups made by different manufacturers. Endohm's symmetrically opposing circular disc electrodes, situated above and beneath the membrane, allow a more-uniform current density to flow across the membrane than is achieved with the STX2 electrodes. The background resistance of a blank insert is reduced from 150 Ω (when using the hand-held STX2 electrodes) to less than 5 Ω, as per

the manufacturer. Resistance measurements can also be made using special sliders for use with Snapwell chambers, e.g., P2302, Physiologic Instruments, USA or CHM5 from World Precision Instruments, USA using the classical Ussing chambers.

5. Any effect of hydrostatic pressure on net ion movement is eliminated by maintaining equal volumes of solutions on both sides of the tissue. Osmotic influences are removed by keeping concentrations of all individual ions equal on both sides of the tissue. Ion movement driven by transmural PD is eliminated by measuring fluxes under short-circuit conditions (a current is passed through the tissues in the direction opposite to the current to bring the transmural potential difference to zero). Removal of these influences will eliminate contributions from leakage, facilitated diffusion, and simple diffusion, and thus the net ion flux will presumably reflect only the active transport mechanisms.

6. It should be noted that while transepithelial resistance (R_t) measurements primarily reflect changes in paracellular conductance in a leaky tissue, ~10 % of the changes in resistance could be due to transcellular conductance of ions (i.e., mediated by apical and basolateral transporters under voltage clamp settings). A final factor to keep in mind when interpreting G_t and R_t measurements is the volume of the lateral paracellular space. Ion conductance through the paracellular space is limited by both the tight junctional complex and the relative apposition of the basolateral membranes of adjacent epithelial cells. These factors determine the volume of the surrounding aqueous column, i.e., the lateral intercellular space. The fluid in the lateral intercellular space can be affected by active chloride secretion, e.g., in cAMP-activated chloride secretion, which decreases the volume of the lateral intercellular space, or by an absorptive state, e.g., in epithelial sodium channel (ENaC)-mediated sodium absorption, which increases the volume of the lateral intracellular space. Since paracellular resistance is the sum of the junctional resistance and the resistance along the paracellular space, a collapse of the paracellular space can decrease paracellular permeability and alter R_t. Hence, it is important that the G_t measurements and the dilution potential measurements described below be carefully correlated with morphological analysis.

7. The dilution potential measurements can also be performed in the normal bicarbonate Ringer's solution, however, in this case the chambers are gassed with 95 % O_2/5 % CO_2. This would also necessitate the use of transport inhibitors, such as bumetanide (100 µM, serosal), and inhibitors of basolateral potassium channels, to reduce the transepithelial PD to close to 0 mV. In complete bicarbonate Ringer's solution there will

also be interference from other permeable cations and ions, such as bicarbonate, depending on how selective the paracellular pathway is for the ions.

8. Dilution potentials should also be measured by apical compartment replacement with the $NaCl_{0.5}$ solution, although in this case the polarity of the diffusion potential and junction potential will be reversed.

9. The paracellular permeability of gastrointestinal epithelia is usually moderately sodium selective with a $NaCl_{0.5}$ dilution potential of around –7 mV ($V_2 - V_1$ with respect to the blood side of the epithelium), which corresponds to a $P_{Cl}/P_{Na} \approx 0.4$, which makes it about 2.5 times more permeable to sodium than chloride (Fig. 6). Experimental modulations (i.e., radiation treatment, drugs, diet, disease) or genetically modified animals may show increased sodium permeability ratio or a switch in the epithelia resulting in the tight junctions being predominately chloride selective, i.e., +3.5 mV, which corresponds to a $P_{Cl}/P_{Na} \approx 1.5$, which makes the paracellular pathway about 1.5 times more permeable to chloride than sodium.

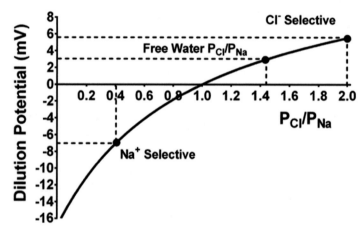

Fig. 6 Theoretical plot of the Goldman–Hodgkin–Katz equation for a 150 mM NaCl/75 mM NaCl dilution potential experiment. Equation $\left(V_D = -\dfrac{RT}{F} \ln\left[\dfrac{(\alpha + \beta)}{1 + \alpha\beta} \right] \right)$ is used to calculate the P_{Cl}/P_{Na} ratios for observed dilution potentials. Three typical data points are plotted, on the left, –7 mV, a usually observed dilution potential for gastrointestinal epithelia, showing about 2.5 times more permeability of sodium than chloride ions. The middle data point, +2.8 mV, would indicate that the paracellular pathways are water filled channels, being about 1.4 times more permeable to chloride than sodium ions, identical to the ratio observed for free water permeability. In the far right data point, a dilution potential of +5.5 mV would indicate a chloride selective transepithelial pathway, being about twice as permeable for chloride as sodium ions

References

1. Farquhar MG, Palade GE (1963) Junctional complexes in various epithelia. J Cell Biol 17:375–412

2. Fromter E, Diamond J (1972) Route of passive ion permeation in epithelia. Nat New Biol 235:9–13

3. Powell DW (1981) Barrier function of epithelia. Am J Physiol 241:G275–G288

4. Frizzell RA, Schultz SG (1972) Ionic conductances of extracellular shunt pathway in rabbit ileum. Influence of shunt on transmural sodium transport and electrical potential differences. J Gen Physiol 59:318–346

5. Eigler FW (1961) Short-circuit current measurements in proximal tubule of Necturus kidney. Am J Physiol 201:157–163

6. Sourisseau T, Georgiadis A, Tsapara A, Ali RR, Pestell R, Matter K, Balda MS (2006) Regulation of PCNA and cyclin D1 expression and epithelial morphogenesis by the ZO-1-regulated transcription factor ZONAB/DbpA. Mol Cell Biol 26:2387–2398

7. Hille B (1984) Ionic channels of excitable membranes. Sinauer, Sunderland, MA

8. Yu AS, Cheng MH, Angelow S, Gunzel D, Kanzawa SA, Schneeberger EE, Fromm M, Coalson RD (2009) Molecular basis for cation selectivity in claudin-2-based paracellular pores: identification of an electrostatic interaction site. J Gen Physiol 133:111–127

9. Berman JM, Awayda MS (2013) Redox artifacts in electrophysiological recordings. Am J Physiol Cell Physiol 304:C604–C613

Chapter 7

HPLC-Based Metabolomic Analysis of Normal and Inflamed Gut

Daniel J. Kao, Jordi M. Lanis, Erica Alexeev, and Douglas J. Kominsky

Abstract

The idiopathic inflammatory bowel diseases, which include Crohn's disease and ulcerative colitis, are multifactorial chronic conditions that result in numerous perturbations of metabolism in the gastrointestinal mucosa. Thus, methodologies for the qualitative and quantitative analysis of small molecule metabolites in mucosal tissues are important for further elucidation of mechanisms driving inflammation and the metabolic consequences of inflammation. High-performance liquid chromatography (HPLC) is a ubiquitous analytical technique that can be adapted for both targeted and non-targeted metabolomic analysis. Here, protocols for reversed-phase (RP) HPLC-based methods using two different detection modalities are presented. Ultraviolet detection is used for the analysis of adenine nucleotide metabolites, whereas electrochemical detection is used for the analysis of multiple amino acid metabolites. These methodologies provide platforms for further characterization of the metabolic changes that occur during gastrointestinal inflammation.

Key words High-performance liquid chromatography, Electrochemical detection, Adenosine, Tryptophan, Inflammatory bowel disease, Intestinal epithelial cell

1 Introduction

While nuclear magnetic resonance (NMR) and liquid chromatography-mass spectrometry (LC/MS) are powerful techniques for metabolic fingerprinting of complex samples, traditional high-performance liquid chromatography (HPLC) continues to be indispensable for profiling of specific metabolites. HPLC retains its advantages as being a rapid, quantitative, adaptable, and readily accessible method for analysis of simple to moderately complex metabolite samples. In some cases, however, HPLC may be inadequate to resolve complex mixtures or may be unable to resolve certain metabolites depending on their physicochemical properties. HPLC separates complex mixtures into individual compounds through partitioning of analytes between mobile and stationary phases. The quality of separation is determined by the differences in interactions of the analytes with each of the phases. Once

Andrei I. Ivanov (ed.), *Gastrointestinal Physiology and Diseases: Methods and Protocols,* Methods in Molecular Biology, vol. 1422, DOI 10.1007/978-1-4939-3603-8_7, © Springer Science+Business Media New York 2016

compounds have been separated, they are then passed through a detector where an electrical signal corresponds to the amount of compound present [1]. The advantage of HPLC lies in its adaptability, as this technique has the ability to separate, identify, and quantify compounds present in any sample that can be dissolved in liquid. HPLC is more versatile than gas chromatography since it is not limited to volatile and thermally stable samples, and the choice of mobile and stationary phases is wider [1]. In this chapter, we will discuss HPLC with ultraviolet (UV) detection and HPLC with electrochemical detection, both of which have been used for metabolomics studies in IBD.

The metabolism of adenine nucleotides has been carefully examined due to the important role of Ado in the inflamed gastrointestinal mucosa [2]. The analysis of the adenine nucleotide metabolites lends itself well to reversed-phase HPLC (RP-HPLC) with UV detection for multiple reasons [3–7]. First, purines are easily detected by UV absorbance and adenosine has an absorbance peak around 260 nm. This allows for UV detection of adenine nucleotides with less interference from protein absorbance peaks, which occur around 190–201 nm and 280 nm for peptide bonds and aromatic amino acids, respectively. Second, adenine nucleotide metabolites can be efficiently separated by a number of chromatographic methods, allowing for quantitative profiling of the metabolites. Finally, the lower complexity of the extracellular metabolites allows for more straightforward species identification of the metabolites, especially in the case of in vitro systems.

While HPLC is an effective analytical technique for the profiling of adenine nucleotide metabolites, there are obstacles that need to be considered before applying this technique. While not a limitation of the technique itself, one consideration in the profiling of adenine nucleotide metabolites is uptake of metabolites by cells. For example, adenosine is actively and passively transported intracellularly by epithelial cells and red blood cells within minutes to seconds by equilibrative and concentrative nucleoside transporters. Therefore, when profiling these metabolites, it may be desirable to either inhibit the transporters using dipyridamole and S-(4-nitrobenzyl)-6-thioinosine (NBTI) [8] or by using derivatized adenine nucleotide analogs that are not substrates for these transporters [7]. Depending on the application, it may be preferable to use the endogenous metabolites and not interfere with the transporters in order to determine a more relevant steady-state concentration of the metabolites.

Derivatization of adenine nucleotide metabolites can be used to both enhance detection of the metabolites as well as block their transport in order to accurately determine their metabolism. Using the technique as described by Barrio et al., chloroacetaldehyde can be used to make $1,N^6$-etheno-derivatized adenine nucleotides [9].

These can be used either as exogenous nucleotides to profile the activities of the nucleotide-metabolizing enzymes or endogenous adenine nucleotide metabolites can be derivatized for analysis. In the method described in this chapter, exogenous $1,N^6$-etheno derivatives are used to monitor the extracellular metabolism of adenine nucleotides by cultured intestinal epithelial cells.

While UV detection of compounds is used in traditional HPLC methods, HPLC coupled with electrochemical detection (ECD) is an alternative detection platform affording significant advantages compared to traditional detection methods. For instance, ECD is much more selective, providing the ability to measure metabolites that are difficult to separate [10]. Additionally, ECD represents an extremely sensitive detection platform with a broad dynamic range [11]. The basic HPLC principles are the same; analytes are separated using a column and eluted in a mobile phase before flowing through a detector. In electrochemical detection, the detector is an electrode that applies a specific voltage to the passing compounds, causing them to be oxidized or reduced. Upon oxidation, compounds release electrons to the transfer of electrons create an electrical current, which is relative to the concentration of the analyte.

Interest in simple quantifiable methods such as ECD-HPLC has increased as the significance of tissue metabolism during inflammation has been realized. This technique enables the detection of a number of compounds including amino acids, biogenic amines (dopamine, epinephrine), hormones and hormone metabolites, caratenoids, retinoids, vitamins (A, B6, folic acid, C, D, E, K), purines, pyridine, and sulfides (glutathione). Metabolic processes through compounds such as tryptophan, creatine, vitamin metabolites, and catecholamines have been found to have a substantial role in IBD. Glover et al. found that the phosphocreatine–creatine kinase energy circuit is crucial for proper barrier function, and becomes dysregulated during inflammation [12], while intestinal increases in the neurotransmitter norepinephrine can lead to inflammation through increased bacterial translocation [13, 14]. Additionally, use of this platform has been utilized to characterize the role of retinoic acid in regulatory T cell (T reg) differentiation [15]. Finally, tryptophan catabolites can affect mucosal immunity through interaction with the aryl hydrocarbon receptor (AHR), which controls the production of the anti-inflammatory cytokine IL-22 [16, 17].

In this chapter, methods for the analysis of adenine nucleotide metabolites using HPLC with UV detection are presented. These are followed by methods utilizing HPLC with electrochemical detection for the analysis of an array of amino acids, biogenic amines including catecholamines, and other amino acid catabolites, using tryptophan and kynurenine as examples.

2 Materials

2.1 Tissue Culture Components

1. Growth media: Maintain T84 human colonic carcinoma cells in Dulbecco's Modified Eagle Medium (DMEM)/Ham's F-12 (1:1) media with l-Glutamine and 15 mM HEPES supplemented with 10 % (v/v) iron-supplemented bovine calf serum (HyClone, Logan, UT, USA), 1 % (v/v) penicillin/streptomycin and 1 % (v/v) GlutaMAX.

2. Cell culture plates: 6-well polystyrene tissue culture-treated cell culture dishes.

2.2 1,N⁶-Ethenoadenosine Assay

1. 1,N^6-Ethenoadenosine (Sigma-Aldrich, St. Louis, MO), 100 mM in DMSO (see **Note 1**).

2. 1,N^6-Etheno-5′-monophosphate (Sigma-Aldrich, 100 mM in DMSO.

3. 1,N^6-Etheno-5′-triphosphate, 5 mM in 50 mM Tris–HCl, pH 7.5.

4. Hank's buffered salt solution (HBSS), modified, without phenol red and sodium bicarbonate with 10 mM HEPES, pH 7.4. Dissolve contents of packet in 900 ml ultrapure water. Add 10 ml 1 M HEPES. Adjust pH to 7.4 with sodium hydroxide. Adjust final volume to 1000 ml. Filter-sterilize using a bottle top 0.2 μm PES filter.

5. Bottle top filter, 500 ml, 0.22 μm PES.

6. 1N Hydrochloric acid.

7. Syringe filter, 0.2 μm PVDF.

2.3 RP-HPLC-UV

1. Instrument: Agilent Technologies liquid chromatography system with Infinity 1260 quaternary pump VL, Infinity 1260 ALS autosampler, Infinity 1290 thermostat, Infinity 1260 thermostated column compartment, and Infinity diode array detector (Agilent Technologies, Santa Clara, CA, USA).

2. Column: Phenomenex Luna C18 [2] 150×4.6 mm ID, 5 μm particle size, 100 Å pore size (see **Note 2**).

3. Method 1 (Isocratic): Buffer: Ultrapure water:Acetonitrile (96:4), 0.5 mM tetrabutylammonium bisulfate (TBAS). Dissolve 170 mg TBAS (FW 339.5 g/mol) in 900 ml ultrapure water. Add 40 ml HPLC grade acetonitrile and then add ultrapure water for a total volume of 1000 ml. Sterile filter using a bottle-top 0.2 μm PES filter.

4. Method 2 (Gradient) [3]: Buffer A: 30 mM KH_2PO_4 0.8 M TBAS, pH 5.45. Dissolve 4.8 g monobasic potassium phosphate (M.W. 136.1 g/mol) and 272 mg TBAS in 900 ml ultrapure water. Adjust pH to 5.45 using 1 M potassium hydroxide (KOH). Add ultrapure water to make 1 L. Sterile

filter using a bottle-top 0.2 μm PES filter. Buffer B: 30 mM KH_2PO_4 0.8 M TBAS, pH 7.0 in equal volume acetonitrile. Dissolve 2.04 g monobasic potassium phosphate and 136 mg TBAS in 450 ml ultrapure water. Adjust pH to 7.0 using 1 M KOH and adjust total volume to 500 ml. Sterile filter using a bottle top 0.2 μm PES filter into a 1 L bottle. Add 500 ml acetonitrile.

5. Software: OpenLAB CDS, Chemstation Edition Rev C.01.06 (Agilent Technologies).

2.4 ECD-HPLC

1. Perchloric acid diluted to 6 % with ultrapure H_2O. Prepare 100 ml.

2. Potassium hydroxide, 3 M. Dissolve 16.83 g of KOH into 100 ml of ultrapure water.

3. MDTM buffer. Dissolve 20.67 g NaH_2PO_4 and 0.735 g 1-octanesulfonic acid into 1.5 L of ultrapure water. Add 200 μl of triethylamine and 200 ml of HPLC grade acetonitrile. Mix and bring to pH 3.0 with phosphoric acid. Bring up to 2 L with H_2O.

4. L-Tryptophan standard, 100 μM. Dissolve 20.423 mg of reagent grade L-tryptophan into 1 ml of 0.5 M HCl.

5. L-kynurenine standard, 100 μM. Dissolve 20.821 mg of L-kynurenine into 1 ml of 0.5 M HCl.

6. Instrument: Coularray Model 5600A, 16-channel electrochemical detector with Coularray Data Station 3.00, ESA 584 solvent delivery system, and ESA 540 autosampler (*see* **Note 3**).

7. Column: Acclaim Polar Advantage II C18 column—5 μm 120Å, 4.6 × 150 mm (*see* **Note 4**).

3 Methods

3.1 Tissue Culture

1. Grow T84 colonic epithelial cells to confluence in T-75 flask. Feed every 2–3 days while in culture.

2. Dissociate cells with 0.25 % trypsin using standard methods and seed in 6-well plate at an initial density of 2×10^5 cells/well.

3. Feed every 3–4 days. Monolayers will be ready for $1,N^6$-ethenoadenosine assay when confluent.

3.2 $1,N^6$-Ethenoaenosine Assay

1. Warm HBSS in 37 °C water bath for 30 min.

2. Prepare 1 ml of working stock of eAMP for each well to be used in the assay and one additional ml to be used as a standard. Dilute 100 mM stock eAMP solution 1:1000 in HBSS for a final concentration of 100 μM. Keep working stock at 37 °C (*see* **Note 5**).

3. Prepare 1 ml of working stock of eATP for each well to be used in the assay. Dilute 5 mM stock eATP solution 1:1000 in HBSS for a final concentration of 100 μM. Keep working stock at 37 °C.

4. Aspirate media from each well of the 6-well plate.

5. Carefully pipette 2 ml of pre-warmed HBSS into each well. Swirl and aspirate wash solution.

6. Repeat wash step one additional time.

7. Pipette 1 ml of the working stock of eAMP, eATP, or HBSS onto monolayer.

8. Return plate to 37 °C incubator for 10 min, or for the desired length of time (*see* **Note 6**).

9. Pipette 1 ml of supernatant in a fresh 1.5 ml microcentrifuge tube.

10. Acidify each sample by adding 10 μl of 1N HCL. Invert to mix and put on ice.

11. Filter each sample with a fresh 0.2 μm syringe filter.

12. Analyze samples on HPLC immediately or store at −80 °C until time of analysis.

3.3 RP-HPLC-UV

1. Place appropriate mobile phase solvents in appropriate buffer compartment and place the appropriate inlet tubing in the bottles.

2. Equilibrate column with at least 20 column volumes of buffer. For a 150×4.6 mm ID column, equilibrate for at least 35 min at a flow rate of 1 ml/min.

3. Create method(s) as follows: Method 1—Flow rate: 1 ml/min. Isocratic method, 100 % buffer A 0–14 min. 5 min post-run. Temperature: 25 °C. Injection volume: 40 μl (*see* **Note 7**); Method 2—Flow rate: 1 ml/min (constant). Gradient: Initial 10 % Buffer B. Starting at 0.5 min, start a linear gradient of 4 % B/min up to 20 % B at 3 min. At 7 min, start a linear gradient of 7.5 % B/min up to 50 % B at 11 min. At 21 min, start a linear gradient of −10 % B/min down to 10 % B at 25 min. Post-run time 20 min at 10 % B, flow rate 1 ml/min. Temperature: 25 °C. Injection volume 40 μl. Detection: 210 and 260 nm (*see* **Note 8**).

4. Run the method without injecting any sample onto the column. If the baseline chromatogram is normal, proceed with analysis of samples.

5. Pipette 60 μl of each sample into sample vials and place in autosampler tray, noting the location and identity of each sample. Include blank samples with acidified HBSS for baseline subtraction later (Fig. 1, *see* **Note 9**).

6. Assign sample locations in the sequence table according to the positions of the samples in the sample tray.

7. Run sequence.

3.4 Analysis of RP-HPLC-UV Chromatograms

1. Using the data analysis software package included with the HPLC system, subtract the chromatogram tracing of the acidified buffer without analytes from each of the chromatograms to generate chromatograms with flat baselines (Fig. 1).

2. Integrate chromatograms according to the data analysis software provided with the system (*see* **Note 10**).

3. Using chromatograms of known standards, calculate concentrations of analytes for each chromatogram (*see* **Note 11**).

3.5 PCA Extraction of Metabolites from Cells and Tissues for ECD-HPLC Analysis

1. Wash cells in ice-cold PBS and incubate cells with 6 % PCA, enough to just cover the cells, approximately 50–100 μl per cm², on ice for 5 min. Scrape cells and pipette supernatant and debris into microfuge tubes. For tissue, mechanically homogenize tissue in 6 % PCA, followed by sonication (*see* **Note 12**).

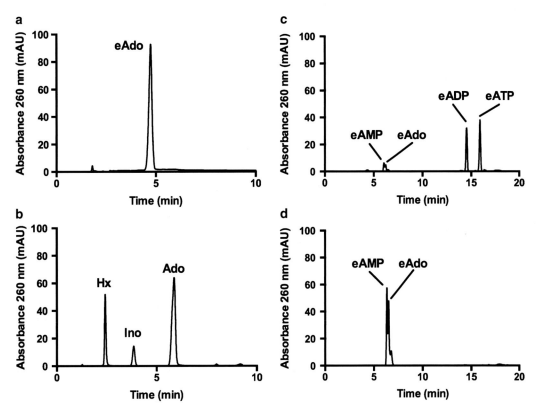

Fig. 1 HPLC-UV chromatograms of adenine nucleotide metabolites. (**a**) Conversion of 1,N⁶-etheno-5′-adenosine monophosphate (eAMP) to 1,N⁶-ethenoadenosine (eAdo) by T84 colonic carcinoma cells. Monolayers were treated with 100 μM eAMP for 10 min at 37 °C. Supernatants were analyzed using Method 1. As shown, eAdo is retained on the C18 column RP-HPLC column, but eAMP is not retained. (**b**) Adenosine and its purine metabolites inosine (Ino) and hypoxanthine (Hx) analyzed using Method 1. (**c**) Conversion of 1,N⁶-etheno-5′-adenosine triphosphate (eATP) to 1,N⁶-etheno-5′-adenosine diphosphate (eADP), eAMP, and eAdo by T84 monolayers. Monolayers were treated with 100 μM eATP for 10 min at 37 °C. Supernatants were analyzed using Method 2. (**d**) Conversion of 1,N⁶-etheno-5′-adenosine monophosphate (eAMP) to 1,N⁶-etheno-5′-adenosine diphosphate (eADP), eAMP, and eAdo by T84 monolayers. Monolayers were treated with 100 μM eAMP for 10 min at 37 °C. Supernatants were analyzed using Method 2

2. Centrifuge samples for 5 min at $12{,}000 \times g$ at 4 °C in order to remove the precipitated proteins and lipids (*see* **Note 13**).

3. Add supernatant to 1/3 volume of 3M KOH (e.g. 100 μl of KOH added to 300 μl of PCA) and mix tubes by inverting gently. Centrifuge at $12{,}000 \times g$ for 5 min to separate the potassium salts and transfer the supernatants to fresh tubes.

3.6 Separation and Identification of Metabolites via ECD-HPLC

1. Power on all instruments and manually set the minimum pressure on the pumps to 10 Bar. This will help prevent any damage to the electrodes in case of flow rate issues. Set the maximum pressure to 300 Bar using the Control tab within the data station program (*see* **Note 14**).

2. Manually flush the pumps with MDTM buffer to ensure there are no bubbles in the system, then turn on the pumps using the data station and equilibrate the C18 column for 30 min at a flow rate of 1 ml/min.

3. Set the voltages on the electrodes from 0 mV, with every channel 50 mV higher than the previous channel, up to 750 mV on channel 16. Let the electrodes equilibrate until the voltages stabilize, at least 1 h (*see* **Note 15**).

4. Meanwhile, prepare combined standards of tryptophan and kynurenine by serial dilution of the 100 mM stocks 1:1000 in MDTM to equal 100 μM, then serial 1:10 dilutions to 10, 1, 0.1, and 0.01 μM.

5. Set up and name your new "Study" in the data station. Create a new sample batch for the standards using an isocratic flow of MDTM buffer over 20 min at 1.5 ml/min, and detect the peaks with the electrodes set at the previously described voltages (Fig. 2).

6. Using the data station analysis software, open up the chromatogram from the middle concentration, analyze, and label the peaks. Kynurenine is the first peak at approximately 6 min, and tryptophan toward the end around 16 min (Fig. 2).

7. Using the data station analysis software, create a calibration table and indicate the concentrations of the standards. Analyze the remaining standards. The standard peaks should now all be identified and used to create the calibration curve. This information will be used by the program to identify and measure the concentration of tryptophan and kynurenine in the unknown samples (Fig. 3).

8. Prepare to analyze the unknown samples by diluting the extracts 1:2 in MDTM buffer. Set up a new sample batch within the same study that contains the calibrated standards and label the unknowns as appropriate. Collect data from the unknowns using the same isocratic method: 1.5 ml/min over 20 min, 0–750 mV in 50 mV increments (Fig. 4).

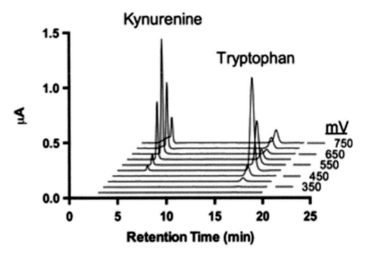

Fig. 2 ECD-HPLC chromatograms of tryptophan and kynurenine. 10 μM mixed standards of tryptophan and kynurenine in MDTM mobile phase were separated and identified by ECD-HPLC. Each metabolite has a different retention time (*x*-axis) on the column and has a different redox potential from the electrodes. The chromatograms from channels 7–16 are shown and represent 300–750 mV, and the resulting current from the eluting compounds is reported as μAmps (*y*-axis). The peak representing kynurenine has the most specific peak at 650 mV with a retention time of 6 min and tryptophan has the strongest signal at 500 mV and is labeled at 16 min

9. Once the data have been recorded, use the data analysis software to analyze the unknown samples. Any peaks matching the retention time and voltage of the standards indicated in the calibration table will be labeled and the concentrations interpolated based on peak size (Fig. 5, *see* **Note 16**).

4   Notes

1. If $1,N^6$-ethenoadenine compounds are not commercially available, they may be synthesized from adenine nucleotides and adenosine through a reaction with chloroacetaldehyde [3, 9].

2. The method described is not dependent on the use of this specific RP-HPLC column. If another column is readily available, a set of standard etheno-derivatized compounds can be analyzed to test its suitability.

3. This method is not dependent on the use of this specific solvent delivery system and detector, however, optimization may be necessary for use with other systems.

4. This protocol has been optimized for use with this column. Use of a different C18 column could affect back pressure, flow rates, and retention time.

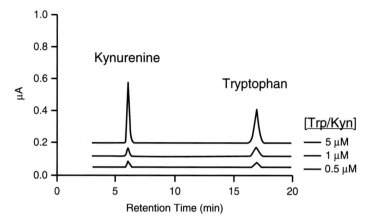

Analyte	Conc. (μM)	Retention (min)	Area (μC)	Height nA
Kynurenine	5	6.175	9.41	777
	1	6.05	1.47	125
	0.5	6.083	0.654	57.0
Tryptophan	5	16.617	17.1	623
	1	17.0	4.24	144
	0.5	16.675	1.93	81.4

Fig. 3 Standard curve analysis of tryptophan and kynurenine by ECD-HPLC. 0.5, 1, and 5 μM mixed standards of tryptophan and kynurenine in MDTM mobile phase were separated and identified by ECD-HPLC. A simplified chromatogram combining each compound's peak detection potential (500, 650 mV) is shown for simplicity (x-axis: retention time, y-axis: μA). The Coularray data station calculates the peak height and area from these chromatograms, and uses the known concentrations of the standards to create a standard curve

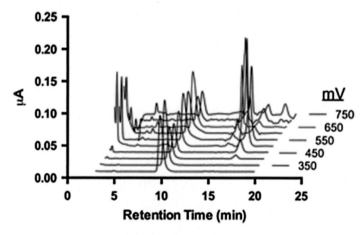

Fig. 4 ECD-HPLC chromatograms of T84 IEC extracts. PCA extracts from T84 cells were separated and identified by ECD-HPLC. The chromatograms from electrode potentials of 300 mV through 750 mV are shown from *bottom to top*, with the retention time on the x-axis and current in μA on the y-axis. Many more peaks are detected in the cell extract compared to the standards in Fig. 3. The peak detection potentials of 500 mV for tryptophan and 650 mV for kynurenine are highlighted in *blue* and *red* respectively

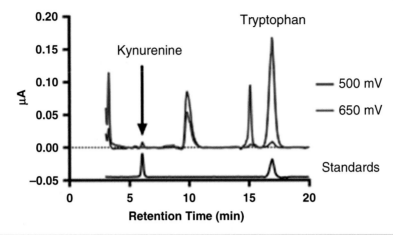

Analyte	Retention (min.)	Area (µC)	Height (nA)	*Extrapolated Concentration (µM)
Kynurenine	6.0	0.117	8.65	0.09
Tryptophan	16.54	5.37	147	2.7

Fig. 5 Overlay ECD-HPLC chromatograms of T84 IEC extracts and standards. The chromatogram from this figure is simplified to show only the peak detection potentials of tryptophan at 500 mV in *blue* and kynurenine at 650 mV in *red*. A chromatogram from the standard curve has also been aligned, and the matching peaks for kynurenine and tryptophan in the T84 sample are labeled. The Coularray data station uses the peak potentials and retention time to match compounds in the unknown samples to those in the standard curve. The peak height and area are used to extrapolate the concentrations of analytes that correspond to the standards

5. Derivatized 5′-AMP or 5′-ATP (or both) can be used in this assay depending on the aspect of the metabolic pathway of interest.

6. The time course of metabolism of adenine nucleotides will vary depending on the particular system and conditions being used. Typically, the reaction proceeds to a significant degree within 5–30 min.

7. While Method 1 is a rapid, isocratic method, there are limitations under the described conditions. 1,N^6-Etheno-5′-monophosphate and 1,N^6-Etheno-5′-triphosphate are not retained on the column. If quantification of 1,N^6-Ethenoadenosine alone is adequate, this method will suffice. If quantification of adenine nucleotides is desired, Method 2 is recommended.

8. In the described methods, a standard in-line UV diode array detector is used. To increase sensitivity and for complex samples, a fluorescence detector is recommended with an excitation wavelength of 280 nm and emission wavelength of 410 nm.

9. Depending on the sensitivity of the HPLC and detector and the amount of reactants and products being analyzed, injection volume can be increased or decreased to produce chromatograms with adequate resolution and quantifiable detection.

10. Details of the integration procedure will vary from one HPLC system to another. Typically, default parameters for integration will produce satisfactory results.

11. Typically, the amount of nucleotide is normalized to protein content of the cell monolayer. This can be determined by harvesting the monolayer after completion of the assay and quantifying total protein using a Bradford assay or bicinchoninic acid assay.

12. Tissue samples may be flash frozen in liquid nitrogen and stored at −80 °C before processing. Be sure to weigh the tissue before processing to ensure standardization.

13. It is important to remove the supernatant quickly, as the proteins and lipids may begin to resolubilize.

14. These maximum pressures are based off of the ESA solvent delivery system and Acclaim column. If using different equipment, adjust the max pressure accordingly.

15. Always ensure that the pumps are running and mobile phase is flowing through the cells when the electrodes are turned on. Operation of the cells without buffer flow can cause permanent damage to the electrodes.

16. Occasionally, small peak size, drift, or other interferences can prevent the unknown peaks from being matched with the standards. In order to overcome this, the peak detection parameters can be adjusted, or in some cases it may be necessary to manually fit the peak.

References

1. Hagan RL (1994) High-performance liquid chromatography for small-scale studies of drug stability. Am J Hosp Pharm 51:2162–2175

2. Colgan SP, Fennimore B, Ehrentraut SF (2013) Adenosine and gastrointestinal inflammation. J Mol Med 91:157–164

3. Bhatt DP, Chen X, Geiger JD, Rosenberger TA (2012) A sensitive HPLC-based method to quantify adenine nucleotides in primary astrocyte cell cultures. J Chromatogr B Analyt Technol Biomed Life Sci 889–890:110–115

4. Chunn JL, Young HWJ, Banerjee SK, Colasurdo GN, Blackburn MR (2001) Adenosine-dependent airway inflammation and hyperresponsiveness in partially adenosine deaminase-deficient mice. J Immunol 167:4676–4685

5. Knudsen TB, Winters RS, Otey SK, Blackburn MR, Airhart MJ, Church JK, Skalko RG (1992) Effects of (R)-deoxycoformycin (pentostatin) on intrauterine nucleoside catabolism and embryo viability in the pregnant mouse. Teratology 45:91–103

6. Lennon PF, Taylor CT, Stahl GL, Colgan SP (1998) Neutrophil-derived 5′-adenosine monophosphate promotes endothelial barrier function via CD73-mediated conversion to adenosine and endothelial A2B receptor activation. J Exp Med 188:1433–1443

7. Synnestvedt K, Furuta GT, Comerford KM, Louis N et al (2002) Ecto-5′-nucleotidase

(CD73) regulation by hypoxia-inducible factor-1 mediates permeability changes in intestinal epithelia. J Clin Invest 110:993–1002

8. Ward JL, Tse CM (1999) Nucleoside transport in human colonic epithelial cell lines: evidence for two Na$^+$-independent transport systems in T84 and Caco-2 cells. Biochim Biophys Acta 1419:15–22

9. Barrio JR, Secrist JA III, Leonard NJ (1972) Fluorescent adenosine and cytidine derivatives. Biochem Biophys Res Commun 46:597–604

10. Tsunoda M (2006) Recent advances in methods for the analysis of catecholamines and their metabolites. Anal Bioanal Chem 386: 506–514

11. Rozet E, Morello R, Lecomte F, Martin GB, Chiap P, Crommen J, Boos KS, Hubert P (2006) Performances of a multidimensional on-line SPE-LC-ECD method for the determination of three major catecholamines in native human urine: validation, risk and uncertainty assessments. J Chromatogr B Analyt Technol Biomed Life Sci 844:251–260

12. Glover LE, Bowers BE, Saeedi B, Ehrentraut SF et al (2013) Control of creatine metabolism by HIF is an endogenous mechanism of barrier regulation in colitis. Proc Natl Acad Sci U S A 110:19820–19825

13. Green BT, Lyte M, Kulkarni-Narla A, Brown DR (2003) Neuromodulation of enteropathogen internalization in Peyer's patches from porcine jejunum. J Neuroimmunol 141:74–82

14. Brown DR, Price LD (2008) Catecholamines and sympathomimetic drugs decrease early Salmonella Typhimurium uptake into porcine Peyer's patches. FEMS Immunol Med Microbiol 52:29–35

15. Collins CB, Aherne CM, Kominsky D, McNamee EN et al (2011) Retinoic acid attenuates ileitis by restoring the balance between T-helper 17 and T regulatory cells. Gastroenterology 141:1821–1831

16. Monteleone I, Rizzo A, Sarra M, Sica G et al (2011) Aryl hydrocarbon receptor-induced signals up-regulate IL-22 production and inhibit inflammation in the gastrointestinal tract. Gastroenterology 141:237–248

17. Zelante T, Iannitti Rossana G, Cunha C, De Luca A et al (2013) Tryptophan catabolites from microbiota engage aryl hydrocarbon receptor and balance mucosal reactivity via interleukin-22. Immunity 39:372–385

Chapter 8

NMR-Based Metabolomic Analysis of Normal and Inflamed Gut

Daniel J. Kao, Jordi M. Lanis, Erica Alexeev, and Douglas J. Kominsky

Abstract

Crohn's disease and ulcerative colitis, the two major forms of idiopathic inflammatory bowel disease (IBD), are thought to occur through a loss of intestinal barrier leading to an inappropriate immune response toward intestinal microbiota. While genome-wide association studies (GWAS) have provided much information about susceptibility loci associated with these diseases, the etiology of IBD is still unknown. Metabolomic analysis allows for the comprehensive measurement of multiple small molecule metabolites in biological samples. During the past decade, metabolomic techniques have yielded novel and potentially important findings, revealing insight into metabolic perturbations associated with these diseases. This chapter provides metabolomic methodologies describing a nuclear magnetic resonance (NMR)-based non-targeted approach that has been utilized to make important contributions toward a better understanding of IBD.

Key words Metabolomics, Nuclear magnetic resonance, Inflammatory bowel disease, Intestinal epithelium, Mucosa

1 Introduction

Chronic inflammatory and immune responses are characterized by significant shifts in metabolic activity. These changes are exemplified in inflammatory bowel diseases (IBD). The two main subtypes of IBD, Crohn's disease (CD) and ulcerative colitis (UC), are currently areas of intense investigation. Though they may overlap in symptoms and disease location, both are distinct diseases with variations in clinical, pathological, and immunological features [1, 2]. These debilitating disorders of unknown etiology [3] are characterized by chronic inflammation of the human gastrointestinal tract, driven by aberrant interactions between the immune system and the intestinal microbiota [4, 5].

IBD is characterized by the breakdown of the intestinal epithelial barrier leading to increased exposure of the mucosal immune system to luminal antigens. It is thought that this exposure

Andrei I. Ivanov (ed.), *Gastrointestinal Physiology and Diseases: Methods and Protocols*, Methods in Molecular Biology, vol. 1422, DOI 10.1007/978-1-4939-3603-8_8, © Springer Science+Business Media New York 2016

promotes inflammation and continued mucosal injury [1, 6]. Sites of ongoing inflammation are associated with substantial changes in tissue metabolism that result in part from perturbations of vasculature [7, 8], leading to the decreased supply of oxygen and depletion of nutrients [9–11]. Further, continued inflammation leads to the generation of large quantities of reactive nitrogen and oxygen intermediates [12].

Several lines of evidence have implicated the involvement of metabolic changes in the pathogenesis of IBD. For instance, insults initiated by inflammatory response result in decreased O_2 supply, decreased ATP generation, and activation of intestinal epithelial HIF-1α, a transcription factor that functions as a global regulator of oxygen homeostasis and is a central orchestrator of all cellular metabolism [13]. During mucosal inflammation, HIF plays a protective role [14]. Cultured epithelial cells subjected to hypoxia and animal models of inflammation have shown that the HIF-regulated transcriptional profile promotes intestinal epithelial barrier function [15].

Genetic information concerning IBD has grown rapidly, as there have been numerous associations between IBD and specific gene variants. Using genome-wide association studies (GWAS), more than 163 independent autosomal genetic risk loci have been identified for IBD, specifically for the two major IBD subtypes CD and UC [16]. Further, there is considerable evidence of support that the dysregulation of the normally controlled immune response to commensal bacteria in genetically susceptible individuals drives IBD [17]. GWAS have helped support the idea that there are several susceptible genes common in IBD; however, the etiology of the disease remains unknown. Consequently, it is vital to continue to explore more comprehensive analytical tools for studying IBD. Mass spectrometry and nuclear magnetic resonance are methodologies that allow for the analysis of multiple metabolites in biological samples and have become invaluable tools for metabolomics analysis.

Mass spectrometry, in essence, measures the molecular mass of chemical compounds and their fragmentation products. This technique entails the ionization and fragmentation of compounds into smaller molecules that are then detected and quantified [18]. Further, gas chromatography (GC) and liquid chromatography (LC) are commonly used interfaces for MS to enable the physical separation of metabolites prior to MS analysis and enhance the detection of individual analytes [19]. Technological advances have dramatically improved analyte detection and reduced overall cost, making LC/MS a widely available approach for metabolomic analysis that does not necessarily require dedicated expertise.

NMR measures the magnetic resonance of atomic nuclei in molecules. The resonance frequencies of nuclei are influenced by the nature and number of surrounding nuclei, and as a result, the frequency and pattern of resonance provides a spectroscopic signature for each metabolite based on its chemical structure [20]. Unlike MS, identification of metabolites can be performed without prior separation of compounds in the sample. However, not all metabolites can be individually identified if their signals overlap and signals of metabolites present at lower concentrations may be overwhelmed by signals from more abundant metabolites [21].

While MS and NMR are both useful methodologies for metabolomics analysis, each of these technologies has distinct advantages and disadvantages. To date, NMR has been the dominant platform for metabolomic analysis and thus will be the focus of the methods presented here. The primary advantages of NMR are (1) minimal sample preparation; (2) non-selectivity in metabolite detection and ability to quantify multiple metabolites; (3) the ability to provide highly reproducible results; and (4) the non-destructive nature of the platform, allowing samples to be further analyzed. The most significant drawbacks of the NMR platform are the lower level of sensitivity compared to MS as well as higher costs.

In the past decade, a number of studies have applied NMR-based metabolomics to the study of IBD. The most utilized NMR-active isotope in these studies is proton (^{1}H-NMR), which enables for detection of all proton-containing low molecular weight metabolites with a limit of detection of approximately 10 μM. Additionally, ^{31}P-NMR allows for the detection of metabolites that provide insight into energy or phospholipid metabolism [22–24]. Finally, ^{13}C-NMR following incorporation of a labeled precursor enables the interrogation of cellular pathways such as glycolysis and the TCA cycle [22–24]. These studies include the analysis of in vitro inflammatory models [25] and the interrogation of mouse models of IBD including the analysis of biofluid (serum, urine) [26, 27], feces [26, 28], and intestinal tissue [25, 29]. Furthermore, metabolomic studies have also been carried out utilizing human specimens including biofluids [30, 31], fecal samples [32–34], and human tissue specimens [35–37].

In this chapter, we will review the methodologies for the preparation of in vitro cell extracts and the extraction of intestinal tissue for NMR analysis. Additional extraction protocols for biofluid or fecal samples for NMR analysis are beyond the scope of this endeavor but excellent protocols can be found elsewhere [32, 38]. Likewise, NMR sample run protocols and data analysis methods are only summarized here.

2 Materials

2.1 Sample Collection

1. Ice container.
2. Liquid nitrogen.

2.2 Cell Culture

1. Dulbecco's Modified Eagle's Medium (DMEM) without glucose.
2. [1-^{13}C]-labeled glucose (Cambridge Isotopes Laboratories).
3. Phosphate-Buffered Solution (PBS).
4. 0.9 % sodium chloride (NaCl).
5. Teflon cell scrapers.

2.3 Sample Extraction

1. Methanol
2. Chloroform
3. Perchloric acid ($HClO_4$).
4. Potassium hydroxide (KOH).

2.4 NMR Analysis

1. Deuterium oxide (D_2O) (Cambridge Isotope Laboratories).
2. Deuterated chloroform ($CDCl_3$).
3. Deuterated methanol (CD_3OD).
4. Deuterated DCl and NaOD).
5. Trimethylsilyl propionic-2,2,3,3,-d$_4$ acid (TSP).
6. Methyl diphosphoric acid (MDP, Sigma-Aldrich).
7. Ethylene diamine tetraacetic acid (EDTA).
8. 5 mm NMR tubes (Wilmad).
9. 1 mm NMR capillaries (Bruker Medical).

3 Methods

3.1 Perchloric Acid Cell Extraction

1. This protocol describes a 4 h incubation with 5 mmol/L [1-^{13}C]-labeled glucose for glucose uptake and metabolism studies by ^{13}C-NMR. However, glucose can be replaced with other labeled compounds based on the study focus (*see* **Note 1**).
2. Use 3–4 confluent 15 cm Petri dishes. Aspirate culture medium, wash each dish with 5 ml sterile PBS, add 15 ml of glucose-free serum-free medium supplemented with 13.7 mg of [1-^{13}C] labeled glucose (5 mmol/L).
3. Incubate for 4 h under normal culture conditions.
4. Pre-cool NaCl, 8 % PCA and bidistilled water in an ice bath. Pre-cool the centrifuge to 4 °C.

5. Transfer 5 ml of [1-^{13}C] labeled medium from each Petri dish to a 50 ml conical tube and place in ice bath.

6. Aspirate the remaining media.

7. Wash cells with 6 ml of ice-cold NaCl; aspirate NaCl.

8. Place the dish in liquid nitrogen until all cells are frozen (30–60 s).

9. Add 2 ml ice-cold 8 % PCA to the frozen cells.

10. Scrape the cells from the dish surface and transfer cell suspension into a 50 ml conical tube (ice bath).

11. Add 2 ml of ice-cold water into the dish and transfer remaining cells from the dish in the same tube.

12. Repeat with all four dishes.

13. Place the tube containing cell PCA extract in an ultrasonic ice bath for 5 min.

14. Centrifuge at 1300×g for 20 min at 4 °C.

15. Transfer the supernatant into a new 50 ml conical tube and place in ice bath.

16. Resuspend the pellet in 2 ml ice-cold 8 % PCA in the original 50 ml tube. Vortex well to resuspend the pellet.

17. Place the tube in an ultrasonic ice bath for 5 min.

18. Centrifuge the resulting suspension at 1300×g for 20 min at 4 °C.

19. Collect the supernatant in the same 50 ml tube as before; save the pellet for **step 24** (ice bath).

20. Neutralize the supernatant with KOH (first using 8 M, then 1 M and 0.1 M) so that it is at pH 7.0.

21. Centrifuge at 1300×g for 10 min at 4 °C.

22. Transfer the supernatant into a lyophilizer glass, freeze in liquid nitrogen, and lyophilize it in a freeze-dry system overnight.

23. Re-dissolve the lyophilized sample in 0.5 ml D$_2$O, centrifuge at 1300×g for 10 min, transfer into a 5-mm NMR tube, and adjust the sample pH to 7.2 using DCl and NaOD, for NMR analysis.

24. Re-dissolve the tissue pellet (**step 19**) in 4 ml ice-cold water, vortex well, and place in an ultrasonic ice bath for 10 min.

25. Neutralize the pellet suspension (containing tissue lipids) with KOH so that it is at pH 7.0, transfer into a lyophilizer glass (do not centrifuge), freeze in liquid nitrogen, and lyophilize it in a freeze-dry system overnight.

26. Re-dissolve the lyophilized lipids in 0.6 ml deuterium methanol/chloroform (1:2) solution, centrifuge at 1300×g for 10 min, and transfer into a 5 mm NMR tube for NMR analysis.

27. Centrifuge the collected 20 ml of ^{13}C-labeled medium at $1300 \times g$ for 5 min at 4 °C (to remove cell debris).

28. Lyophilize 10 ml of the labeled medium overnight.

29. Re-dissolve the lyophilized medium in 1 ml of D_2O, centrifuge at $1300 \times g$ for 10 min, and transfer into a 5 mm NMR tube for NMR analysis.

3.2 Perchloric Acid Tissue Extraction

1. Quickly remove 100–300 mg tissue and immediately snap freeze by wrapping sample in pre-cooled aluminum foil and immediately place in liquid nitrogen. Store all frozen tissue samples at –80 °C until extraction (*see* **Note 1**).

2. Pre-cool 8 % PCA in an ice bath. Pre-cool a mortar, pestle, and spatula by immersing them in a liquid nitrogen bath. Set and pre-cool the centrifuge at 4 °C. Transfer 4 ml of 8 % PCA in a 50 ml conical tube and pre-cool it in an ice bath.

3. Weigh tissue and record weight. Place frozen tissue in the pre-cooled pestle and crush with the mortar, while periodically cooling the pestle and mortar in liquid nitrogen.

4. Using the pre-cooled spatula, transfer the crushed tissue into the 50 ml conical tube containing 4 ml of pre-cooled 8 % PCA. Add an additional 2 ml of ice-cold 8 % PCA. Mix well using a vortex mixer and homogenize tissue sample using an electric tissue homogenizer (pre-cool the blade in the ice-cold water bath between the samples).

5. Place the tube containing tissue/PCA extract in an ultrasonic ice bath for 5 min.

6. Centrifuge at $1300 \times g$ for 20 min at 4 °C. Transfer the supernatant into a new 50 ml conical tube.

7. Resuspend the tissue pellet in 2 ml ice-cold 8 % PCA in the original tube. Vortex well to resuspend the pellet. Centrifuge the sample as before (**step 6**). Transfer the supernatant in the same 50 ml conical tube as before, and save the pellet for **step 12**.

8. Neutralize the supernatant with KOH (first using 8 M, then 1 M and 0.1 M) so that it is at pH 7.0.

9. Centrifuge at $1300 \times g$ for 10 min at 4 °C.

10. Transfer the supernatant to a lyophilizer glass, freeze in liquid nitrogen, and lyophilize it in a freeze-dry system overnight.

11. Re-dissolve the lyophilized sample in 0.5 ml D_2O, centrifuge at $1300 \times g$ for 10 min, transfer into a 5 mm NMR tube and adjust its pH to 7.2 using DCl and NaOD, for NMR analysis.

12. Re-dissolve the tissue pellet (**step 7**) in 4 ml ice-cold water, vortex well, and put it in an ultrasound ice bath for 10 min.

13. Neutralize the pellet suspension (containing tissue lipids) with KOH so that it is at pH 7.0, transfer it into a lyophilizer glass (do not centrifuge), freeze in liquid nitrogen, and lyophilize it in a freeze-dry system overnight.

14. Re-dissolve the lyophilized lipids in 1 ml deuterium methanol/ chloroform (1:2) solution, centrifuge at $1300 \times g$ for 10 min, and transfer into a 5 mm NMR tube for NMR analysis.

3.3 NMR Data Acquisition

The described acquisition parameters require the use of superconductive high-resolution NMR spectrometers with the proton resonance frequency of 300–600 MHz (magnetic field strength: 7.0–14.0 T) equipped with inverse (for proton detection) or QNP (for multinuclear detection) 5 mm probes. The described methods describe experimental parameters for one-dimensional NMR metabolite quantification only (*see* **Notes 2–4**).

1. Proton acquisition parameters (^{1}H-NMR). The following are possible acquisition parameters for ^{1}H-NMR spectra, which provide a linear correlation of metabolite concentration and signal integral. In order to calculate metabolite concentrations it is necessary to add the concentration standard TSP into D_2O into thin sealed glass capillaries, which can be placed in the 5 mm NMR tube as external concentration standard. Fully relaxed ^{1}H-NMR spectra of water-soluble metabolites and lipids can be obtained at 500 MHz (11.7 T) using a 5-mm HX (or TXI) inverse probe using solvent presaturation pulses (such as "zgpr" for Bruker) and the following acquisition parameters: 30° flip angle, 6000 Hz sweep width, 12.8 s repetition time, time-domain data points of 32 K, power level for solvent presaturation of 59 dB and 128 transients. *See* Fig. 1 for examples of generated ^{1}H-NMR spectra from both in vitro and murine colon tissue.

2. Carbon acquisition parameters (^{13}C-NMR). For ^{13}C-NMR measurements, composite pulse (WALTZ-16) proton decoupling is required for the acquisition of ^{13}C-NMR spectra. The use of QNP or broad-band probes is recommended. Water-soluble and lipid ^{13}C-NMR spectra can be recorded with 30° flip angle, 29,411 Hz sweep width, 3 s repetition time, time domain data points of 16 K (zero filled to 32 K before Fourier transform), and 20,000 transients for water-soluble metabolites or 6000 transients for lipids.

3. Phosphor acquisition parameters (^{31}P-NMR). Addition of 100 mmol/L EDTA is recommended for cell extract studies to chelate divalent cations which otherwise would cause signal broadening of the nucleoside tri- and diphosphate signals. A thin sealed capillary, containing methyl diphosphoric acid (MDP), is utilized as a concentration standard for ^{31}P-metabolite quantification. The following acquisition parameters can be used to

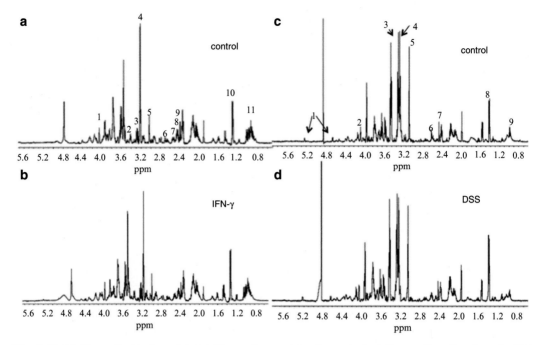

Fig. 1 Metabolic analysis of modeled inflammation and murine colonic inflammation. Panels **a–b**; T84 human colonic epithelial cells were exposed to IFN-γ at 10 ng/ml concentrations for 48 h. Representative proton spectra from control (**a**) and IFN-γ-treated T84 cells (10 ng/ml) (**b**). Numbered peaks are: 1, inositol; 2, taurine; 3, betaine; 4, total cholines; 5, total creatine; 6, methionine; 7, total glutathione; 8, glutamine; 9, succinate; 10, lactate; 11, leucine, isoleucine, and valine. Panels **c–d**; wild type mice received DSS for 5 days. Tissue was extracted and used for NMR analysis of endogenous metabolites. Representative proton spectra from control (**c**) and DSS-treated (3 %, 6 days treatment). (**d**) colonic tissue extracts. Numbered peaks are: 1, glucose; 2, inositol; 3, betaine; 4, total cholines; 5, total creatine; 6, total glutathione; 7, succinate; 8, lactate; 9, leucine, isoleucine, and valine. Reproduced from Ref. 25 with permission from the American Association of Immunologists

acquire fully relaxed ^{31}P-NMR spectra (without saturation effects) of water-soluble metabolites and lipids using the following acquisition parameters: 30° flip angle, 10,000 Hz sweep width, 3.7 s repetition time, time-domain data points of 16 K (zero filled to 32 K before Fourier transform), and 4000–8000 transients.

3.4 Data Analysis

For NMR data processing, various NMR software platforms can be utilized such as TopSpin and XWINNMR. Likewise, a number of statistical analyses may be employed for these purposes. Perform exponential multiplication and Fourier transformation on acquired spectral FID (free induction decay) with the following line broadening: LB = 0.2 Hz (^{1}H-NMR), 2 Hz (^{13}C-NMR), and 1 Hz (^{31}P-NMR). Proceed further with phase and baseline correction. For calibration, set TSP to 0 ppm (^{1}H-NMR), C3-lactate to

21 ppm (^{13}C-NMR), and MDP signal to 18.6 ppm (^{31}P-NMR). Perform peak integration and metabolite quantification using an NMR postprocessing software.

4 Notes

1. It is critical to remember throughout the process of sample collection and extraction that procedures be performed at low temperatures in order to avoid metabolite loss due to sample degradation. Likewise, steps requiring pH adjustment must be performed with care as NMR spectra are pH-sensitive.

2. All metabolites are characterized by their NMR chemical shift. The chemical shift is the normalized absorption frequency and is expressed in parts per million (ppm) (Fig. 1).

3. Due to overlapping chemical shifts of two or more metabolites in one-dimensional NMR spectra, two-dimensional NMR techniques (such as ^{1}H/^{13}C-HSQC) can be used for precise metabolite identification.

4. The integral of a spectral peak is measured in arbitrary units. The value is obtained by computer integration, using NMR software as described above, and is proportional to the number of nuclei contributing to the resonance. The normalized integral values (divided by the number of nuclei) are proportional to the metabolite concentration.

References

1. Baumgart DC, Carding SR (2007) Inflammatory bowel disease: cause and immunobiology. Lancet 369:1627–1640

2. Brand S (2009) Crohn's disease: Th1, Th17 or both? The change of a paradigm: new immunological and genetic insights implicate Th17 cells in the pathogenesis of Crohn's disease. Gut 58:1152–1167

3. Xavier RJ, Podolsky DK (2007) Unravelling the pathogenesis of inflammatory bowel disease. Nature 448:427–434

4. Khor B, Gardet A, Xavier RJ (2011) Genetics and pathogenesis of inflammatory bowel disease. Nature 474:307–317

5. Manichanh C, Borruel N, Casellas F, Guarner F (2012) The gut microbiota in IBD. Nat Rev Gastroenterol Hepatol 9:599–608

6. Sartor RB (1995) Current concepts of the etiology and pathogenesis of ulcerative colitis and Crohn's disease. Gastroenterol Clin North Am 24:475–507

7. Danese S, Dejana E, Fiocchi C (2007) Immune regulation by microvascular endothelial cells: directing innate and adaptive immunity, coagulation, and inflammation. J Immunol 178:6017–6022

8. Hatoum OA, Binion DG, Gutterman DD (2005) Paradox of simultaneous intestinal ischaemia and hyperaemia in inflammatory bowel disease. Eur J Clin Invest 35:599–609

9. Haddad JJ (2003) Science review: redox and oxygen-sensitive transcription factors in the regulation of oxidant-mediated lung injury: role for hypoxia-inducible factor-1alpha. Crit Care 7:47–54

10. Kokura S, Yoshida N, Yoshikawa T (2002) Anoxia/reoxygenation-induced leukocyte-endothelial cell interactions. Free Radic Biol Med 33:427–432

11. Saadi S, Wrenshall LE, Platt JL (2002) Regional manifestations and control of the immune system. FASEB J 16:849–856

12. Cummins EP, Seeballuck F, Keely SJ et al (2008) The hydroxylase inhibitor dimethyloxalylglycine is protective in a murine model of colitis. Gastroenterology 134:156–165

13. Semenza GL (2009) Regulation of oxygen homeostasis by hypoxia-inducible factor 1. Physiology (Bethesda) 24:97–106

14. Furuta GT, Turner JR, Taylor CT et al (2001) Hypoxia-inducible factor 1-dependent induction of intestinal trefoil factor protects barrier function during hypoxia. J Exp Med 193:1027–1034

15. Glover LE, Colgan SP (2011) Hypoxia and metabolic factors that influence inflammatory bowel disease pathogenesis. Gastroenterology 140:1748–1755

16. Jostins L, Ripke S, Weersma RK et al (2012) Host-microbe interactions have shaped the genetic architecture of inflammatory bowel disease. Nature 491:119–124

17. Cho JH (2008) The genetics and immunopathogenesis of inflammatory bowel disease. Nat Rev Immunol 8:458–466

18. Dettmer K, Aronov PA, Hammock BD (2007) Mass spectrometry-based metabolomics. Mass Spectrom Rev 26:51–78

19. Dunn WB, Bailey NJ, Johnson HE (2005) Measuring the metabolome: current analytical technologies. Analyst 130:606–625

20. Beckonert O, Keun HC, Ebbels TM, Bundy J, Holmes E, Lindon JC, Nicholson JK (2007) Metabolic profiling, metabolomic and metabonomic procedures for NMR spectroscopy of urine, plasma, serum and tissue extracts. Nat Protoc 2:2692–2703

21. Andriulli A, Loperfido S, Napolitano G et al (2007) Incidence rates of post-ERCP complications: a systematic survey of prospective studies. Am J Gastroenterol 102:1781–1788

22. Glunde K, Serkova NJ (2006) Therapeutic targets and biomarkers identified in cancer choline phospholipid metabolism. Pharmacogenomics 7:1109–1123

23. Klawitter J, Kominsky DJ, Brown JL et al (2009) Metabolic characteristics of imatinib resistance in chronic myeloid leukaemia cells. Br J Pharmacol 158:588–600

24. Kominsky DJ, Klawitter J, Brown JL, Boros LG, Melo JV, Eckhardt SG, Serkova NJ (2009) Abnormalities in glucose uptake and metabolism in imatinib-resistant human BCR-ABL-positive cells. Clin Cancer Res 15:3442–3450

25. Kominsky DJ, Keely S, MacManus CF et al (2011) An endogenously anti-inflammatory role for methylation in mucosal inflammation identified through metabolite profiling. J Immunol 186:6505–6514

26. Romick-Rosendale LE, Goodpaster AM, Hanwright PJ, Patel NB, Wheeler ET, Chona DL, Kennedy MA (2009) NMR-based metabonomics analysis of mouse urine and fecal extracts following oral treatment with the broad-spectrum antibiotic enrofloxacin (Baytril). Magn Reson Chem 47(Suppl 1):S36–S46

27. Schicho R, Nazyrova A, Shaykhutdinov R, Duggan G, Vogel HJ, Storr M (2010) Quantitative metabolomic profiling of serum and urine in DSS-induced ulcerative colitis of mice by (1)H NMR spectroscopy. J Proteome Res 9:6265–6273

28. Hong YS, Ahn YT, Park JC et al (2010) ^{1}H NMR-based metabonomic assessment of probiotic effects in a colitis mouse model. Arch Pharm Res 33:1091–1101

29. Dong F, Zhang L, Hao F, Tang H, Wang Y (2013) Systemic responses of mice to dextran sulfate sodium-induced acute ulcerative colitis using ^{1}H NMR spectroscopy. J Proteome Res 12:2958–2966

30. Williams HR, Cox IJ, Walker DG et al (2009) Characterization of inflammatory bowel disease with urinary metabolic profiling. Am J Gastroenterol 104:1435–1444

31. Williams HR, Willsmore JD, Cox IJ, Walker DG, Cobbold JF, Taylor-Robinson SD, Orchard TR (2012) Serum metabolic profiling in inflammatory bowel disease. Dig Dis Sci 57:2157–2165

32. Jacobs DM, Deltimple N, van Velzen E, van Dorsten FA, Bingham M, Vaughan EE, van Duynhoven J (2008) (1)H NMR metabolite profiling of feces as a tool to assess the impact of nutrition on the human microbiome. NMR Biomed 21:615–626

33. Marchesi JR, Holmes E, Khan F, Kochhar S, Scanlan P, Shanahan F, Wilson ID, Wang Y (2007) Rapid and noninvasive metabonomic characterization of inflammatory bowel disease. J Proteome Res 6:546–551

34. Bjerrum JT, Wang Y, Hao F, Coskun M, Ludwig C, Gunther U, Nielsen OH (2015) Metabonomics of human fecal extracts characterize ulcerative colitis, Crohn's disease and healthy individuals. Metabolomics 11:122–133

35. Balasubramanian K, Kumar S, Singh RR, Sharma U, Ahuja V, Makharia GK, Jagannathan NR (2009) Metabolism of the colonic mucosa in patients with inflammatory bowel diseases: an in vitro proton magnetic resonance spectroscopy study. Magn Reson Imaging 27:79–86

36. Bjerrum JT, Nielsen OH, Hao F, Tang H, Nicholson JK, Wang Y, Olsen J (2010) Metabonomics in ulcerative colitis:

diagnostics, biomarker identification, and insight into the pathophysiology. J Proteome Res 9:954–962

37. Sharma U, Singh RR, Ahuja V, Makharia GK, Jagannathan NR (2010) Similarity in the metabolic profile in macroscopically involved and un-involved colonic mucosa in patients with inflammatory bowel disease: an in vitro proton ((1)H) MR spectroscopy study. Magn Reson Imaging 28:1022–1029

38. Serkova NJ, Glunde K (2009) Metabolomics of cancer. Methods Mol Biol 520:273–295

Chapter 9

Analysis of microRNA Levels in Intestinal Epithelial Cells

Hang Thi Thu Nguyen

Abstract

The field of microRNA (miRNA) research is expanding rapidly with the crucial role of miRNAs in almost every biological process and their implication in many diseases. The role of miRNAs in modulating inflammatory responses in the gut has attracted many research groups including us. Here, we first briefly summarize our current understanding of the role of miRNAs in maintaining and regulating gut physiopathology and in inflammatory bowel diseases. We then describe in detail our techniques to analyze miRNA levels with notes that we have collected and summarized during our experiments.

Key words microRNA, Inflammatory bowel disease, Total RNA extraction, cDNA analysis, qRT-PCR

1 Introduction

MicroRNAs (miRNAs, miR) are small (approximately 20–22 nucleotides), non-coding RNAs that post-transcriptionally regulate gene expression by binding to the 3′-untranslated region (3′-UTR) of target mRNAs, leading to mRNA destabilization and/or inhibition of translation (1).

Each miRNA can target hundreds of mRNAs, and a single mRNA is often the target of multiple miRNAs. It is estimated that miRNAs regulate more than 60 % of protein-coding mRNAs [1], incorporating this post-transcriptional control pathway within various important cellular functions such as differentiation, proliferation, signal transduction, and apoptosis [2].

MiRNAs play a crucial role in regulation of the innate and adaptive immune systems, which has evolved to maintain self-tolerance and to recognize efficiently specific pathogens. The innate immune system acts as a first defense providing an immediate response to pathogens. It is activated via recognition of pathogen-associated molecular patterns (PAMPs) by extracellular receptor termed toll-like receptors (TLRs) or cytoplasmic nucleotide binding oligomerization domain-containing protein (NOD)-like receptors, promoting downstream signaling cascades through

Andrei I. Ivanov (ed.), *Gastrointestinal Physiology and Diseases: Methods and Protocols*, Methods in Molecular Biology, vol. 1422, DOI 10.1007/978-1-4939-3603-8_9, © Springer Science+Business Media New York 2016

pathways including nuclear factor-kappa B (NF-κB), mitogen-activated protein kinase, and interferon (IFN) regulatory factors. These processes are highly regulated by miRNAs. In addition to their role in regulating the innate immune system, miRNAs have been implicated in the complex network of signaling involved in maturation and activation of the adaptive immune system. This is mostly mediated via their role in controlling the development and activation of T and B cells. Key findings of the role of miRNAs in the innate and adaptive immune systems have been reviewed elsewhere [3–5].

As miRNAs play a critical role in almost every biological process, aberrant miRNA regulation has been associated with several human disorders including inflammatory bowel disease (IBD). IBD, comprised of Crohn's disease (CD) and ulcerative colitis (UC), is a chronic inflammatory gastrointestinal disorder, of which the etiology is multifactorial and involves environmental, genetic, and microbial factors [6]. Recent studies using quantitative RT-PCR (qRT-PCR) and miRNA arrays have identified distinct miRNA expression profiles in tissues and peripheral blood of patients with IBD versus control subjects and even UC versus CD [3, 7]. The results from these studies have been somewhat conflicting, due to at least in part differences in housekeeping genes and methods used to normalize miRNA levels, controls used (healthy individuals or "symptomatic control" patients), and disease duration and therapy which could influence miRNA levels [5]. The functional studies of miRNAs in IBD pathogenesis have only recently begun, however they have explored the link between miRNAs and defect in autophagy, epithelial barrier dysfunction, and inflammatory responses [3, 5, 7], which are the hallmarks of IBD pathogenesis. The increasing evidence and knowledge of miRNA function and dysregulation in IBD have raised miRNAs as potential diagnostic and therapeutic targets for future clinical applications. To read more about the role of miRNAs in IBD, please refer to the recent reviews [3, 5, 7].

In this chapter, we aim to describe our techniques with materials required to analyze miRNA levels [3, 8–12]. Notes that we have collected and summarized during experiments will be discussed. We would like to emphasize that this is only one example of methods used to analyze miRNA levels which we routinely use for our research.

2 Materials

2.1 Isolation of Total RNA

1. Phosphate-buffered saline (PBS).
2. Chloroform.
3. Ethanol 100 %.

4. DEPC-treated water: Add 0.1 mL of diethylpyrocarbonate (DEPC, Sigma) to 100 mL of water and shake vigorously. Incubate the solution for 12 h at 37 °C, then autoclave to remove the DEPC and filter.

5. MiRNeasy mini kit (Qiagen) including:

   (a) QIAzol Lysis Reagent, store at 4 °C.

   (b) Buffer RWT and Buffer RPE are supplied as a concentrate. Add 2 and 4 volumes of 100 % ethanol to Buffer RWT and Buffer RPE, respectively, to obtain a working solution.

   (c) RNeasy MinElute spin column.

2.2 First-Strand cDNA Synthesis

1. 1 mM Tris, pH 8.0 prepared in DEPC-treated water (*see* **Note 1**).

2. DEPC-treated water.

3. NCode™ miRNA first-strand cDNA synthesis (Invitrogen) inclu ding:

   5× miRNA reaction buffer.

   25 mM MnCl$_2$.

   Poly A polymerase.

   Annealing buffer.

   Universal RT primer 25 µM.

   2× First-strand reaction mix.

   SuperScriptR III RT/RNaseOUT™ enzyme mix.

4. 0.2 mL RNAse-free individual PCR tubes.

5. Thermocycler.

2.3 Real-Time Quantitative RT-PCR Amplification of miRNAs

1. Universal qRT-PCR reverse primer (10 µM) provided in the NCode™ miRNA first-strand cDNA synthesis kit.

2. Forward primer designed for each miRNA (*see* **Note 2**).
   The sequence of forward primer is identical to the entire mature miRNA sequence, except that U is replaced by T. For example:

   Mature hsa-miR-130a sequence: CAGUGCAAUGUUAAAA GGGCAU.

   Forward primer sequence: CAGTGCAATGTTAAAAGGGCAT.

3. SYBR Premix Ex Taq II (Takara).

4. MilliQ water or nuclease-free water.

5. 96-well qRT-PCR plates (4titude).

6. qRT-PCR optically clear adhesive film for 96-well plate (4titude).

7. Eppendorf Mastercycler® RealPlex^2 (Eppendorf).

3 Methods

3.1 Isolation of Total RNA Including miRNAs

1. Cell or tissue lysate preparation (*see* **Note 3**):

   (a) For cells grown in a monolayer, cells are washed with PBS, and 700 μL QIAzol Lysis Reagent is then added to the cell-culture dish (up to 1×10^7 cells). Agitate the dish for 10 min at room temperature (RT). Pipet the lysate into an eppendorf. Vortex or pipet to mix and ensure that no cell clumps are visible.

   (b) For tissues (flash-frozen in liquid nitrogen and stored at –70 °C), ≤30 mg of tissue is homogenized in 700 μL of QIAzol Lysis Reagent.

2. Incubate the homogenate at RT for 5 min. This step promotes dissociation of nucleoprotein complexes.

3. Add 140 μL chloroform to the tube containing the homogenate. Shake the tube vigorously for 15 s.

4. Incubate at RT for 2–3 min.

5. Centrifuge for 15 min at $12,000 \times g$ at 4 °C.

6. After centrifugation, the sample separates into 3 phases: an upper, colorless, aqueous phase containing RNA; a white interphase; and a lower, red, organic phase. Transfer the upper aqueous phase to a new collection tube (*see* **Note 4**).

7. Add 1.5 volumes of 100 % ethanol and mix thoroughly by pipetting.

8. Pipet up to 700 μL of the sample including any precipitate into an RNeasy MinElute spin column placed in a 2 mL collection tube. Centrifuge for 15 s at $10,000 \times g$ at RT. Discard the flow-through.

9. Repeat **step 8** until the whole sample has been pipetted into the spin column. Discard the flow-through each time.

10. Add 700 μL Buffer RWT to the RNeasy Mini spin column. Centrifuge for 15 s at $10,000 \times g$ at RT to wash the column. Discard the flow-through.

11. Add 500 μL Buffer RPE onto the RNeasy Mini spin column. Centrifuge for 15 s at $10,000 \times g$ at RT to wash the column. Discard the flow-through.

12. Add another 500 μL Buffer RPE to the RNeasy Mini spin column. Centrifuge for 2 min at $10,000 \times g$ at RT to dry the RNeasy Mini spin column membrane.

13. Place the RNeasy Mini spin column into a new 2 mL collection tube, and centrifuge for 1 min at $10,000 \times g$ at RT to eliminate

any possible carryover of Buffer RPE or if residual flow-through remains on the outside of the RNeasy Mini spin column after **step 12**. Transfer the RNeasy Mini spin column to a new 1.5 mL RNA-free collection tube.

14. Pipet 30–50 μL DEPC-treated water directly onto the RNeasy Mini spin column membrane. Centrifuge for 1 min at $10,000 \times g$ at RT to elute the RNA (*see* **Notes 5** and **6**).

15. Assess RNA quality and quantity using NanoDrop (*see* **Note 7**).

16. Check the integrity of purified total RNA by electrophoresis on 1.2 % agarose gel (*see* **Note 8**).

17. Store at –80 °C for further uses (*see* **Note 9**).

3.2   Generation of cDNA

3.2.1   Poly(A) Tailing Procedure

1. Use 100 ng to 1 μg of total RNA. Based on the quantity of total RNA, dilute a volume of 10 mM ATP in 1 mM Tris (pH 8.0) according to the following formula:
    ATP dilution factor = 5000/____ ng of total RNA
    Example: If 1 μg of total RNA is used, the ATP dilution factor is 5000/1000 ng = 5. Dilute the ATP 1:5 by adding 10 μL of 10 mM ATP to 40 μL of 1 mM Tris, pH 8.0.

2. Dilute total RNA in 0.2 mL individual PCR tubes to 100 ng–1 μg in 16 μL (*see* **Note 10**) using DEPC-treated water and place the tubes on ice.

3. Prepare a master mix on ice (*see* **Note 10**).

Component	Volume (for 1 reaction)
5× miRNA Reaction Buffer	5 μL
25 mM MnCl$_2$	2.5 μL
Diluted ATP (from **step 1**)	1 μL
Poly A Polymerase	0.5 μL
Total	9 μL

4. Mix master mix by flicking, centrifuge briefly, and add 9 mL of master mix to 16 μL of total RNA (containing 100 ng–1 μg RNA) from **step 2**.

5. Mix gently and spin the tube briefly to collect the contents.

6. Place the tubes in a thermocycler and run the following cycle: incubation at 37 °C for 15 min.

7. After incubation, places the tubes on ice

8. Pipet 4 μL from each tube (25 μL in total) to a new 0.2 mL RNase-free PCR tube, and use this for first-strand cDNA synthesis.

9. Store the rest of polyadenylated RNA at −20 °C for further uses.

3.2.2 First-Strand cDNA Synthesis

1. Prepare mix A on ice.

Component	Volume (for 1 reaction)
Annealing Buffer	1 μL
Universal RT Primer (25 μM)	3 μL
Total volume	4 μL

2. Vortex mix A, centrifuge briefly, and add 4 μL of mix A to 4 μL of polyadenylated RNA from **step 10** of Poly(A) tailing procedure.

3. Mix gently and centrifuge the tubes briefly to collect the contents.

4. Place the tubes in a thermocycler and run the following cycle: incubation at 65 °C for 5 min with heated lid on PCR-block (*see* **Note 11**).

5. After incubation, place the tube on ice for 1 min, centrifuge the tubes briefly to collect the contents, and replace them on ice.

6. Prepare mix B on ice:

Component	Volume (for 1 reaction)
2× First-Strand Reaction Mix	10 μL
SuperScriptR III RT/ RNaseOUT™ Enzyme Mix	2 μL

7. Vortex mix B, centrifuge briefly, and add 12 μL of mix B to the content in the tubes from **step 5**, for a final volume of 20 μL.

8. Mix gently and spin the tubes briefly to collect the contents.

9. Place the tubes in a thermocycler and run the following cycle with heated lid on PCR-block: 50 °C for 50 min, 85 °C for 5 min, and hold at 4 °C.

10. Store aliquots at −20 °C or proceed directly to quantitative RT-PCR.

3.3 Real-Time Quantitative RT-PCR Amplification of miRNAs

1. Dilute the cDNA (from **step 7** of B-First-strand cDNA synthesis) 1:10 in nuclease-free water.

2. Prepare the qRT-PCR mix:

Component	Volume (for 1 reaction)
2× SYBR Premix Ex Taq II (*see* **Note 12**)	5 μL
Forward primer (10 μM)	0.2 μL (200 nM final concentration)
Universal qPCR Primer (10 μM)	0.2 μL (200 nM final concentration)
MiliQ water	3.6 μL
Total	9 μL

3. Mix gently and spin briefly.

4. Add 9 μL of the mix to each well of a qRT-PCR 96-well plate.

5. Add 1 μL of cDNA (diluted 1:10 from **step 1**) to each well. For negative qRT-PCR control, use 1 μL of total RNA (the same amount as used for cDNA synthesis) to ensure that the qRT-PCR does not amplify genomic DNA contaminants.

6. Seal the plate with the adhesive film, and spin gently so that all components are at the bottom of the plate.

7. Place the plate in the Eppendorf Mastercycler® RealPlex and program the instrument.

   We usually perform a first gradient qRT-PCR run to identify the optimal annealing temperature for the primers. For this purpose, the program is set up as below:

   95 °C for 30 s.

   40 cycles of: 95 °C—5 s; Gradient from 55 to 65 °C (*see* **Note 13**)—60 s.

   Melting curve (See instrument documentation).

   Hold the reaction at 4 °C

   (a) For the Eppendorf Mastercycler® RealPlex, if a gradient 50–60 °C is programmed for the annealing temperature, there will be 12 different temperatures for 12 columns of the qRT-PCR plate in the same run (Fig. 1). This means that samples in different columns will be run at different temperatures.

   (b) We usually use one cDNA sample synthesized from total RNA extracted from untreated human/mouse cultured cells for this test. When there are several miRNAs to be analyzed, up to 8 forward primers specific to 8 mature miRNAs can be tested. Each row of the qRT-PCR plate will contain a different qRT-PCR mix with a different forward primer to analyze a different miRNA (Fig. 1).

   (c) From melting curve analysis (*see* **Note 14**), determine the optimal annealing temperature, which gives one specific product.

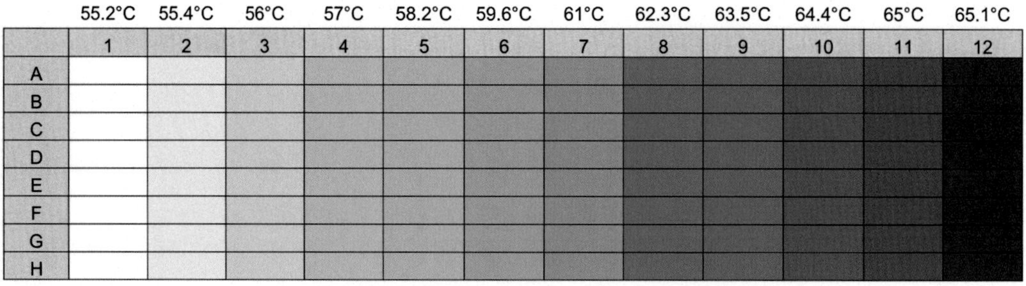

	55.2°C	55.4°C	56°C	57°C	58.2°C	59.6°C	61°C	62.3°C	63.5°C	64.4°C	65°C	65.1°C
	1	2	3	4	5	6	7	8	9	10	11	12
A												
B												
C												
D												
E												
F												
G												
H												

Fig. 1 Example of a qRT-PCR plate with a gradient annealing temperature 55–65 °C

8. Perform qRT-PCR to analyze expression of each miRNA at optimal annealing temperature (*see* **Note 15**).

9. Perform qRT-PCR to analyze expression of the non-coding small nuclear U6 RNA, used as an internal control, using the forward primer GCAAGGATGACACGCAAATTCGT and annealing temperature = 59 °C.

10. Calculate fold-induction of miRNA levels using the Ct method as follows: $\Delta\Delta Ct = (Ct_{\mathrm{miRNA}} - Ct_{\mathrm{U6}})_{\mathrm{treatment}} - (Ct_{\mathrm{miRNA}} - Ct_{\mathrm{U6}})_{\mathrm{nontreatment}}$, and the final data were derived from $2^{-\Delta\Delta Ct}$.

One example of our analysis of miRNA levels was presented in Fig. 3, indicating changes in the levels of mature homo sapiens (has)-miR-30c and hsa-miR-130a in human intestinal epithelial T84 cells upon infection with Crohn's disease-associated adherent-invasive *Escherichia coli* reference strain LF82 [8].

4   Notes

1. DEPC reacts with primary amines and cannot be used directly to treat Tris buffers. DEPC is highly unstable in the presence of Tris buffers and decomposes rapidly into ethanol and CO_2. When preparing Tris buffers, treat water with DEPC first, and then dissolve Tris to make the appropriate buffer.

2. In most cases, we can amplify the miRNA of interest using the forward primer of which the sequence is identical to that of the entire mature miRNA except that U is replaced by T. However, for some GC-rich miRNA sequences, it may be necessary to design a truncated forward primer to ensure that its GC content is in the range of 40–60 %.

3. When processing samples, do not exceed the maximum amount of starting material (up to 10^7 cells or 30 mg of tissues in 700 μL of QIAzol Lysis Reagent. Otherwise, homogenization will be incomplete and cellular debris may interfere with the binding of RNA to the RNeasy Mini spin column membrane, resulting in lower RNA yield and purity.

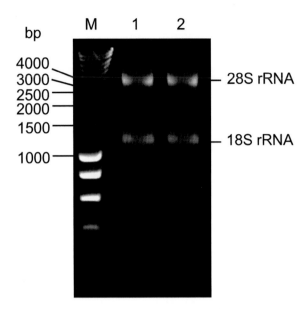

Fig. 2 Agarose gel electrophoresis analysis of total RNA isolated from human intestinal epithelial T84. Total RNA were isolated from T84 cells using the miRNeasy kit (Qiagen), and 500 μg of total RNA were migrated on 1.2 % agarose gel (lanes 1 and 2). M: DNA SmartLadder (Eurogentec)

4. When removing the upper aqueous phase containing RNA, avoid withdrawing contaminants. Leaving a little of the upper aqueous phase containing RNA can minimize the risk of carrying over contaminants.

5. DNase digestion during extraction of total RNA is not required for detecting mature miRNAs.

6. After RNA elution, do not put RNA samples into a vacuum dryer that may have RNase contamination.

7. RNA purity is assessed by the ratio of the readings at 260 nm and 280 nm (A260/A280). A ratio of approximately 2.0 indicates a good quality RNA sample. A lower ratio may indicate the presence of other contaminants potentially carried over from the RNA isolation procedure.

8. Analysis of RNA integrity by electrophoresis on agarose gel: The respective ribosomal bands should appear as sharp bands. The 28S ribosomal RNA band should be present at approximately twice the amount of the 18S RNA band (Fig. 2). If the ribosomal bands appear as a smear toward smaller RNAs, it is likely that the RNAs are degraded during isolation.

9. The quality of the total RNA isolated using miRNeasy kit is suitable for our analyses of miRNA levels by qRT-PCR [8–12] or by miRNA array [9, 11].

10. To economize the NCode™ miRNA first-strand cDNA synthesis kit, the amount of all components required for the

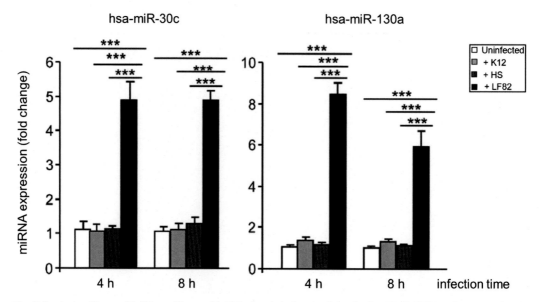

Fig. 3 Analysis of hsa-miR-30c and hsa-miR-130a levels in human intestinal epithelial T84 cells upon infection with Crohn's disease-associated adherent-invasive *Escherichia coli* (AIEC). T84 cells were infected with the AIEC LF82 reference strain, the non-pathogenic *E. coli* K12 MG1655 or the commensal *E. coli* HS strain. MiRNA levels in uninfected and infected cells analyzed by qRT-PCR were shown. Data are means ± SEM of 6 replicates and are representative of 3 independent experiments. *$P < 0.05$; **$P < 0.005$; ***$P < 0.001$ (Reproduced from Ref. 8 with permission from Elsevier)

master mix in Poly(A) tailing procedure could be divided by 2. In this case, dilute total RNA to 100 ng–1 μg in 8 μL using DEPC-treated water.

11. When using the thermocycler, we usually select to heat lid to avoid the evaporation of samples at high temperature.

12. Minimize exposure of 2X SYBR Premix Ex Taq II to direct light. Exposure to direct light for an extended period of time may result in loss of fluorescent signal intensity.

13. We have obtained optimal qRT-PCR results with the annealing temperature in the range of 55–65 °C. If this range of annealing temperature does not give good and specific qRT-PCR results, another gradient for annealing temperature (for example 50–60 °C or 55–68 °C) could be tested. Raising the annealing temperature may result in better discrimination of closely related miRNA sequences, but may decrease the sensitivity of the reaction. If this still does not give good results, forward primer redesign needs to be considered.

14. Melting curve analysis should always be performed after qRT-PCR to analyze the specificity of the reaction. For more information, visit www.lifetechnologies.com/qpcr and type in the search box "melting curve analysis".

15. When performing qRT-PCR, prepare duplicate wells per sample per miRNA. We usually perform 6 replicates per condition for in vitro studies and ≥8 mice/group for in vivo studies, and the experiments are repeated at least 3 times.

References

1. Friedman RC, Farh KK, Burge CB, Bartel DP (2009) Most mammalian mRNAs are conserved targets of microRNAs. Genome Res 19:92–105

2. Ambros V (2004) The functions of animal microRNAs. Nature 431:350–355

3. Raisch J, Darfeuille-Michaud A, Nguyen HT (2013) Role of microRNAs in the immune system, inflammation and cancer. World J Gastroenterol 19:2985–2996

4. Lu LF, Liston A (2009) microRNA in the immune system, microRNA as an immune system. Immunology 127:291–298

5. Kalla R, Ventham NT, Kennedy NA, Quintana JF, Nimmo ER, Buck AH, Satsangi J (2015) microRNAs: new players in IBD. Gut 64:504–517

6. Carriere J, Darfeuille-Michaud A, Nguyen HT (2014) Infectious etiopathogenesis of Crohn's disease. World J Gastroenterol 20:12102–12117

7. Chapman CG, Pekow J (2015) The emerging role of miRNAs in inflammatory bowel disease: a review. Therap Adv Gastroenterol 8:4–22

8. Nguyen HT, Dalmasso G, Muller S, Carriere J, Seibold F, Darfeuille-Michaud A (2014) Crohn's disease-associated adherent invasive Escherichia coli modulate levels of microRNAs in intestinal epithelial cells to reduce autophagy. Gastroenterology 146:508–519

9. Dalmasso G, Nguyen HT, Yan Y, Laroui H, Charania MA, Ayyadurai S, Sitaraman SV, Merlin D (2011) Microbiota modulate host gene expression via microRNAs. PLoS One 6:e19293

10. Dalmasso G, Nguyen HT, Yan Y, Laroui H, Charania MA, Obertone TS, Sitaraman SV, Merlin D (2011) microRNA-92b regulates expression of the oligopeptide transporter PepT1 in intestinal epithelial cells. Am J Physiol Gastrointest Liver Physiol 300:G52–G59

11. Dalmasso G, Nguyen HT, Yan Y, Laroui H, Srinivasan S, Sitaraman SV, Merlin D (2010) microRNAs determine human intestinal epithelial cell fate. Differentiation 80:147–154

12. Nguyen HT, Dalmasso G, Yan Y, Laroui H, Dahan S, Mayer L, Sitaraman SV, Merlin D (2010) microRNA-7 modulates CD98 expression during intestinal epithelial cell differentiation. J Biol Chem 285:1479–1489

Part II

Imaging Analysis of the Gastrointestinal System In Vivo

Chapter 10

Detecting Reactive Oxygen Species Generation and Stem Cell Proliferation in the Drosophila Intestine

Liping Luo, April R. Reedy, and Rheinallt M. Jones

Abstract

The conservation of intestinal stem cell crypt dynamics between *Drosophila melanogaster* and mammals allows for the genetically tractable fly model to be used for analyses of intestinal development, homeostasis, and renewal in relation to microbiota. The invertebrate fly model is advantageous for genetic research due to its anatomical and genetic simplicity and short lifespan. Accordingly, experimental resources such as large numbers of mutant and genetically modified flies have been developed. We have developed techniques to generate germ-free *Drosophila*, monoassociate them with candidate bacteria, and assess ensuing physiological responses within the gut tissue that include the generation of reactive oxygen species and cell proliferation.

Key words *Drosophila melanogaster*, Reactive oxygen species, ROS, Symbiosis, Commensal bacteria, Cell proliferation, Probiotics, Germ-free, Monoassociation

1 Introduction

The *Drosophila melanogaster* model organism has been integral to biological research for more than a century. Its enduring attraction to researchers can be attributed to its ease of breeding and genetic tractability. Indeed, *Drosophila* has been used to study developmental, cell, neuro, and behavioral biology. The *Drosophila* intestine has been recently recognized to possess remarkable homology with the mammalian intestinal crypt-villus unit, permitting the fly to be utilized as a model to study gut development. Comprehensive review articles comparing the development of the fly and mammalian gut, and detailing the fly's relevance as a model for human intestinal physiology have been recently published [1–4]. The *Drosophila* midgut is comprised of an epithelial monolayer, interspersed with hormone-producing enteroendocrine cells. Adult *Drosophila* midgut cells are continuously replenished by a population of pluripotent intestinal stem cells (ISCs) which adjoin the intestinal basement membrane [5, 6]. Similar to mammalian ISCs,

Andrei I. Ivanov (ed.), *Gastrointestinal Physiology and Diseases: Methods and Protocols*, Methods in Molecular Biology, vol. 1422, DOI 10.1007/978-1-4939-3603-8_10, © Springer Science+Business Media New York 2016

Notch and WNT signaling has been shown to be required for the differentiation of *Drosophila* ISC daughter cells into enterocytes or enteroendocrine cells [5, 7, 8]. Other signaling pathways such as EGFR signaling [9, 10], BMP ligand-mediated JAK/Stat signaling [11], JNK signaling [12], and the Warts/Yorkie/Hippo pathway [13] have been shown to maintain *Drosophila* ISC homeostasis.

Physiological levels of reactive oxygen species (ROS) are increasingly recognized in the transduction of intracellular signaling events in a wide variety of cell types including epithelial cells [14–18]. Interestingly, ROS has been shown to be required for cell proliferation and differentiation of *Drosophila* hematopoietic progenitors [19], in the control of the transition from proliferation to differentiation in the plant root [20], and the regeneration of an amputated *Xenopus* tadpole tail [21]. At the molecular level, ROS-regulated signaling is mediated by "sensor" regulatory proteins, whose activity can be influenced by ROS. These redox-sensitive proteins are modified by reversible H_2O_2-mediated oxidation of active site cysteine residues [22]. However, the extent to which ROS participate as signaling molecules in bacterial-induced gut developmental and homeostatic processes in the intestine remains understudied.

We previously used the *Drosophila* model to study host–pathogen interactions in epithelial tissues [23, 24]. In recent investigations, we employed the *Drosophila* model to investigate host–commensal bacterial interactions. We discovered that lactobacilli colonization of germ-free *Drosophila* can induce the NADPH oxidase-mediated cellular ROS generation, and induce ROS-dependent cell proliferation in the fly intestine without any outward negative pathological influences on cell health [25]. Here, we describe the methods we developed to generate germ-free *Drosophila*, and to detect ROS generation and cell proliferation in the intestine following colonization with lactobacilli. We also describe the use of a novel ROS-sensitive dye known as hydrocyanines which can function as fluorescent sensors for imaging ROS generation in vivo [26].

2   Materials

1. Alcohol lamp.
2. Aluminum foil.
3. Bleach.
4. Bovine serum albumin (BSA) in phosphate-buffered saline (PBS): 3 % BSA in PBS, pH 7.4.
5. Cell strainer, 100 μm.
6. Cell culture plate, 12 well.
7. Click-iT® EdU Alexa Fluor® 488 Imaging Kit (Life Technologies).

8. 18 × 18-mm Coverslips.

9. Forceps.

10. *Lactobacillus plantarum* strain ATCC 14917.

11. *Lactobacillus plantarum Lactobacilli* growth medium (Difco).

12. 4 % Paraformaldehyde solution (PFA).

13. Sterile PBS 1X diluted from 10X PBS.

14. Permeabilization solution: 0.5 % Triton X-100 in PBS.

15. Sterile Petri dishes.

16. phospho-Histone H3 antibody (Cell Signaling Technology).

17. PYREX® 3 Depression Glass Spot Plates (Corning).

18. ROSstar 550 reactive oxygen species-sensitive dye (LI-COR Biosciences).

19. Scalpels (sterile).

20. TO-PRO-3 (Life Technologies).

21. VectaShield (Vector Laboratories).

22. Autoclavable Vials for *Drosophila* husbandry (Genesee Scientific).

23. Fly Food: Weight dextrose 50 g/L, sucrose 25 g/L, yeast extract 15 g/L, cornmeal 60 g/L, agar 6.5 g/L, tryptone 30 g/L, and molasses 65 g/L in a glass beaker, then add 1 L water and mix. Microwave for 3 min. Observe the beaker contents regularly. When food comes to a boil, stop the microwave and mix the food thoroughly. The liquid should resemble the consistency of broth/soup. Aliquot 10 mL of the microwaved food to *Drosophila* husbandry vials, apply plugs to the vials, and cover the plugs with aluminum foil. Autoclave vials as liquid cycle (121 °C for 20 min). Store vials at 4 °C until required.

24. DAPI nuclear stain.

25. SYTO24 nuclear stain (Life Technologies).

3 Methods

3.1 Generation of Germ-Free *Drosophila*

1. Synchronize animals for egg collection by transferring (tapping) adult flies into a bottle with fresh fly food (without yeast). Incubate the bottle at 25 °C for about 12 h or overnight.

2. Dissociate eggs collected in the bottle with a steady stream of PBS from a wash bottle (*see* **Note 1**) and transfer them into a sterile cell strainer. Wash collected eggs with PBS (*see* **Note 2**).

3. Working in a sterile hood, place the entire cell strainer into a petri dish containing 50 % bleach solution (*see* **Note 3**). Bleach eggs for 5 min. Thoroughly rinse the sides of the cell strainer

with bleach. With sterile forceps agitate the cell strainer up and down a few times to expose all eggs to bleach.

4. Move the entire cell strainer into a petri dish containing sterile PBS solution. Agitate the cell strainer in sterile PBS multiple times to wash the eggs.

5. Repeat **step 4** with a new petri dish with fresh sterile PBS solution three times.

6. Firmly grasp the top plastic lip of the cell strainer and cut the mesh out of strainer with a sterile scalpel (*see* **Note 4**).

7. Grab the mesh along the edge with sterile forceps (*see* **Note 5**) and place it into an autoclaved vial containing autoclaved *Drosophila* food.

8. Once the mesh harboring germ-free embryos (**step 7**) has been transferred to the vial containing autoclaved *Drosophila* food, incubate the vial at 25 °C.

9. After 5 days incubation and the food is churned, work under a hood to test the food and larvae mixture for axenia by undertaking standard 16S rDNA PCR analysis [27].

3.2 Monoassociation of Germ-Free Third Instar Drosophila with Lactobacillus plantarum

1. The night before the experiment, inoculate 100 mL sterile MRS medium with the *Lactobacillus plantarum* (taken directly from glycerol stock). Culture *L. plantarum* in bottle with tightly closed lid and with no shaking at 37 °C overnight.

2. Centrifuge the culture suspension ($3200 \times g$, 15 min). Discard the supernatant. Wash the pellet twice by suspending in 25 mL of sterile PBS, centrifuging (3200 g, 15 min) then discarding the supernatant.

3. Suspend the pellet in 1–2 mL of sterile PBS.

4. Measure the colony forming units CFU/mL (*see* **Note 6**) then adjust the *L. plantarum* culture suspension volume with sterile PBS to 2×10^{10} CFU/mL.

5. Using sterile aseptic technique, puree/smash 1 g of solidified germ-free food (*see* **Note 7**) in a petri dish.

6. Mix 1 mL of the *L. plantarum* (2×10^{10} CFU/mL), or 1 mL PBS control with the 1 g of pureed food. Final concentration of assayed bacteria should be 1×10^{10} CFU/mL.

7. Pick about 30 germ-free early third instar larvae (*see* **Notes 8 and 9**) and place into the 12-well cell culture plate containing food/*L. plantarum* mixture.

8. Incubate the plate at 25 °C for 4 h (*see* **Note 10**).

9. Dissect gut (Subheading 3.4) and perform immunofluorescence analysis (Subheading 3.5).

3.3 Detection of Bacterial-Induced ROS Generation in the Intestine of Third Instar Drosophila Larvae

Carry out all procedures at room temperature unless otherwise specified. Maintain sterile aseptic technique throughout manipulations.

1. Follow **steps 1–6** from Subheading 3.2 for the monoassociation of germ-free larvae with *L. plantarum*.

2. Place ~400 µL autoclaved and pureed fly food with *L. plantarum* into a sterile 12-well plate (*see* **Note 11**).

3. Add ROSstar 550 probe at 100 µM final concentration to each well and mix.

4. Select about 30 early germ-free third instar larvae (*see* **Notes 8 and 9**) and place into wells.

5. Keep the 12-well plate at 25 °C in the dark for 4 h (*see* **Note 10**).

6. Dissect out the gut of the third instar larvae (Subheading 3.4).

7. Transfer the guts into a depression of a glass spot plate containing 4 % PFA solution and fix for 20 min (*see* **Note 12**).

8. Wash the intestines three times (5 min each wash) by sequentially transferring them to glass spot plate depressions filled with PBS.

9. For DNA counter stain, dilute TO-PRO-3 solution 1:2000 in PBS. Place 1 mL into a depression. Incubate the intestines in TO-PRO-3 solution for 5 min at room temperature.

10. Orient the intestines linearly on a slide in a small drop of PBS with forceps (*see* **Note 13**). Let the PBS drop completely evaporate, then place 1 small drop of glycerol or VectaShield on top of the intestines. Gently orient and drop a coverslip onto the slide. Seal around the coverslip with nail polish if necessary (for inverted microscopy).

11. Visualize reacted ROSstar 550 probe by confocal microscopy with the Cy3 channel (Fig. 1).

3.4 Dissection of Drosophila Larval and Adult Intestine

3.4.1 For Larvae

1. Fill the concave depressions of a glass spot plate with PBS.

2. Pick 10–15 larvae into the first depression/well with forceps (*see* **Note 14**).

3. Transfer one larvae to a new depression/well.

4. Orient the larvae (anterior to the left, posterior to the right) by grasping the posterior spiracles with forceps held in your right hand and the pharyngeal apparatus with forceps held in your left hand.

5. Move the left forceps within a few millimeters anterior to the right forceps at the posterior of the larvae.

6. Firmly clamp the larvae with the left forceps, then move the right forceps laterally to forcibly remove the tail (*see* **Note 15**). The internal organs including the intestine should extrude from torn posterior (*see* **Note 16**).

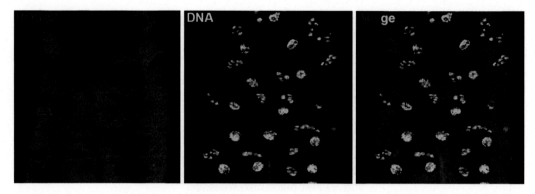

Fig. 1 Detection of ROS generation in the posterior intestine of third instar *Drosophila* larvae following colonization of germ-free third instar larvae with *Lactobacillus plantarum*. ROS was detected by oxidation of the ROSstar 550 ROS-sensitive dye included in the media and ingested by the larvae. Note ROS generation within the cytoplasm but not within the nucleus of enterocytes

7. Hold the pharyngeal apparatus with the left forceps and run the right forceps down the length of the larvae to squeeze out the internal organs from the torn posterior.

8. Dissect the fat bodies away from the intestine (*see* **Note 17**).

9. Transfer the dissected gut into a new depression containing PBS and perform fixation step immediately.

3.4.2 For Adult Drosophila

1. Fill the concave depressions of a glass spot plate with 1X PBS.

2. Remove 10–15 abdomens from anesthetized adult flies (*see* **Note 18**) and place into the first depression/well.

3. Transfer one abdomen at a time to a new depression/well for dissection.

4. Dissect away a little of the extreme posterior of the abdomen. The internal organs including the intestine should extrude from the large hole in the anterior abdomen.

5. Holding the posterior region of the cuticle with the right forceps, run the left forceps down the length of the abdomen to squeeze out the internal organs from the torn anterior.

6. Dissect the fat bodies and eggs away from the intestine (*see* **Note 17**).

7. Transfer the dissected gut into a new depression containing PBS and perform fixation step immediately.

3.5 Immunofluorescence Detection of Phospho-Histone H3-Positive Cells in the Drosophila Intestine

1. Previously dissected intestines (*see* Subheading 3.4) are sequentially transferred to depressions in glass spot plates containing sequential solutions to fix, wash and antibody stain as described below (*see* **Note 19**).

2. Transfer 10 dissected fly intestines (*see* Subheading 3.4) from PBS into 4 % PFA solution. Fix at room temperature for 20 min (*see* **Note 12**).

3. Wash the intestines twice by sequentially transferring them to depressions containing PBS.

4. Transfer the intestines into a depression with PBS + 0.3 % Triton X-100 solution and permeate for 15 min.

5. Wash the intestines twice with PBS + 0.1 % Triton X-100.

6. Transfer the intestines into 50 μL of primary antibody solution (phospho-Histone H3 antibody is suspended at 1:100 in PBS + 0.1 % Triton X-100).

7. Wrap the entire glass spot plate in parafilm then incubate overnight at 4 °C (*see* **Note 21**).

8. Wash twice with PBS + 0.1 % Triton X-100.

9. Transfer the intestines into 50 μL secondary antibody (suspended 1:100 in PBS + 0.1 % Triton X-100) at room temperature for 1 h.

10. Wash twice with PBS + 0.1 % Triton X-100.

11. For DNA counter stain, dilute TO-PRO-3 solution 1:2000 in PBS. Place 1 mL into a depression. Incubate the intestines in TO-PRO-3 solution for 5 min at room temperature (*see* **Note 20**).

12. Wash the intestines twice with PBS (as done in **step 3**).

13. Orient the intestines linearly on a slide in a small drop of PBS (*see* **Note 13**). Place a small drop of glycerol or VectaShield on to each intestine. Gently orient and drop a coverslip onto the slide. Seal around the coverslip with nail polish if necessary (for inverted confocal microscope).

14. Visualize the phospho-Histone H3-positive cells using confocal microscopy.

3.6 Detection of the Incorporation of EdU into Proliferating Cells in the Drosophila Intestine

1. For larvae, place 400 μL pureed food in a 12-well plate (*see* **Note 11**). Add EdU to food to a final concentration of 0.4 mM. Transfer larvae into the mixture and allow to feed on EdU-containing food for 4 h.

2. For adults, starve the flies for 4–6 h in empty vials (*see* **Note 23**). Then transfer adult flies to a vial containing pureed food containing 1 mM EdU final concentration. Let animals feed for 24 h.

3. Follow the dissection procedure for 10 larvae or adults (Subheading 3.4). All dissected intestines are sequentially transferred to depressions in glass spot plates containing sequential solutions to fix, wash, and immunolabel the tissues as described below.

4. Transfer 10 dissected fly intestines (*see* Subheading 3.4) from PBS into 4 % PFA solution. Fix for 20 min (*see* **Note 12**).

5. Wash the intestines in a depression containing PBS.

6. Transfer the intestines into a depression with PBS + 3 % BSA.

7. Wash the intestines twice with PBS + 3 % BSA.

8. Permeabilize the intestines by incubating them in PBS + 0.5 % Triton X-100 for 20 min.

9. Prepare 1X Click-iT® EdU buffer (according to the manufacturer's protocol) by diluting the 10X solution 1:10 in deionized water. Also prepare the Click-iT® reaction cocktail (according to the manufacturer's instructions). Prepare enough volume of cocktail for 50 μL of solution per assayed intestine. Prepare both solutions fresh and use the solution on the same day as assay (*see* **Notes 24 and 25**).

10. Wash the intestines twice with 1 mL PBS + 3 % BSA.

11. Add 50 μL of Click-iT® reaction cocktail to each depression well that will contain an assayed intestine.

12. Transfer the intestines into the Click-iT® reaction cocktail then incubate the plate for 30 min at room temperature (*see* **Note 22**).

13. Wash the intestines twice with 3 % BSA in PBS.

14. Perform DNA counter staining and mount intestines on slides as described in Subheading 3.4. Detect EdU incorporation by confocal microscopy at 488 nm (Fig. 2).

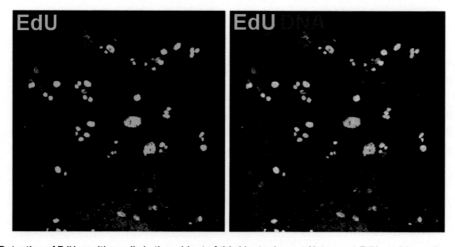

Fig. 2 Detection of EdU-positive cells in the midgut of third instar larvae. Note most EdU-positive cells are adult midgut progenitor (AMP) cells in clusters with smaller nuclei. Note also a few EdU-positive larger nuclei which are a result of endonuclear replication of enterocyte DNA

4   Notes

1. Do not dislodge a large amount of agar.

2. If the animals are not well synchronized it may be necessary to pick out the larvae underneath the dissecting scope.

3. You can label the cell strainers to keep track of different fly lines being made germ-free.

4. Balance the cut mesh carefully in the cell strainer to not let it drop.

5. Flame the forceps with an alcohol lamp inside the hood.

6. One can correlate the turbidometric measure of the cells (measured by optical density (OD) at 550λ) with CFU/mL concentration and use this for future CFU/mL approximations.

7. Coarsely smash the germ-free food with a sterile spatula, then pipette the mixture up and down with a sterile 1 mL pipette tip to further homogenize.

8. Remove early third instar larvae by scooping food out from a populated vial/bottle on to a petri plate with PBS, then pick third instar larvae from the food/PBS solution. Third instar larvae that are crawling up the walls of the vial/bottle do not feed as well.

9. Picking the larvae to a KimWipe first can remove any excess PBS before exposure to experimental conditions.

10. Cover the petri plate or 12-well plate with plastic wrap to prevent the larvae from escaping.

11. The pureed food should be just enough to cover the larvae without drowning them.

12. Perform this procedure under the hood.

13. Orient the intestine in a small drop (3–5 µL) of PBS. Let the drop dry before adding mounting media.

14. Carefully pick the larvae up by the posterior spiracles to minimize damage to the body.

15. The hindgut is attached to the tail so it is necessary to forcefully remove the most posterior section of the larvae to separate the intestine.

16. If the internal organs do not extrude from the posterior it may suggest tail removal was incomplete or inefficient. Reorient the larvae and remove the posterior again.

17. The fat bodies look granulated and white in comparison to the more opaque/yellowish intestine.

18. Remove all abdomens from flies anesthetized by carbon dioxide (CO_2) before proceeding further.

19. The same 3-well glass spot plate is used for all solutions and processing of the same genotype. The wells are just washed and dried.

20. Alternative DNA counter stains include DAPI (emission maximum at 461 nm) or SYTO24 (emission maximum at 515 nm).

21. Intestines should remain stationary to reduce shearing.

22. Wrap in tin foil to keep out light.

23. Tap flies into empty vials or bottles to starve.

24. This protocol uses 500 μL of Click-iT® reaction cocktail per glass depression. A smaller volume can be used as long as the remaining reaction components are maintained at the same ratios.

25. It is important to add the ingredients in the order listed in the table; otherwise, the reaction will not proceed optimally. Use the Click-iT® reaction cocktail within 15 min of preparation.

References

1. Jiang H, Edgar BA (2012) Intestinal stem cell function in *Drosophila* and mice. Curr Opin Genet Dev 22:354–360

2. Apidianakis Y, Rahme LG (2011) *Drosophila melanogaster* as a model for human intestinal infection and pathology. Dis Model Mech 4:21–30

3. Karpowicz P, Perrimon N (2010) All for one, and one for all: the clonality of the intestinal stem cell niche. F1000 Biol Rep 2:73

4. Wang P, Hou SX (2010) Regulation of intestinal stem cells in mammals and *Drosophila*. J Cell Physiol 222:33–37

5. Micchelli CA, Perrimon N (2006) Evidence that stem cells reside in the adult *Drosophila* midgut epithelium. Nature 439:475–479

6. Ohlstein B, Spradling A (2006) The adult *Drosophila* posterior midgut is maintained by pluripotent stem cells. Nature 439:470–474

7. Lin G, Xu N, Xi R (2008) Paracrine Wingless signalling controls self-renewal of *Drosophila* intestinal stem cells. Nature 455:1119–1123

8. Ohlstein B, Spradling A (2007) Multipotent *Drosophila* intestinal stem cells specify daughter cell fates by differential notch signaling. Science 315:988–992

9. Jiang H, Edgar BA (2009) EGFR signaling regulates the proliferation of *Drosophila* adult midgut progenitors. Development 136:483–493

10. Jiang H, Grenley MO, Bravo MJ, Blumhagen RZ, Edgar BA (2011) EGFR/Ras/MAPK signaling mediates adult midgut epithelial homeostasis and regeneration in *Drosophila*. Cell Stem Cell 8:84–95

11. Jiang H, Patel PH, Kohlmaier A, Grenley MO, McEwen DG, Edgar BA (2009) Cytokine/Jak/Stat signaling mediates regeneration and homeostasis in the *Drosophila* midgut. Cell 137:1343–1355

12. Biteau B, Hochmuth CE, Jasper H (2008) JNK activity in somatic stem cells causes loss of tissue homeostasis in the aging *Drosophila* gut. Cell Stem Cell 3:442–455

13. Staley BK, Irvine KD (2010) Warts and Yorkie mediate intestinal regeneration by influencing stem cell proliferation. Curr Biol 20:1580–1587

14. Kumar A, Wu H, Collier-Hyams LS, Hansen JM, Li T, Yamoah K, Pan Z-Q, Jones DP, Neish AS (2007) Commensal bacteria modulate cullin-dependent signaling via generation of reactive oxygen species. EMBO J 26:4457–4466

15. Kumar A, Wu H, Collier-Hyams LS, Kwon YM, Hanson JM, Neish AS (2009) The bacterial fermentation product butyrate influences epithelial signaling via reactive oxygen species-mediated changes in cullin-1 neddylation. J Immunol 182:538–546

16. Swanson PA 2nd, Kumar A, Samarin S, Vijay-Kumar M et al (2011) Enteric commensal bacteria potentiate epithelial restitution via reactive oxygen species-mediated inactivation of focal adhesion kinase phosphatases. Proc Natl Acad Sci U S A 108:8803–8808

17. Wentworth CC, Alam A, Jones RM, Nusrat A, Neish AS (2011) Enteric commensal bacteria induce ERK pathway signaling via formyl peptide receptor (FPR)-dependent redox modulation of Dual specific phosphatase 3 (DUSP3). J Biol Chem 286:38448–38455

18. Alam A, Leoni G, Wentworth CC, Kwal JM et al (2013) Redox signaling regulates commensal-mediated mucosal homeostasis and restitution and requires formyl peptide receptor 1. Mucosal Immunol 17:645–655

19. Owusu-Ansah E, Banerjee U (2009) Reactive oxygen species prime *Drosophila* haematopoietic progenitors for differentiation. Nature 461:537–541

20. Takashima S, Mkrtchyan M, Younossi-Hartenstein A, Merriam JR, Hartenstein V (2008) The behaviour of Drosophila adult hindgut stem cells is controlled by Wnt and Hh signalling. Nature 454:651–655

21. Love NR, Chen Y, Ishibashi S, Kritsiligkou P et al (2013) Amputation-induced reactive oxygen species are required for successful Xenopus tadpole tail regeneration. Nat Cell Biol 15:222–228

22. Ray PD, Huang BW, Tsuji Y (2012) Reactive oxygen species (ROS) homeostasis and redox regulation in cellular signaling. Cell Signal 24:981–990

23. Jones RM, Wu H, Wentworth C, Luo L, Collier-Hyams L, Neish AS (2008) *Salmonella* AvrA coordinates suppression of host immune and apoptotic defenses via JNK pathway blockade. Cell Host Microbe 3:233–244

24. Jones RM, Luo L, Moberg KH (2012) Aeromonas salmonicida secreted protein AopP is a potent inducer of apoptosis in a mammalian and a *Drosophila* model. Cell Microbiol 14:274–285

25. Jones RM, Luo L, Ardita CS, Richardson AN et al (2013) Symbiotic lactobacilli stimulate gut epithelial proliferation via Nox-mediated generation of reactive oxygen species. EMBO J 32:3017–3028

26. Kundu K, Knight SF, Willett N, Lee S, Taylor WR, Murthy N (2009) Hydrocyanines: a class of fluorescent sensors that can image reactive oxygen species in cell culture, tissue, and in vivo. Angew Chem 48:299–303

27. Weisburg WG, Barns SM, Pelletier DA, Lane DJ (1991) 16S ribosomal DNA amplification for phylogenetic study. J Bacteriol 173:697–703

Chapter 11

Imaging Inflammatory Hypoxia in the Murine Gut

Alyssa K. Whitney and Eric L. Campbell

Abstract

The inflammatory bowel diseases (IBD), including Crohn's disease and ulcerative colitis, result in chronic inflammation to the gastrointestinal tract. In ulcerative colitis, inflammation tends to be more superficial and restricted to the colon; contrastingly, Crohn's disease presents as patchy, more penetrative inflammation that can occur throughout the gastrointestinal tract. Other differences between these diseases include the nature of their respective immune responses—Crohn's disease presents as a Th1 and ulcerative colitis as a Th2-type inflammation. During any inflammatory episode, metabolic demand on the tissue increases accompanying the influx of inflammatory cells, increasing the demand for ATP and oxygen. When availability of oxygen is limiting, tissues become hypoxic, which results in adaptive pathways to enable survival of hypoxic episodes. The primary pathway activated is the HIF (hypoxia inducible factor) transcription factor, which regulates adaptive pathways including genes controlling glycolytic metabolism and angiogenesis. In adequately oxygenated tissues (i.e. normoxia), the HIF protein is constantly produced, but oxygen-dependent enzymes called prolyl-hydroxylases utilize available oxygen to hydroxylate HIF on proline residues, targeting it for ubiquitination and subsequent degradation. Here we describe methods for inducing, visualizing, and quantifying in vivo "inflammatory hypoxia," using the murine gut as a model system.

Key words Hypoxia, HIF, Hypoxyprobe, ODD-Luc, Inflammation

1 Introduction

The observation that inflammation and hypoxia appear to coincide in the murine gut was described in a landmark publication by Karhausen et al. [1]. Since then, the concept that hypoxia and inflammation are inextricably linked has gained much momentum [2]. Considering the steep oxygen gradient between the anoxic lumen of the gut and the highly vascularized mucosa, it would appear that not only during inflammation is hypoxia evident—indeed "physiologic hypoxia" has been reported in the epithelial cells adjacent to the colonic lumen [3, 4].

A variety of methods have been employed to detect hypoxia in vivo. Varghese et al. first reported that 2-nitroimidazole compounds form adducts in hypoxic cells in vivo [5]. These adducts only formed in hypoxic cells exhibiting a $pO_2 < 10$ mmHg.

Andrei I. Ivanov (ed.), *Gastrointestinal Physiology and Diseases: Methods and Protocols*, Methods in Molecular Biology, vol. 1422, DOI 10.1007/978-1-4939-3603-8_11, © Springer Science+Business Media New York 2016

In normoxic cells, the nitroimidazole/pimonidazole compounds passively diffuse both into and out of cells without forming adducts. Later, antibodies were raised against these adducts [6]. The pimonidazole compounds, including Pimonidazole-HCl (Hypoxyprobe-1), exhibit widespread bioavailability, with low toxicity and allow specific staining of regions of hypoxia in vivo.

In 2005, Bill Kaelin's group developed a hypoxia reporter mouse by linking the oxygen-dependent degradation domain (ODD) of HIF to Firefly Luciferase on the ROSA26 locus [7]. Thus, Luciferase is rapidly synthesized in all cells of the mouse's body; however, in normoxic tissues, Luciferase is targeted for degradation with similar kinetics to HIF (due to the HIF-ODD motif). Prolyl hydroxylase enzymes target the Luciferase for hydroxylation on proline residues and it is subsequently degraded with similar kinetics to HIF regulation. Following exposure to inflammatory conditions, hypoxia–ischemia experiments, etc., Luciferase activity can be both quantified and visualized by intraperitoneal injection of substrate (d-Luciferin) followed by induction of anesthesia and imaging on an animal imager. A strong hypoxic signal appears in the kidneys, which can obscure visualization of colonic hypoxia. In order to get reliable visualization of colonic hypoxia, it is preferable to sacrifice the mouse, remove the colon, and transfer to a petri dish [8]. Other means of quantification include homogenization of the tissues in passive lysis buffer (Promega) followed by in vitro addition of Luciferin substrate and data acquisition by Luminometer; this method is more convenient for high-throughput analysis. ODD-Luc mice can also be used to visualize tissue hypoxia by microscopy, similar to Hypoxyprobe-1. However, the antibody for detecting Firefly Luciferase works better on frozen sections [8]. Here we describe methods for inducing murine colitis, processing tissue (frozen and formalin fixed paraffin embedded) and visualizing colonic hypoxia in ODD-Luc mice (IVIS imaging and anti-Firefly Luciferase staining) and wild-type mice (Hypoxyprobe-1 staining).

2 Materials

2.1 Murine Colitis Models

1. Appropriate institutional approval should be sought; delineating the proposed breeding, housing, experimental conditions and expected outcomes, prior to experimentation.

2. Carefully select the appropriate murine model, based on the specific scientific question (*see* **Notes 1–8**).

3. Dextran sulfate sodium (DSS) solution: dissolve DSS (MW 36,000–50,000, MP Biomedicals) to the desired concentration (typically 2–3.5 % w/v) in tap water (*see* **Notes 9** and **10**).

4. 2,4,6-trinitrobenzenesulfonic acid (TNBS) skin-paint solution: dilute 5 % TNBS stock solution 1:5 v/v in 200 proof

EtOH (both Sigma) [final concentration 1 % TNBS in 80 % EtOH] (*see* **Note 11**).

5. TNBS enema solution: dilute 5 % TNBS stock solution 1:2 v/v in 200 proof EtOH to obtain a final concentration 2.5 % TNBS in 50 % EtOH.

6. Small gavage/feeding needle (~20 gauge).

7. IVIS imager (Perkin Elmer) or similar bioimaging system.

8. Luciferin solution: Prepare 30 mg/ml of d-luciferin (Caliper Lifesciences) in PBS, pH 7.2.

9. Hypoxyprobe-1 stock solution: Dissolve Pimonidazole-HCl (Hypoxyprobe Inc.) to 10 mg/ml in PBS or normal saline. Stock solution is stable for 2–4 years at 4 °C, if protected from light.

10. Dissection tools (scissors, forceps, fine scissors, fine forceps).

11. Syringe (5 ml) with 20 g × 1.5″ blunt dispensing needle attached (Brico Medical Supplies) for flushing colonic contents.

12. Scales and 500 ml plastic beaker to monitor animal weights.

13. Institutional approved anesthetic (e.g. Isoflurane, Pentobarbital, etc.).

14. Biopsy foam pads.

15. Neutral buffered formalin.

16. Tru-flow tissue cassettes.

17. 70 % EtOH in distilled water (diH$_2$O).

2.2 Hypoxyprobe Immunohisto-chemistry

1. Oven, hybridization chamber or microwave to melt paraffin.

2. Tissue-Tek slide staining system with baths set up for Xylene, 100 % EtOH, 95 % EtOH, 85 % EtOH, 70 % EtOH, and PBS.

3. Citric acid-based Antigen Unmasking Solution (Vector Laboratories).

4. Decloaking chamber (Biocare Medical).

5. Dylight™549-conjugated or FITC-conjugated monoclonal 4.3.11.3 monoclonal antibody (Hypoxyprobe Inc.).

6. Tween permeabilization solution: 0.05 % Tween-20 in PBS, pH 7.2–7.4.

7. PAP pen (Cosmo Bio).

8. Blocking solution: 1 % bovine serum albumin (BSA) in PBS, pH 7.2–7.4.

9. DAPI nuclear counterstain solution: Prepare 5 mg/ml stock solution of DAPI (Life Technologies) in diH$_2$O and dilute it 1:50,000 in PBS to make a working solution.

10. ProLong Gold antifade reagent (Life Technologies).

11. Coverslip glass and jewelers forceps to lower onto section.

12. Epifluorescent microscope with appropriate filter sets to visualize fluorescent conjugate.

13. OPTIONAL: Sodium borohydride.

14. OPTIMAL: ImageIT (Life Technologies).

2.3 Anti-Firefly Luciferase Immuno-histochemistry

1. 4 % paraformaldehyde (PFA) solution in PBS: mix equal volumes of 8 % PFA and 2× PBS. To prepare 2× PBS: dissolve 10.9 g Na_2HPO_4, and 3.2 g NaH_2PO_4 in 400 ml of diH_2O, adjust pH to 7.4 and bring up to 500 ml with diH_2O. To prepare 8 % PFA: weigh 40 g of prilled PFA, transfer to a beaker with magnetic stir bar (*see* **Note 12**). Add 500 ml of diH_2O to the beaker and transfer to a hotplate with magnetic stirrer function in a fume hood. Monitor the temperature using a thermometer placed in the beaker. Maintain the temperature at 60–65 °C (*see* **Note 13**). Incubate the solution on hotplate with continuous stirring and add 2 N NaOH, dropwise until PFA suspension is transformed into a clear solution (~5 drops of 2 N NaOH).

   Remove 8 % PFA from hotplate and allow to cool. Add 2× PBS (500 ml) and adjust pH to 7.4 with diluted HCl if necessary. When cooled to room temperature, filter, and aliquot the final PFA solution, protect it from light and freeze aliquots at −20 °C.

2. Cork tiles cut into approximately ¼″ by 3″ sections, to enable submersion in 15 ml falcon tube.

3. Insect pins (or 27-gauge needles, bent in the middle and with the plastic cautiously removed by pliers).

4. 30 % Sucrose/PBS solution: dissolve 30 g of sucrose in 100 ml of PBS.

5. Optimal cutting temperature compound (OCT) and plastic tissue molds (Sakura).

6. OCT:Sucrose/PBS solution: combine equal volumes (e.g. 25 ml) of 30 % Sucrose/PBS solution with 25 ml of OCT. Vortex until solution is homogenous.

7. Normal serum in PBS (Jackson Immunological).

8. Anti-Firefly Luciferase polyclonal antibody (MBL; PM016).

9. Fluorescently conjugated secondary antibodies (raised in the same animal as used for normal serum), we prefer AlexaFluor 488 (green) or AlexaFluor 568 (red) conjugated anti-rabbit, IgG (H + L) (ThermoFisher).

3 Methods

3.1 Induction of TNBS Colitis

1. One week prior to administration of TNBS, anesthetize and shave the bellies of the mice and skin paint with TNBS or vehicle solutions. You will need ~200 μl of either TNBS skin-paint solution or vehicle (80 % EtOH) per mouse.

2. On the day of the experiment, anesthetize the mice and administer an enema of 100 μl of TNBS solution or vehicle. Use either a flexible catheter or a small gavage needle affixed to a 1 ml syringe. To facilitate a more easy entry, lubricate the anus with a drop of solution, prior to insertion. You will need ~100 μl of either TNBS solution or vehicle (50 % EtOH) per mouse.

3. Try to make all of your instillations a similar length of time. Following instillation, suspend the mouse by the tail for 30–60 s. Before the mouse wakes take a baseline body weight measurement. Return the mouse to its cage and observe for recovery from anesthesia.

4. Weigh the mice daily to plot a weight loss curve, which is a clinical correlate to disease severity. If the mice lose more than 15 % of their body weight, they should be sacrificed.

3.2 Induction of DSS Colitis

1. Weigh mice prior to administration of DSS (day 0).

2. Provide between 2.5 and 3.5 % DSS to the drinking water of mice (*see* **Note 14**).

3. Change the DSS/drinking water every two days during the experiment. Return the mice to drinking water on Day 5. Continue to measure body weight daily. It may also be beneficial to track other signs of disease severity to compile a DAI (disease activity index). DAIs are typically arbitrary scores assigned by the investigator to aid in statistical discrimination of disease severity between experimental groups. For example: 0 = no disease evident; 1 = decreased activity; 2 = soft stool; 3 = trace of blood in stool; 4 = bloody stool; 5 = hypothermia/matted fur/severe rectal bleeding.

4. Typically peak disease does not occur until days 6–9, with resolution occurring between days 12 and 15.

5. Should any mouse lose >15 % body weight, sacrifice the animal.

3.3 Administration of Hypoxyprobe-1

1. Administer 60 mg/kg of Hypoxyprobe-1 solution to mice at least 60 min before euthanasia.

2. Harvest the entire colon in euthanized animals. Remove fecal material by flushing the lumen with PBS, using a 5 ml syringe with blunt dispensing needle attached.

3. Pre-wet a biopsy foam pad with neutral buffered formalin, place in the bottom of a labeled histology tissue cassette.

4. Transect colon and lay flat into histology cassette. Cover with another pre-wetted biopsy foam pad and close the cassette.

5. Fix in neutral buffered formalin overnight at 4 °C.

6. After 24 h remove cassettes from formalin and transfer to 70 % EtOH.

7. Process cassettes for embedding in paraffin and sectioning.

3.4 Hypoxyprobe-1 Staining of Formalin-Fixed Paraffin Embedded Tissue Sections

1. Transfer paraffin-embedded slides to dipping apparatus from Tissue-Tek staining bath system and warm to 60 °C in an oven (for <20 min).

2. Immediately immerse in Xylene bath for 10 min. After 10 min transfer to second Xylene bath (*see* **Note 15**).

3. Gradually rehydrate the tissue by cycling down through 100 % EtOH, 95 % EtOH, 85 % EtOH, 70 % EtOH, and finally PBS. Dip slides at least 10 times in each solution.

4. Replace PBS with 1× Antigen Unmasking Solution. Perform antigen retrieval by heating to 125 °C for 20 min in a decloaking chamber.

5. When chamber cools sufficiently, remove slides and allow to cool on benchtop.

6. OPTIONAL STEP: Following antigen retrieval, incubate slides in 1 % sodium borohydride in PBS (w/v) to remove unreacted aldehydes (*see* **Note 16**).

7. Perform 3× washes in PBS (5 min each).

8. Permeabilize sections with PBS-Tween for 5 min (*see* **Note 17**).

9. Perform 3× washes in PBS (5 min each).

10. Circle tissue sections with PAP pen.

11. OPTIONAL STEP: Apply drop of imageIT to slides for 30 min at room temp.

12. Block with ~50 μl of blocking buffer for 20 min at room temp.

13. Incubate with fluorescently conjugated anti-Hypoxyprobe antibody (1:100 diluted in blocking buffer overnight at 4 °C).

Ensure slides remain hydrated (*see* **Note 18**) and protect from light for the remainder of the procedure.

14. Wash slides 3 times by applying PBS for 5 min, followed by blotting on paper towels.

15. Apply DAPI counterstain and incubate for 5 min.

16. Wash slides 3 times with PBS for 5 min each.

17. IMPORTANT: Wash slides once with diH$_2$O [this prevents the formation of salt crystals during curing which can make images hazy].

15. Air dry slides in the dark [do not allow to over-dry].

16. Mount slides with a drop of antifade reagent and slowly lower a coverslip onto the slide using jeweler's forceps.

17. Allow slides to cure at room temperature in the dark for 24 h before visualization (Fig. 1, left panel).

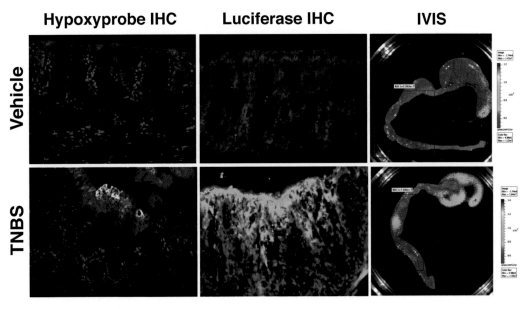

Fig. 1 Examples of different methods of in vivo imaging for "inflammatory hypoxia." *Top panels* were from vehicle (80 % EtOH)-treated mice and *bottom panels* were harvested from TNBS-treated mice 24 h post-induction. *Left panels* are from formalin-fixed paraffin embedded tissue processed for Hypoxyprobe-1 staining (*red*) and nuclear DAPI (*blue*). *Middle panels* are from OCT-embedded frozen tissue stained with anti-Firefly Luciferase (*green*) and DAPI (*blue*). *Right panels* depict Luciferase activity in d-luciferin-treated mice, colons were removed prior to imaging to accurately visualize colitis-induced hypoxia. *Red box* indicates user-defined ROI (region of interest) for subsequent quantification

3.5 Luciferase Staining in Frozen Tissue Sections

1. For each experiment, prepare fresh 4 % PFA in PBS.

2. Remove colons from sacrificed animals, flush the lumen with PBS and transect along the mesenteric line with scissors.

3. Using insect pins and thin slices of cork board, pin tissue to cork board and immerse in 15 ml Falcon tube containing 7 ml of 4 % PFA in PBS.

4. Fix overnight at 4 °C.

5. Next day dispose of 4 % PFA according to facility chemical waste guidelines. Remove pins from cork board and float tissues in 30 % Sucrose/PBS. This step is essential to prevent cell rupture, the sucrose acts to cryoprotect the tissue. When tissues become infiltrated with sucrose they will sink to the bottom of the tube.

6. Transfer tissues to 50:50 solution of OCT:30 % Sucrose/PBS and incubate prior to embedding.

7. Tissues can now be embedded in 100 % OCT.

8. [Optional step] To aid with tissue orientation: first fill the bottom of a labeled plastic mold with OCT. Place mold briefly onto a block of dry ice. Do not wait for it to freeze completely, but remove after it becomes tacky. Lay tissue in the desired orientation on the sticky OCT.

9. Place tissue in plastic mold and fill with OCT until the mold is full and the tissue completely surrounded.

10. Place mold on a block of dry ice until it is completely frozen (allows for slow and even freezing).

11. Store frozen tissue blocks at −80 °C until ready for cryosectioning. Blocks should be wrapped in foil or plastic to protect from drying out.

12. Warm tissue blocks up slowly either by placing in a pre-cooled cyrotstat or a −20 °C freezer (at least 15–20 min). Affix the block to the chuck with a drop of fresh OCT. Cut 5 μm sections and transfer to positively charged microscopy slides.

13. Warm slides to room temperature and air dry (helps sections to adhere to slides).

14. Fix in 100 % MeOH (at −20 °C) for 10 min.

15. Wash sections in 0.05 % Tween in PBS for 5 min.

16. Wash sections twice in PBS for 5 min.

17. Block in 5 % normal serum in PBS for 20 min. (CRITICAL FACTOR: the source of normal serum depends on the source of the secondary antibody used. For example, if you use a goat-derived anti-rabbit IgG secondary antibody, normal goat serum should be used.)

18. Incubate with 1:200 dilution of anti-Firefly Luciferase antibody in blocking solution at 4 °C in a humidified chamber.

19. Wash sections three times with PBS (5 min for each wash).

20. Incubate in the dark with 1:1000 dilution of fluorescently conjugated secondary anti-rabbit secondary in blocking solution, at room temperature for 1 h.

21. Wash sections three times with PBS (5 min for each wash).

22. OPTIONAL: Additional primary antibodies followed by appropriate secondary antibodies can be applied if desired.

23. Counterstain with DAPI.

24. Wash twice with PBS.

25. Wash once with diH$_2$O.

26. Air dry and mount with anti-fade reagent. Allow to cure in the dark and seal edges with nail varnish.

27. Visualize staining on epifluorescent microscope (Fig. 1, middle panel).

3.6 Quantitative In Vivo Imaging

1. Initialize IVIS or similar luminescent bioimager. Many systems may take up to 5 min to warm up.

2. Inject each mouse twice by bilateral intraperitoneal injection with 100 μl of d-luciferin using a 26-gauge needle.

3. Wait 5 min (*see* **Note 19**).

4. Sacrifice mouse by institutionally approved methods and harvest the cecum and colon.

5. **Steps 2–4** should take ~10 min, which is the optimal time for Luciferase activity.

6. Place tissue in a petri dish in the center of the imaging chamber and close the door.

7. Adjustment of exposure settings and binning may be necessary, however, once set all mice in the experiment should be recorded using the same settings.

8. Utilize bioimager software to define ROI (regions of interest) and quantify luminescence (Fig. 1, right panel).

4 Notes

1. Pimonidazole staining has the advantage that it can be used in any genetic background of mouse (i.e. wild-type or transgenic), however, staining is not quantitative.

2. ODD-Luc mice can be used to quantify Firefly Luciferase activity, however, these mice are commonly and incorrectly

referred to as "HIF reporter mice." ODD-Luc mice stabilize Luciferase in an oxygen-dependent fashion, *similar to the post-translational* regulation of HIF. It is important to note that alterations in HIF levels due to negative/positive feedback loops (such as altered mRNA synthesis, microRNA-mediated regulation, etc.) can lead to differences between tissue Luciferase and tissue HIF levels.

3. ODD-Luc mice are also raised on an FVB background, which limits immunological studies such as generation of bone marrow chimeras, adoptive transfer, etc. into the more commonly used C57/B6. Should such studies be necessary, backcrossing of ODD-Luc onto the B6 background should be performed.

4. Experimental considerations: Mice should be age- and sex-matched before undertaking a colitis experiment. Typically mice respond most consistently to colitis between 8 and 12 weeks. If possible, test groups should have littermate controls to minimize differences based on the microbiome. If purchasing mice from a commercial vendor (e.g. Jackson laboratories or Charles River), we recommend that animals be housed in the facility for at least 2 weeks for their microbiome to adjust to the new facility.

5. Experimental considerations: The choice of mouse line used will depend on the desired endpoints of the experiment. It is important to note that various commercially available mouse strains exhibit varying degrees of susceptibility to colitis [9]. For instance, if the investigators aim to elucidate the role of a transgenic animal, backcrossed to a C57/B6J background, wild-type C57/B6J mice should be used as a control for the transgene (e.g. BALBc mice would not be a suitable control).

6. Choice of colitis-inducing agent: DSS colitis is the most commonly utilized chemically induced colitis model, supposedly modeling human ulcerative colitis. It relies on erosion of colonic mucus [10] which allows direct physical interactions between the luminal bacteria and the colonic epithelial cells. The resulting inflammation is primarily granulocytic, with large numbers of neutrophils, eosinophils, and monocytes. Widespread damage to the crypt architecture occurs, with shedding or damage to the epithelium. Damage extends proximally from the anus, approximately half-way up the colon. The DSS model is not self-resolving, removal of DSS from the drinking water is necessary to allow resolution of disease.

7. Choice of colitis-inducing agent: DSS has the advantage of being easy to perform by administration in drinking water.

8. Choice of colitis-inducing agent: TNBS colitis is technically more challenging but results in a more Th1-type inflammation,

with eventual T-cell involvement, supposedly modeling human Crohn's disease. It relies on prior exposure to a hapten-forming agent on the skin, one week prior to induction of colitis. Subsequently, rectal instillation of TNBS in ethanol results in haptenization of luminal contents, accompanied by a transient barrier breach. The profile of immune cell infiltration differs from DSS and mucosal damage occurring from TNBS is not continuous, like DSS, and recapitulates skip lesions seen in human Crohn's disease. Under normal conditions TNBS colitis should be self-resolving.

9. The percentage of DSS must be worked out empirically for each animal room and facility. Variability is thought to exist due to differing microflora between different locations.

10. Use tap water, not distilled or ultrapure water, as these can have a lower pH.

11. TNBS is explosive when dry. Consult your Environmental Health & Safety office for guidelines on disposal of empty or old material.

12. Prilled PFA minimizes health risk due to inhalation hazard while weighing. Non-prilled PFA can be used, however, ensure to wear appropriate respiratory protection.

13. In order to dissolve, PFA needs to be heated to at least 60 °C and needs to be at an optimal pH of ~8. PFA will break down at temperatures exceeding 70 °C.

14. Other sources and molecular weights are available, the only form that works for colitis is DSS MW 36,000–50,000 purchased from MP Biomedical.

15. Official Hypoxyprobe website recommends Clear-Rite 3 as a less toxic alternative to Xylenes.

16. Warning: if mixing sodium borohydride in a closed container, be aware that hydrogen gas is rapidly released.

17. Official protocol recommends the use of PBS-Brij (0.2 % Brij 35 in PBS).

18. We use a 20-slide tray from Fisher scientific (cat#: 50-367-057). Place Kimwipes or blotting paper in the troughs and soak with water.

19. Bioimagers can be used to visualize live anesthetized mice, however because parts of the GI tract (including the colon, cecum, stomach, etc.) are physically obscured by other organs and because of high background from hypoxic organs, such as the kidney; imaging hypoxia in the GI tract in a living mouse is difficult.

Acknowledgement

This work was supported by a career development award from the Crohn's and Colitis Foundation of America and an NIH Grant R01-DK103639 both to ELC.

References

1. Karhausen J, Furuta GT, Tomaszewski JE, Johnson RS, Colgan SP, Haase VH (2004) Epithelial hypoxia-inducible factor-1 is protective in murine experimental colitis. J Clin Invest 114:1098–1106

2. Eltzschig HK, Carmeliet P (2011) Hypoxia and inflammation. N Engl J Med 364:656–665

3. Colgan SP, Taylor CT (2010) Hypoxia: an alarm signal during intestinal inflammation. Nat Rev Gastroenterol Hepatol 7:281–287

4. Kelly CJ, Zheng L, Campbell EL, Saeedi B et al (2015) Crosstalk between microbiota-derived short-chain fatty acids and intestinal epithelial HIF augments tissue barrier function. Cell Host Microbe 17:662–671

5. Varghese AJ, Gulyas S, Mohindra JK (1976) Hypoxia-dependent reduction of 1-(2-nitro-1-imidazolyl)-3-methoxy-2-propanol by Chinese hamster ovary cells and KHT tumor cells in vitro and in vivo. Cancer Res 36:3761–3765

6. Raleigh JA, Miller GG, Franko AJ, Koch CJ, Fuciarelli AF, Kelly DA (1987) Fluorescence immunohistochemical detection of hypoxic cells in spheroids and tumours. Br J Cancer 56:395–400

7. Safran M, Kim WY, O'Connell F, Flippin L, Gunzler V, Horner JW, Depinho RA, Kaelin WG Jr (2006) Mouse model for noninvasive imaging of HIF prolyl hydroxylase activity: assessment of an oral agent that stimulates erythropoietin production. Proc Natl Acad Sci U S A 103:105–110

8. Campbell EL, Bruyninckx WJ, Kelly CJ, Glover LE et al (2014) Transmigrating neutrophils shape the mucosal microenvironment through localized oxygen depletion to influence resolution of inflammation. Immunity 40:66–77

9. Mahler M, Bristol IJ, Leiter EH, Workman AE, Birkenmeier EH, Elson CO, Sundberg JP (1998) Differential susceptibility of inbred mouse strains to dextran sulfate sodium-induced colitis. Am J Physiol 274:G544–G551

10. Johansson ME, Gustafsson JK, Sjoberg KE, Petersson J, Holm L, Sjovall H, Hansson GC (2010) Bacteria penetrate the inner mucus layer before inflammation in the dextran sulfate colitis model. PLoS One 5:e12238

Chapter 12

Label-Free Imaging of Eosinophilic Esophagitis Mouse Models Using Optical Coherence Tomography

Aneesh Alex, Elia D. Tait Wojno, David Artis, and Chao Zhou

Abstract

Eosinophilic esophagitis (EoE) is an immune-mediated disorder characterized by esophageal inflammation and related structural changes causing symptoms such as feeding difficulties and food impaction. The pathophysiological mechanisms underlying EoE remain poorly understood. Preclinical studies using mouse models have been critical in comprehending human disease mechanisms and associated pathways. In this chapter, we describe an experimental method using a noninvasive label-free optical imaging technique, optical coherence tomography, to characterize the pathophysiological changes in the esophagus of mice with EoE-like disease ex vivo.

Key words Label-free imaging, Optical coherence tomography, Eosinophilic esophagitis, EoE mouse models, Optical imaging, Gastrointestinal imaging

1 Introduction

Eosinophilic esophagitis (EoE) is a food allergy-associated chronic immune disease that affects children and adults [1, 2]. Hallmark features of EoE include esophageal eosinophilia and inflammation, and structural esophageal changes such as strictures leading to feeding difficulties, dysphagia, and food impaction [1, 3]. The incidence of EoE has dramatically increased in the past two decades [4]. However, the current treatment strategies are nonspecific and impose significant dietary restrictions on patients [5, 6]. As a complex immune-mediated disorder, the underlying mechanisms involved in EoE pathogenesis are not clearly understood. Hence, there is an urgent need to understand signaling pathways contributing to EoE and pathophysiological changes associated with this disease in order to improve EoE diagnosis and management.

Current EoE diagnosis is based on three criteria: (1) clinical symptoms of esophageal dysfunction; (2) an esophageal biopsy with a maximum eosinophil count ≥15 eosinophils per

Andrei I. Ivanov (ed.), *Gastrointestinal Physiology and Diseases: Methods and Protocols*, Methods in Molecular Biology, vol. 1422, DOI 10.1007/978-1-4939-3603-8_12, © Springer Science+Business Media New York 2016

high power field of view, with few exceptions; and (3) exclusion of other possible causes of esophageal eosinophilia [7]. However, no single symptom is a definitive indicator of EoE, and there are broad variations in the endoscopic appearance of the esophagus of EoE patients [8, 9]. These factors make EoE diagnosis very challenging for clinicians. As per current clinical guidelines, multiple biopsies are required to be taken from patients with EoE symptoms to confirm diagnosis and to assess response to treatment [10]. Hence, identification of more definitive indicators of EoE can tremendously improve the efficacy of EoE diagnosis.

Preclinical studies using experimental animal models that mimic human disorders have been instrumental in gaining deeper understanding of disease mechanisms. For investigating immune-mediated disorders, mouse models are of particular interest due to the similarities between murine and human immunological processes. The role of specific functional pathways can also be determined using transgenic mouse strains [11, 12]. EoE pathogenesis has been linked to various immune cell types including eosinophils, mast cells and lymphocytes, and cytokines [1, 13]. Recent studies using mouse models have shown that EoE pathogenesis is linked to a gain-of-function polymorphism in the gene that encodes the primarily epithelial cell-derived cytokine, thymic stromal lymphopoietin (TSLP) [14, 15]. A mouse model of EoE-like disease associated with enhanced TSLP production exhibited pathophysiological changes in esophageal tissues similar to those observed in human EoE patients, when mice were challenged repeatedly with a model food antigen, ovalbumin (OVA) [16–19].

Such characteristic EoE-associated pathophysiological changes in the esophagus of mouse models can be evaluated using label-free imaging tools. In this chapter, we describe the application of an optical imaging technique known as optical coherence tomography (OCT) for visualizing and characterizing microstructural changes in the esophagus of mice with EoE-like disease. OCT is a promising noninvasive in vivo imaging modality capable of providing 3D structural information with micron-scale resolutions and 1–2 mm penetration depth in biological tissues [20–22]. It is a label-free technique obtaining structural images based on intrinsic scattering contrast of biological samples. OCT has been successfully used to demarcate different layers of mouse gastrointestinal (GI) tract [23–25]. We used OCT to study structural changes in the esophagus of mice with EoE-like disease ex vivo [16, 26]. Using OCT, we were able to accurately quantify structural parameters such as thickness of different esophageal layers in mice with EoE-like disease.

2   Materials

2.1 Reagents for Development of EoE-Like Disease in Mouse Models

1. Phosphate-buffered saline (PBS), pH 7.2.
2. Vitamin D analog, MC903 (Calcipotriol, Tocris Bioscience, Bristol, UK).
3. 100 % ethanol (EtOH).
4. Vitamin D analog sensitization solution: 2 nmol of MC903 in 20 μL of 100 % EtOH.
5. Ovalbumin (OVA). To make the OVA solution, dissolve 100 μg of OVA in 20 μL of PBS for application on ears, and 50 mg of OVA in 100 μL of PBS for oral delivery.

2.2 Mouse Models

1. Wild-type (WT) BALB/c mice (*Tslpr+/+* mice).
2. Mice deficient in the TSLP receptor (*Tslpr-/-* mice).
   Tslpr+/+ mice and *Tslpr-/-* mice were used in order to determine the role of the TSLP pathway in EoE pathogenesis. All mice were provided by Amgen Inc. (California, USA) through Charles River Laboratories Inc. (Massachusetts, USA). Mice were 8–12 weeks old, and all experiments used age- and gender-matched controls (*see* **Note 1**). As shown in Table 1, 4 groups of animals were used in our experiments.

2.3 OCT Imaging Components

1. Plastic tube with an outer diameter of 0.75 mm.
2. Petri dish.
3. A custom-made OCT imaging system (*see* **Note 2** for detailed system specifications).

2.4 Histology Components

1. 4 % paraformaldehyde (PFA).
2. Paraffin wax.
3. A microtome for sectioning.
4. Hematoxylin and eosin (H & E).

Table 1
Mice group classification and sample size in each group

	Tslpr+/+	*Tslpr-/-*
Ethanol + OVA	*Tslpr+/+* control (*n*=6)	*Tslpr-/-* control (*n*=4)
MC903 + OVA	*Tslpr+/+* treatment (*n*=9)	*Tslpr-/-* treatment (*n*=6)

3 Methods

3.1 Murine Models with EoE-Like Disease

1. Every day co-apply the vitamin D analog sensitization solution and OVA solution on the mouse ear. Continue this procedure for 14 days (*see* **Note 3**).

2. In parallel, treat control mice with same volumes of 100 % EtOH and OVA solution.

3. Challenge the mice intragastrically (IG) with 50 mg OVA dissolved in 100 μL PBS on days 14, 17.5, 18, 20, 22, 24, and 26. Perform the IG challenge with OVA using standard oral gavage needles (Kent Scientific, Connecticut, USA) suitable for mice.

4. After the first IG OVA challenge, give the animals continuous access to water containing 1.5 g/L OVA.

5. All the *Tslpr*$^{+/+}$ mice should develop EoE-like disease following epicutaneous sensitization and repeated challenge with OVA by day 27.

6. On day 27, fast each animal for at least 30 min prior to euthanasia. Mice can be euthanized using carbon dioxide (using a protocol approved by the University of Pennsylvania or other Institutional Animal Care and Use Committee (IACUC) or regulatory body). Animals should be washed down with ethanol. The esophagus can be found under the trachea and is dissected out using sharp surgical scissors and fine forceps. Remove the esophagus from the mouse (*see* **Note 4**) and examine the esophagus for the presence of impacted food.

3.2 Sample Preparation and OCT Imaging

1. Carefully insert a plastic tube with an outer diameter of 0.75 mm into the esophagus that has been removed from the mouse (Fig. 1a; *see* **Notes 5** and **6**). The inserted plastic tube helps to identify the esophageal lumen surface in the OCT cross-sections.

2. Place the esophagus in a petri dish. Immerse the esophagus in saline solution to keep the sample hydrated and to reduce light back-reflections from the esophageal surface (*see* **Note 7**).

3. Preview OCT cross-sectional images of the esophagus using a custom-written program in LabVIEW (National Instruments, Austin, USA) and select an appropriate imaging location. Obtain 3D OCT images from different locations along the esophagus: anterior, middle, and posterior regions (Fig. 1b). Each OCT volume covered approximately $3 \times 1.5 \times 1.5$ mm^3 (Fig. 2; *see* **Notes 8** and **9**).

4. After OCT imaging, fix the esophagus in 4 % PFA. Subsequently, embed the fixed esophageal tissues into paraffin, cut into 5 μm sections and stain with H & E for histopathology (*see* **Note 10**).

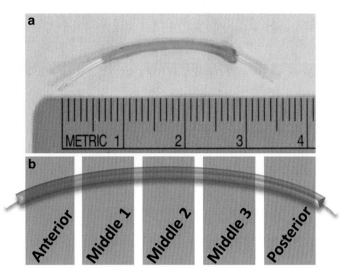

Fig. 1 (**a**) Photograph of esophagus isolated from mouse with plastic tube inserted. (**b**) Schematic indicating various OCT imaging locations along the esophagus (reproduced from Ref. 26 with permission from Optical Society of America)

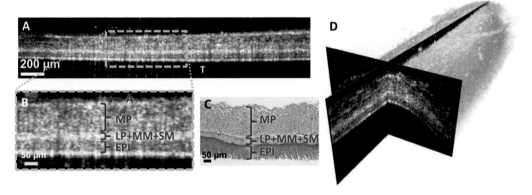

Fig. 2 (**a**) A representative OCT cross-sectional image and (**b**) its magnified view obtained from a wild-type (*Tslpr*+/+ control) mouse. Different esophageal layers can be clearly differentiated (EPI-epithelium, LP-lamina propria, MM-muscularis mucosae, SM-sub-mucosa, MP-muscularis propria, A-adventitia, T-plastic tube). (**c**) Representative histology of the normal esophagus from a wild-type mouse. (**d**) 3D rendering of an OCT volumetric dataset ($3 \times 1.5 \times 1.5$ mm^3) obtained from the middle region of the esophagus of a control mouse (reproduced from Ref. 26 with permission from Optical Society of America)

3.3 OCT Image Analysis

1. Measure the thickness values of different layers of esophagus from cross-sectional OCT images every 200 μm within each 3D OCT dataset. As the thickness of different esophageal layers in the anterior and posterior regions showed significant variations, only use the thickness values obtained from the middle regions of the esophagus (Fig. 3; *see* **Note 11**).

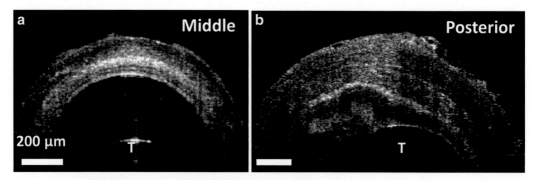

Fig. 3 Axial cross-sections showing structural differences between OCT images obtained from the (**a**) middle and (**b**) posterior regions of the esophagus in a wild-type (*Tslpr*[+/+] control) mouse. Compared to the middle region of the esophagus, the layer thickness values at anterior and posterior regions appeared to be significantly different (reproduced from Ref. 26 with permission from Optical Society of America)

2. Calculate average thickness values for individual epithelial layers from each specimen (Fig. 4).

3. Compare layer thickness values from different mice groups using Student's *t*-test.

4   Notes

1. Animals were bred and housed in specific pathogen-free conditions at the University of Pennsylvania. A licensed veterinarian provided care to any animals requiring medical attention. All experiments were performed under protocols approved by the University of Pennsylvania IACUC.

2. We developed a spectral domain OCT system, which employs a supercontinuum light source (SC-400-4, Fianium Ltd., UK) centered around 1300 nm with a bandwidth of ~200 nm, for imaging mouse esophagus ex vivo [26]. The OCT system was capable of providing axial and transverse resolutions of ~5 μm (in tissue) and ~10 μm, respectively. The OCT system consisted of a Michelson interferometer in which light was split into reference and sample arm beams, and the returning light was detected using a spectrometer. The spectrometer was comprised of an 1145 lines/mm transmission grating (Wasatch Photonics, Logan, UT, USA) and a 1024-pixel line-scan camera (SUI Goodrich, Princeton, NJ, USA). The OCT system was operated at an axial-scan (A-scan) rate of 47 kHz. The incident power on the sample was ~12 mW. The sensitivity of the OCT system was determined to be 92 dB close to the zero delay position. It was possible to image the entire thickness of the esophagus using the OCT system, thus enabling the quantification of the thickness of different esophageal layers noninvasively.

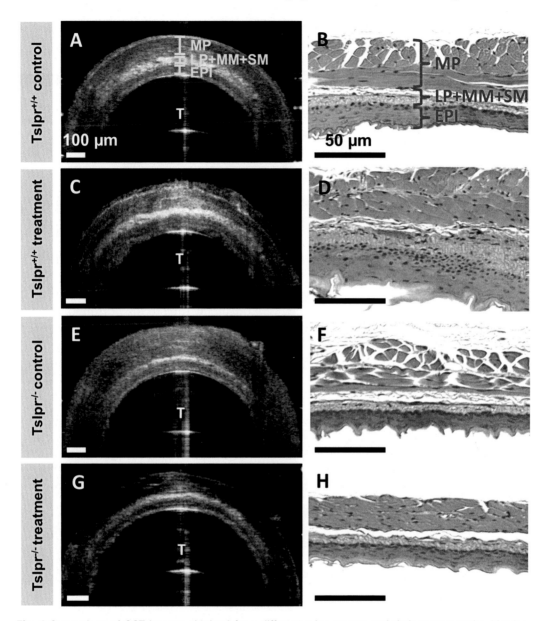

Fig. 4 Comparison of OCT images obtained from different mice groups and their representative histology. Representative OCT cross-section and histology obtained from (**a, b**) *Tslpr⁺/⁺* control, (**c, d**) *Tslpr⁺/⁺* treatment, (**e, f**) *Tslpr⁻/⁻* control, and (**g, h**) *Tslpr⁻/⁻* treatment mice groups (EPI-epithelium, LP-lamina propria, MM-muscularis mucosae, SM-sub-mucosa, MP-muscularis propria). Scale bars denote 100 μm in OCT images and 50 μm in histology. T denotes the location of the inserted plastic tube (reproduced from Ref. 26 with permission from Optical Society of America)

3. First, 2 nmol MC903 in 20 μL of 100 % EtOH is applied on the ears and allowed to dry. Subsequently, 100 μg OVA in 20 μL PBS is painted on the ears and allowed to dry.

4. The esophagus should be cut as far up as possible on the top and at the juncture with the stomach on the bottom.

5. The plastic tube is inserted in order to differentiate the esophageal lumen surface while imaging. The outer diameter (OD) of the plastic tube is a critical parameter. It should be narrow enough so that it goes through the esophagus comfortably. If the tube is too tight, it will apply pressure on the esophageal surface, which might compress the esophagus and affect the layer thickness measurements.

6. Mice with EoE-like disease exhibit symptoms such as food impaction and esophageal strictures. Hence, additional care must be taken while inserting plastic tube in those esophagi.

7. The esophagus must be placed in a stable position before imaging. Fix the edges of the plastic tube to the bottom of the petri dish to prevent them from floating in the solution/moving around while imaging.

8. It is better not use the layer thickness values obtained around the food impaction sites (Fig. 5). The impacted food will com-

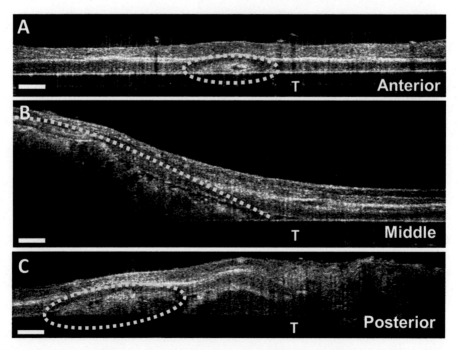

Fig. 5 Food impaction in *Tslpr*+/+ treatment group (mice with EoE-like disease). (**a–c**) OCT cross-sectional images obtained from the food impaction sites that occurred at different regions of esophagus in *Tslpr*+/+ treatment mice. *Yellow markings* show trapped food particles. Scale bars denote 200 μm and T denotes the location of the inserted plastic tube (reproduced from Ref. 26 with permission from Optical Society of America)

press the esophageal layers. Therefore, layer thickness values obtained from these regions may not be accurate.

9. In some mice, especially the diseased ones, food particles might get stuck between the inserted tube and the esophageal surface. If the impacted food is too large, it might limit the OCT beam from reaching the outer esophageal layers due to signal attenuation.

10. Eosinophils on H & E stained histological sections can be readily identified by their characteristic multi-lobular nuclei and faint pink staining in the cytoplasm. Eosinophils identified in this manner can be enumerated, and counts per high-powered field (light microscope) can be obtained to assess disease severity. Eosinophil clusters and infiltration of other immune cells tend to occur in defined regions, rather than being diffuse throughout the esophagus, so it is necessary to obtain the counts per high-powered field from each mouse by counting at least ten fields and taking an average. Some edema may also be present in the epithelial layer of the esophagus.

11. Omit the layer thickness values obtained from the anterior and posterior regions of the esophagus, as these values tend to show a large variation with respect to the thickness values obtained from the middle region. However, we imaged these regions to check the presence of impacted food and to examine the microstructural features in these regions.

Acknowledgement

This work was supported by the Lehigh University Start-up Fund and the National Institute of Health/National Institute of Biomedical Imaging and Bioengineering (NIH/NIBIB) *Pathway to Independence Award* (R00EB010071 to C. Z.). Research in the Artis lab is supported by the US National Institutes of Health (AI061570, AI087990, AI074878, AI095776, AI102942, AI095466, AI095608, and AI097333 to D.A.), T32-AI060516 and F32-AI098365 to E.D.T.W., and the Burroughs Wellcome Fund Investigator in Pathogenesis of Infectious Disease Award to D.A.

References

1. Abonia JP, Rothenberg ME (2012) Eosinophilic esophagitis: rapidly advancing insights. Annu Rev Med 63:421–434

2. Noel RJ, Putnam PE, Rothenberg ME (2004) Eosinophilic esophagitis. N Engl J Med 351:940–941

3. Straumann A, Spichtin H-P, Grize L, Bucher KA, Beglinge RC, Simon H-U (2003) Natural history of primary eosinophilic esophagitis: a follow-up of 30 adult patients for up to 11.5 years. Gastroenterology 125:1660–1669

4. Straumann A, Simon H-U (2005) Eosinophilic esophagitis: escalating epidemiology? J Allergy Clin Immun 115:418–419

5. Arora A, Weiler C, Katzka D (2012) Eosinophilic esophagitis: allergic contribution,

testing, and management. Curr Gastroenterol Rep 14:206–215

6. Straumann A, Schoepfer AM (2012) Therapeutic concepts in adult and paediatric eosinophilic oesophagitis. Nat Rev Gastroenterol Hepatol 9:697–704

7. Collins MH (2008) Histopathologic features of eosinophilic esophagitis. Gastrointest Endosc Clin N Am 18:59–71

8. Dellon ES (2012) Eosinophilic esophagitis: diagnostic tests and criteria. Curr Opin Gastroen 28:382–388

9. Diniz LO, Putnum PE, Towbin AJ (2012) Fluoroscopic findings in pediatric eosinophilic esophagitis. Pediatr Radiol 42:721–727

10. Furuta GT, Liacouras CA, Collins MH, Gupta SK et al (2007) Eosinophilic esophagitis in children and adults: a systematic review and consensus recommendations for diagnosis and treatment. Gastroenterology 133:1342–1363

11. Dinarello CA (2009) Immunological and inflammatory functions of the interleukin-1 family. Annu Rev Immunol 27:519–550

12. Aidinis V, Chandras C, Manoloukos M, Thanassopoulou A et al (2008) MUGEN mouse database; animal models of human immunological diseases. Nucleic Acids Res 36(suppl 1):D1048–D1054

13. Mulder DJ, Justinich CJ (2010) B cells, IgE and mechanisms of type I hypersensitivity in eosinophilic oesophagitis. Gut 59:6–7

14. Rothenberg ME, Spergel JM, Sherrill JD, Annaiah K et al (2010) Common variants at 5q22 associate with pediatric eosinophilic esophagitis. Nat Genet 42:289–291

15. Sherrill JD, Gao P-S, Stucke EM, Blanchard C et al (2010) Variants of thymic stromal lymphopoietin and its receptor associate with eosinophilic esophagitis. J Allergy Clin Immun 126:160–165

16. Noti M, Wojno EDT, Kim BS, Siracusa MC et al (2013) Thymic stromal lymphopoietin-elicited basophil responses promote eosinophilic esophagitis. Nat Med 19:1005–1013

17. DeBrosse CW, Franciosi JP, King EC, Butz BKB et al (2011) Long-term outcomes in pediatric-onset esophageal eosinophilia. J Allergy Clin Immun 128:132–138

18. Bhattacharya B, Carlsten J, Sabo E, Kethu S et al (2007) Increased expression of eotaxin-3 distinguishes between eosinophilic esophagitis and gastroesophageal reflux disease. Hum Pathol 38:1744–1753

19. Hsu Blatman KS, Gonsalves N, Hirano I, Bryce PJ (2011) Expression of mast cell-associated genes is upregulated in adult eosinophilic esophagitis and responds to steroid or dietary therapy. J Allergy Clin Immun 127:1307–1308

20. Huang D, Swanson EA, Lin CP, Schuman JS et al (1991) Optical coherence tomography. Science 254:1178–1181

21. Fercher AF, Hitzenberger CK, Kamp G, El-Zaiat SY (1995) Measurement of intraocular distances by backscattering spectral interferometry. Opt Commun 117:43–48

22. Drexler W, Fujimoto JG (2008) Optical coherence tomography: technology and applications. Springer, Berlin, p 1376

23. Hariri LP, Tumlinson AR, Wade NH, Besselsen DG, Utzinger U, Gerner EW, Barton JK (2007) Ex vivo optical coherence tomography and laser-induced fluorescence spectroscopy imaging of murine gastrointestinal tract. Comp Med 57:175–185

24. Winkler AM, Rice PFS, Drezek RA, Barton JK (2010) Quantitative tool for rapid disease mapping using optical coherence tomography images of azoxymethane-treated mouse colon. J Biomed Opt 15:041512

25. Iftimia N, Iyer AK, Hammer DX, Lue N et al (2012) Fluorescence-guided optical coherence tomography imaging for colon cancer screening: a preliminary mouse study. Biomed Opt Express 3:178–191

26. Alex A, Noti M, Wojno EDT, Artis D, Zhou C (2014) Characterization of eosinophilic esophagitis murine models using optical coherence tomography. Biomed Opt Express 5:609–620

Chapter 13

Near-Infrared Fluorescence Endoscopy to Detect Dysplastic Lesions in the Mouse Colon

Elias Gounaris, Yasushige Ishihara, Manisha Shrivastrav, David Bentrem, and Terrence A. Barrett

Abstract

Near-infrared fluorescence (NIRF) endoscopy has a great potential for efficient early detection of dysplastic lesions in the colon. For preclinical studies, we developed a small animal NIRF endoscope and successfully used this device to identify dysplastic lesions in a murine model of chronic colitis. In this chapter, we present a step-by-step protocol for using NIRF endoscopy to examine the location, the size, and the borders of the dysplastic lesions developed in murine colitis. Our studies suggest that NIRF endoscopy is a specific and sensitive technique that provides a unique opportunity to analyze early stages of tumorigenesis in animal models of colon cancer and to perform surveillance colonoscopy in patients with colitis-associated colon cancer.

Key words Near-infrared probes, Endoscopy, Cathepsin activity, Mucosal inflammation, Mouse colon dysplasia, Colon cancer

1 Introduction

Early detection of dysplastic lesions is the cornerstone for effective colon cancer diagnosis and treatment [1, 2]. Over the past few decades, improvements in endoscope technology have enabled earlier therapeutic interventions [3] thereby achieving a marked reduction in the overall incidence of colon cancer [4]. However, a risk of colon cancer development in patients with a history of inflammatory bowel diseases remains high and this requires development of new efficient imaging procedures for early cancer diagnosis and monitoring.

Conventional wide-angle high-definition (HD) white light endoscopy can detect dysplastic and/or neoplastic lesions based on observed structural changes to the lumen [5]. These structural changes most commonly correspond to the formation of dysplastic polyps; however, many structural aberrations, such as benign pseudo-polyps, remain difficult to accurately characterize with

Andrei I. Ivanov (ed.), *Gastrointestinal Physiology and Diseases: Methods and Protocols*, Methods in Molecular Biology, vol. 1422, DOI 10.1007/978-1-4939-3603-8_13, © Springer Science+Business Media New York 2016

white light endoscopy methods. Pseudo-polyps commonly result in false positives with no dysplastic lesions detected histologically after colectomy. It also remains particularly difficult to detect intra-epithelial neoplasia (flat lesions) with white light endoscopy. These lesions do not produce gross structural changes to the lumen and thus evade detection with conventional endoscopy methods (detection frequency between 40 to 60 %) [6].

Chromo-endoscopy involves topical intra-luminal application of either a methylene blue or indigo carmine solution to differentially alter the optical characteristics of normal and dysplastic tissues. Chromo-endoscopy can be more effective for detection of flat lesions compared to white-light approaches. Random and/or the educated collection of biopsies can improve sensitivity for flat lesion detection. However, recent studies suggest that >45 biopsy samples must be collected to achieve an 85 % probability of detecting a flat lesion with a size on the order of 12 cm^2. Additional studies comparing the sensitivity and specificity of chromo-endoscopy and random biopsy collection demonstrate that many lesions are left undetected by these individual modalities (i.e. some lesions detected with chromo-endoscopy but not random biopsy, and vice versa) [7].

Confocal endoscopes were developed to improve specificity during endoscopic surveillance of the colon. For confocal endoscopy procedures, lesions continue to be detected based on the structural alterations observed with HD white light endoscopy methods; detected lesions are then examined within a small field-of-view with confocal endo-microscopy [8]. The use of fluorescein is advocated as this alters the optical properties of the lumen to improve sensitivity and specificity. More recent studies suggest that topical application of indigo carmine solutions may permit the detection of intra-epithelial neoplasia with confocal endoscopy [9]. However, despite these developments, confocal endo-microscopy has failed to be clinically adopted for endoscopic surveillance of the colon because this technology a) remains principally dependent upon conventional HD white light endoscopy methods for lesion detection; b) requires much longer examination time; c) requires physical contact between the confocal endoscope tip and the lumen, which could distort luminal structures and complicate data interpretation [10].

Recent efforts have been made to develop more sensitive and specific endoscopic approaches involving detection of the lesions after introduction of exogenous fluorescent probes with either specific affinity to colon cancer cells (peptides tagged with fluorescein for human studies or with Cy5.5 dye for murine models [11]), or selective activation by biological activities that are specifically elevated within dysplastic and/or neoplastic tissues [12]. However, this approach suffers from poor tissue penetration of the probes, which limits their binding to the cells directly exposed into the gut lumen. Such poor availability of fluorescence probes limits the overall sensitivity of the technique. Furthermore, the described approaches significantly increase the examination time due to the requirement to wash out colonic mucus before spraying the fluorescent probes and to wash out the unbound probe prior to imaging.

Cysteine cathepsins are lysosomal proteolytic enzymes. Cathepsins are more active in myeloid cells, which are the main cause of innate inflammatory reactions against newly formed dysplastic lesions [13, 14]. In vivo endoscopic detection of cathepsin activity with intravenously administered NIRF probes may offer a highly sensitive and specific technique for detection of dysplastic lesions.

Cathepsin activity can be detected in situ using substrate-based probes (SBP). The intact probe does not emit fluorescence, but their derivatives, produced by enzymatic reactions are characterized by strong fluorescence [15]. The imaging of SBP involves monitoring accumulation of their products generated by specific enzymatic processes. For the cathepsin probes, the longer the substrate is exposed to the specific cathepsins, the stronger the fluorescence signal during NIRF imaging studies. The cathepsin-targeting SBP are tagged with fluorophores that emit at either the 680 or 750 nm wavelengths. ProSense 680 or 750 cathepsin probes are commercially available, and they represent rather nonspecific substrates for a broad range of different cathepsins. However, alternative SBP have also been developed with increased affinity toward specific cathepsins, such as cathepsins B or K [16].

Our studies demonstrated that NIRF cathepsin activity probes can be used to detect polyps in transgenic mouse models by using an invasive intravital microscopy technique [13]. Most recently, we developed a single-channel NIRF endoscopic system that is able to discriminate colitis-induced dysplasia from the inflamed mucosa in a murine model of chronic colitis [14]. For this study, we used a NIRF endoscope with excitation light produced from a Xe light source (excitation at 680 nm after a spectral separation) and the ProSense 680 cathepsin probe. This endoscopic system was able to detect polypoid lesions as well as intraepithelial flat lesions with sensitivity and specificity of 92 % [14]. In the present chapter, we report the development and systemic evaluation of this new NIRF endoscope.

2 Materials

2.1 Chemicals

1. Phosphate Buffered Saline (PBS), pH 7.4 (*see* **Note 1**).

2. Oxygen tank.

3. OCT cryo-embedding medium.

4. Fluorescent cathepsin activity probes: ProSence 680, CathB 680, and CathB 750 (Perkin Elmer).

5. Vascular contrast enhancer, AngioSense 750 (Perkin Elmer).

2.2 Animals and Surgical Tools	1. APC$^{\Delta468}$—Center for Comparative Medicine (CCM); Northwestern University (*see* **Note 2**).
	2. TS4cre/APC$^{lox/+}$ mice CCM, Northwestern University (*see* **Note 3**).
	3. IL-10$^{-/-}$ mice treated with piroxicam, CCM, Northwestern University (*see* **Note 4**).
	4. Tweezers.
	5. Blunt-end scissors.
	6. Gavage needles.
2.3 Endoscopy	1. Xenon light source (*see* **Note 5**).
	2. Laser light source (*see* **Note 6**).
	3. Fiberscopes (*see* **Note 7**).
	4. Isoflurane vaporizer (*see* **Note 8**).
	5. Temperature-regulated plate (*see* **Note 9**).
	6. A windows computer with the appropriate software for the image registration (*see* **Note 10**). A simplified schematic representation of the endoscope is shown in Fig. 1.

3 Methods

According to the Animal Study Protocol approved by Northwestern University's Center of Comparative Medicine, working with NIRF endoscope requires two trained scientists. One person should control animal anesthesia and operates the registration software, whereas the other person should operate the endoscope, controls the path of the fiberscope and the insufflation of the animal's colon.

3.1 Pre-endoscopy Preparation	1. Inject the mice with 2 nmoles of the fluorescent probes before the imaging session. We routinely perform a retro-orbital injection of the probes (*see* **Note 11&12**).
	2. Deeply anesthetize the mice by exposing to 3 % isoflurane and maintain anesthesia by inhaling 2 % isoflurane in 1.5 L/h oxygen delivered by the nozzle (*see* **Note 13**).
	3. Wash the mouse colon using 5 ml of PBS (*see* **Note 14**). PBS additionally acts as lubricant for the insertion of the fiberscope.
3.2 Endoscopy	1. Place the mouse on the thermal regulated plate on its back.
	2. Secure the anesthesia nozzle.
	3. Switch on the BF-XP60 fiberscope (*see* **Note 15**).
	4. Observe both bright field and fluorescent images on the screen. The images are recorded with a rate of 10 frames per second.

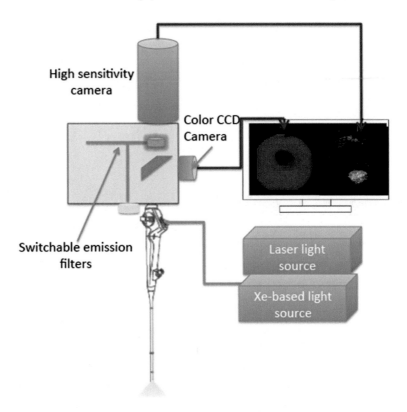

Fig. 1 A simplified diagram of the NIRF endoscopic system. The excitation light going through a set of filters, passes through the fiberscope and the fluorescent and reflectance light is split through a dichroic filter. The color CCD camera detects the white or reflectance light and the fluorescence emission light is directed to the high-sensitivity CCD camera. Both images are projected simultaneously in the computer screen and registered in the computer at the rate of 10 frames/s

5. Insert the fiberscope 6–7 cm (close to cecum).

6. Record fluorescence signal continuously from cecum to rectum by carefully pulling out the fiberscope (Figs. 2 and 3).

7. Collect biopsies as required.

3.3   Data Analysis

1. Use the plugin LOCI of Image J software to open either white light (WL) or fluorescent (FL) images

2. Use a square region of interest (ROI; 30×30 pixels).

3. Use the function "measure" to calculate the mean fluorescence intensity (MFI) of the high emissions part of the image, the tissue noise, and electronic noise. A low emission of the tissue area in the same frame can serve as tissue noise. Since the images we collect are circular and are placed in a square photo, the MFI of the electronic noise can be calculated by selecting the ROI of the non-tissue area.

white light   fluorescence light   pseudocolor

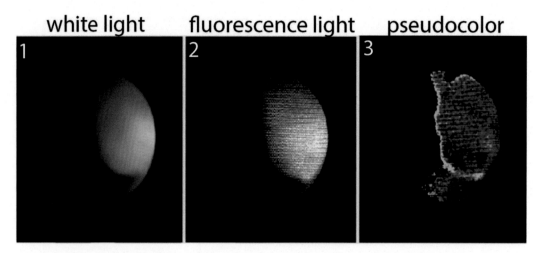

Fig. 2 Different examination modes of NIRF endoscopy. The screenshot demonstrates: (**a**) a white light image; (**b**) a fluorescence image; and (**c**) a fluorescence image with pseudo color

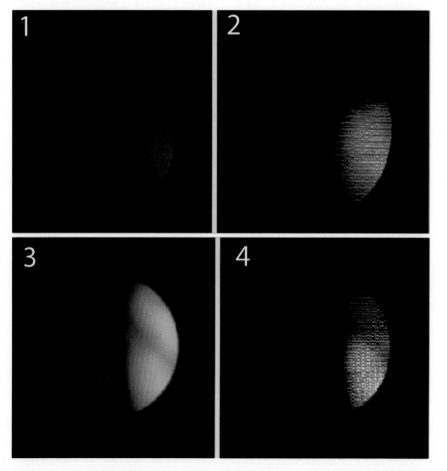

Fig. 3 A typical image of the polyp obtained with laser NIRF endoscope (**a**, **b**) and Xenon light endoscope (**c**, **d**). The polyp was imaged with reflectance excitation (**a**, **c**) and fluorescence excitation and detection (**b**, **d**). Note that the same polyp was used to produce all images

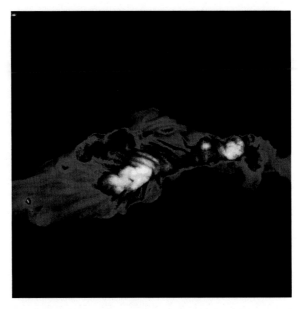

Fig. 4 A typical reflectance fluorescence of a flayed opened colon. The image is pseudo colored using a smart color set of Image J

4. Calculate the signal-to-noise ratio using the formula: [Signal-to-noise ratio = {MFI of high emission-MFI of tissue noise}/ MFI of electronic noise.

3.4 Post-mortem Analysis

1. Flay open the excised colon.

2. Wash the colon with ice-cold PBS.

3. Image reflectance fluorescence of the colon (Fig. 4 and *see* **Note 16**).

4. Compare the size and the numbers of the lesions with the ones observed with the endoscope.

5. Swiss roll the colon, fix it in OCT and section it (5 μm) until completion (*see* **Note 17**).

6. Stain every tenth sections with hematoxylin and eosin (H&E) to verify that each area with high emissions correlated to a dysplastic lesion (Fig. 5).

7. Alternatively, collect biopsies using a suitable biopsy scissor. Section the biopsies mounted in OCT and stain them with H&E for observation (Fig. 6).

4 Notes

1. PBS is not required to be sterile although most of the companies offer it only in sterile form.

2. APC$^{\Delta 468}$ mice develop polyps mostly in the small bowel (~80 polyps per mice). Additionally, 4–8 polyps could be found in

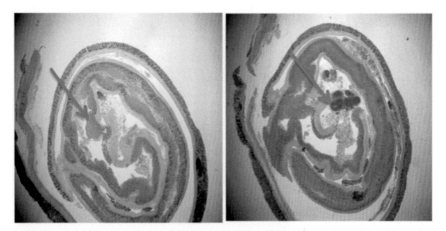

Fig. 5 A typical hematoxylin and eosin staining of murine intestinal Swiss rolls. *Arrows* indicate dysplastic lesions

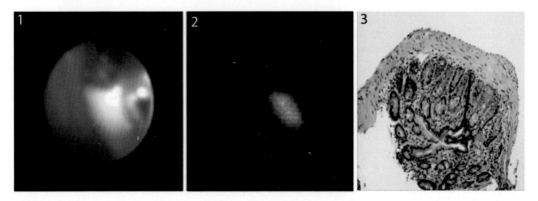

Fig. 6 The NIRF endoscopy can direct a targeted biopsy. (**a**) White light image of a polyp; (**b**) NIRF endoscopy of the same tissue; (**c**) hematoxylin and eosin staining of the biopsy collected from the imaged site that histologically confirms the existence of dysplastic polyp

the colon. The useful age of these mice is 4 months, which is the animal age for the majority of our experiments.

3. TS4cre/APC$^{lox/+}$ mice represent the conditional equivalent of APC$^{\Delta468}$ mice. Since the Cre recombinase is expressing in the colon and in the terminal part of ileum though mice develop polyps in the colon only. We imaged the mice at the age of 4–6 months.

4. IL 10$^{-/-}$ mice are characterized by the constitutive deletion of an anti-inflammatory cytokine IL-10. Piroxicam treatment triggers an acute inflammatory reaction in the colon. In short, 6–8 weeks old IL 10$^{-/-}$ mice were fed with a piroxicam containing chow for 2 weeks. The dose is 60 mg/kg of the chow during the first week and 80 mg/kg during the second. After these 2 weeks the mice were fed with regular chow. Due to the

loss of function of IL-10 this acute inflammation at day 14 becomes chronic at day 28 from the onset of the treatment. At day 56 more than 60 % of the mice develop dysplastic colonic lesions. Histological analysis showed that some of the lesions are invasive and intraepithelial [14].

5. We modified a conventional light source with a Xe lamp followed by spectral separation by filters. Due to the spectral separation the intensity of the light at the end of the fiberscope is sufficient to activate the NIRF probes at least at the 680 nm, but also reasonably low to avoid bleaching the probe.

6. The laser source consists of two lasers, with 633 and 740 nm excitation lines. This laser source offers a superb spectral separation and minimal cross talk between the channels. We used this light source to compare various probe in vivo in the same lesions. The light intensity at the end of the fiberscope is at least five fold higher than the spectrally separated Xe light. The laser source is accompanied with two laser controllers, one for each laser, and two thermal controllers. For our experiments we used the 2/3 of the laser power (~10,000/15,000), while the two laser thermal controllers were operating in maximum power.

7. During our studies, we used three fiberscopes that were linked with the Olympus Universal link. In short, these fiberscopes have a wide spectrum of insertion tube diameters varying from 0.8 to 2.8 mm to provide a wide field of view. Some fiberscopes have bending sections to facilitate their insertion into bending colon sections. We mostly used the BF-XP60 fiberscope, which although has larger tube diameter, also offers an instrument channel (1.2 mm diameter) for biopsy scissors, and insufflating the colon.

8. We used a typical vaporizer with a mixture of 2–3 % of isoflurane in oxygen. The flow of oxygen is 1.5 L/h. There are two isoflurane collectors that are close to the tip of the vaporizer. The weight of the collectors is measured frequently as a measure of isoflurane collected.

9. Mice subjected to anesthesia and endoscopy session should be kept on a heated plate with temperature regulation (37 °C), since anesthesia impairs the ability of the animal to regulate body temperature.

10. The software can operate in various modes:

    (a) *Registration* to register the newly obtained images. Images from both cameras are registered with a frame rate 10 frames/s and stored as single files simultaneously

    (b) *Examination* to examine old files (Fig. 2).

    (c) *Calibration* to calibrate the two cameras of the endoscope.

11. Inject the probes following the manufacturer specifications. In short:

    (a) ProSense 680, 24 h prior the endoscopy session,

    (b) six hours for CathB 680 fast, CathK 680 fast, and

    (c) 10 min for AngioSense 750.

    All the SBPs we used have a peptide backbone comprising poly-l-Lysine and can be diluted in sterile PBS (dilution 1:10, according to the manufacturer's specifications). Other NIRF probes have a different scaffold and different mode of action as they act as suicide inhibitors. These probes require solvents of DMSO varying from 10 to 40 %.

12. Retro-orbital injections are useful for water-soluble probes. For the probe solutions with high DMSO concentrations, we use a tail vein injection.

13. A good method to check the anesthesia level is to apply pressure on the mouse paw. Anesthetized animal would not react to this pressure. Mice can remain anesthetized for more than 2 h.

14. We use gavage needles for the colon washing. The needle can be inserted up to 3–4 cm in the colon.

15. The BF-XP60 fiberscope has an instrument channel, which serves as a channel to insufflate the colon. To do that, we typically use a 5 ml syringe pressing air.

16. For reflectance fluorescence, we use an Olympus OV100 instrument modified to produce NIRF excitation light and detect NIRF emissions. In general, the instrument operates like a low magnification microscope. The whole mount of flayed open colon is placed on a glass plate and exposed to the excitation light. The emissions of the colon are recorded as photographs (tif files). With this reflectance fluorescence, we can detect areas of high emissions ex vivo and correlate them with our in vivo observation with the NIRF endoscope.

17. Usually, more than 150 sections can be collected from an individual Swiss roll.

References

1. Rutter MD (2011) Surveillance programmes for neoplasia in colitis. J Gastroenterol 46:1–5

2. Mayinger B, Neumann F, Kastner C et al (2008) Early detection of premalignant conditions in the colon by fluorescence endoscopy using local sensitization with hexaminolevulinate. Endoscopy 40:106–109

3. Parikh ND, Perl D, Lee MH et al (2014) In vivo diagnostic accuracy of high-resolution microendoscopy in differentiating neoplastic from non-neoplastic colorectal polyps: a prospective study. Am J Gastroenterol 109:68–75

4. Choi PM, Nugent FW, Schoetz DJ et al (1993) Colonoscopic surveillance reduces mortality from colorectal cancer in ulcerative colitis. Gastroenterology 105:418–424

5. Pineton de Chambrun G, Peyrin-Biroulet L, Lémann M et al (2010) Clinical implications of mucosal healing for the management of IBD. Nat Rev Gastroenterol Hepatol 7:15–29

6. Rubin DT, Rothe JA, Hetzel JT et al (2007) Are dysplasia and colorectal cancer endoscopically visible in patients with ulcerative colitis? Gastrointest Endosc 65:998–1004

7. Marion JF, Waye JD, Present DH et al (2008) Chromoendoscopy-targeted biopsies are superior to standard colonoscopic surveillance for detecting dysplasia in inflammatory bowel disease patients: a prospective endoscopic trial. Am J Gastroenterol 103:2342–2349

8. Kiesslich R, Goetz M, Lammersdorf K et al (2007) Chromoscopy-guided endomicroscopy increases the diagnostic yield of intraepithelial neoplasia in ulcerative colitis. Gastroenterology 132:874–882

9. Rispo A, Castiglione F, Staibano S et al (2012) Diagnostic accuracy of confocal laser endomicroscopy in diagnosing dysplasia in patients affected by long-standing ulcerative colitis. World J Gastrointest Endosc 4:414–420

10. Rutter MD, Saunders BP, Wilkinson KH et al (2004) Most dysplasia in ulcerative colitis is visible at colonoscopy. Gastrointest Endosc 60:334–339

11. Wang TD, Friedland S, Sahbaie P et al (2007) Functional imaging of colonic mucosa with a fibered confocal microscope for real-time in vivo pathology. Clin Gastroenterol Hepatol 5:1300–1305

12. Li M, Anastassiades CP, Joshi B et al (2010) Affinity peptide for targeted detection of dysplasia in Barrett's esophagus. Gastroenterology 139:1472–1480

13. Gounaris E, Tung CH, Restaino C et al (2008) Live imaging of cysteine-cathepsin activity reveals dynamics of focal inflammation, angiogenesis, and polyp growth. PLoS One 3, e2916

14. Gounaris E, Martin J, Ishihara Y et al (2013) Fluorescence endoscopy of cathepsin activity discriminates dysplasia from colitis. Inflamm Bowel Dis 19:1339–1345

15. Weissleder R, Tung CH, Mahmood U et al (1999) In vivo imaging of tumors with protease-activated near-infrared fluorescent probes. Nat Biotechnol 17:375–378

16. Leuschner F, Panizzi P, Chico-Calero I et al (2010) Angiotensin-converting enzyme inhibition prevents the release of monocytes from their splenic reservoir in mice with myocardial infarction. Circ Res 107:1364–1373

Chapter 14

Visualization of Signaling Molecules During Neutrophil Recruitment in Transgenic Mice Expressing FRET Biosensors

Rei Mizuno, Yuji Kamioka, Yoshiharu Sakai, and Michiyuki Matsuda

Abstract

A number of chemical mediators regulate neutrophil recruitment to inflammatory sites either positively or negatively. Although the actions of each chemical mediator on the intracellular signaling networks controlling cell migration have been studied with neutrophils cultured in vitro, how such chemical mediators act cooperatively or counteractively in vivo remains largely unknown. To understand the mechanisms regulating neutrophil recruitment to the inflamed intestine in vivo, we recently generated transgenic mice expressing biosensors based on FRET (Förster resonance energy transfer) and set up two-photon excitation microscopy to observe the gastrointestinal tract in living mice. By measuring FRET in neutrophils, we showed activity changes of protein kinases in the neutrophils recruited to inflamed intestines. In this chapter, we describe the protocol used to visualize the protein kinase activities in neutrophils of the inflamed intestine of transgenic mice expressing the FRET biosensors.

Key words FRET, Two-photon microscopy, Intravital imaging, Neutrophil recruitment, Intestinal inflammation

1 Introduction

The hallmark of acute inflammation is the recruitment of neutrophils circulating in the blood vessels to the damaged tissues. This process, called the neutrophil recruitment cascade, is subdivided into the following five steps: rolling, adhesion, crawling, transmigration, and chemotaxis to the damaged tissues [1–5]. Previous in vitro studies have revealed the molecular mechanisms regulating neutrophil recruitment by specific chemokines or cytokines; however, it has not been demonstrated that similar mechanisms operate in vivo. Although the recent advent of in vivo microscopy has enabled us to visualize the neutrophil recruitment to inflammatory sites in living animals [2–4], the activity change of signaling molecules has not been examined because of technical constraints. To overcome this problem, we recently generated transgenic mice expressing FRET

Andrei I. Ivanov (ed.), *Gastrointestinal Physiology and Diseases: Methods and Protocols*, Methods in Molecular Biology, vol. 1422, DOI 10.1007/978-1-4939-3603-8_14, © Springer Science+Business Media New York 2016

biosensors [6] and revealed the role of p42/44 extracellular signal-regulated kinase (ERK) and protein kinase A (PKA) in neutrophil recruitment in vivo [7]. We used two-photon excitation microscopy (TPEM) for the intravital imaging for the following reasons: The infrared light used for TPEM enables us to image to a depth of at least 100 μm in most organs. Furthermore, in comparison to the conventional fluorescent microscopy, two-photon excitation causes less photo-bleaching, allowing long-term time-lapse imaging.

Although blood vessels in the mesentery or cremaster have frequently been used for the observation of neutrophil recruitment in vivo [8–11], we prefer small intestines for the following reasons. First, the intestinal wall is rich in postcapillary venules, where extravasation of neutrophils occurs. Second, intestinal inflammation, such as that caused by lipopolysaccharide (LPS), is more relevant to human diseases. Third, drugs or chemoattractants can be easily administered to the lumen of the gastrointestinal tract. However, intravital imaging of the gastrointestinal tract has been a difficult task because of drifts caused by peristalsis, heart beat, and breathing. To minimize such drift, several researchers have reported methods for intravital imaging of the gastrointestinal tract using custom-made equipment [12–16]. Nonetheless, setting up such equipment requires much time and effort, and therefore intravital imaging of the gastrointestinal tract remains a challenge. In this chapter, we introduce a simple and easy method for observation of the mouse small intestine. We also introduce a method to visualize the activities of signaling molecules in the small intestine by intravital TPEM with transgenic mice expressing FRET biosensors. Specifically, the intravital imaging procedure described in this chapter is for imaging of the inflamed mouse small intestine and is a modification of a previously published method [7]. In conclusion, this chapter describes a method that can be used for analysis of the activities of signaling molecules during neutrophil recruitment in vivo using transgenic mice expressing FRET biosensors (*see* Fig. 1 for an overview).

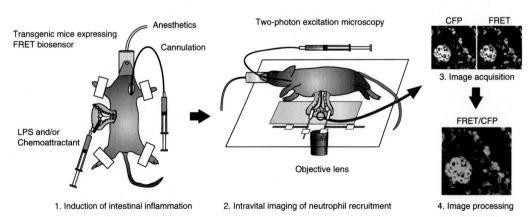

Fig. 1 Schematic view of the experimental setup of intravital imaging of molecular activities during neutrophil recruitment to the inflamed intestines

2 Materials

2.1 Reagents

All reagents and equipment can be substituted with appropriate alternatives from other manufacturers.

1. FRET Mice: Transgenic mice expressing the ERK FRET biosensor EKAREVnes, called Eisuke mice, the PKA FRET biosensor AKAR3EVnes, called PKAchu mice, and other FRET biosensors, collectively called FRET mice, have already been reported [6] and are distributed by the JCRB Laboratory Animal Resource Bank at the National Institute of Biomedical Innovation (Osaka, Japan). Adult FRET mice of either sex were used to visualize the activity of signaling molecules during neutrophil recruitment. Mice were housed in a specific pathogen-free facility and received a routine diet of chow and water ad libitum. All animal studies must be reviewed and approved by the institutional committees.

2. Isoflurane: 100 % solution for inhalation anesthesia (Abbott Laboratories, Abbott Park, IL, USA).

3. Disinfectant: 70 % (vol/vol) ethanol in water.

4. Qtracker 655 vascular labels (Life Technologies, Grand Island, NY, USA).

5. Lipopolysaccharide (LPS).

6. N-formyl-Met-Leu-Phe (fMLP).

7. Phosphate-buffered saline (PBS).

8. Intravenous flush solution: Heparin sodium (100 IU/ml; TERUMO, Tokyo, Japan).

9. PE-conjugated anti-Gr1 antibody (Biolegend, San Diego, CA, USA).

2.2 Equipment and Supplies

1. An inverted two-photon excitation microscope (e.g., IX83/FV1200MPE inverted microscope; Olympus).

2. Silicon oil-immersion objective lens (UPLSAPO 30×/1.05 NA; Olympus).

3. Optical filters used for dual emission imaging: BA685RIF-3 IR-cut filter, two dichroic mirrors DM505 and DM570, and three emission filters, BA460-500 for CFP, BA520-560 for YFP, and 645/60 (Chroma) for Qtracker 655 (all from Olympus).

4. Mercury arc lamp and power supply.

5. InSight DeepSee Laser (Spectra Physics, Santa Clara, CA, USA).

6. MetaMorph software (Molecular Devices LLC, Sunnyvale, CA, USA).

7. Small animal clippers.

8. OPSITE film dressings (Smith & Nephew).

9. Scissors.

10. Forceps.

11. Surgical needle with suture (6/0 silk, Ethicon).

12. Needle holder.

13. Ligature (5/0 silk).

14. Disposal syringes (1 ml) and insulin syringes with a permanently attached 29-gauge needle.

15. General lab supplies: sterile centrifuge tubes, pipettes, alcohol wipes, Kimwipes.

16. Gas anesthesia vaporizer.

17. Cotton swabs.

18. Stereomicroscope (e.g., SZX16; Olympus).

19. Heating pad.

20. Coverslips.

21. Adhesive tape.

22. VASCU-STATT single-use bulldog clamps; straight mini (Scanlan International Inc., Saint Paul, MN, USA).

23. Plastic wrap.

24. Catheter for mice (PUFC-C20-10; Solomon Scientifics, San Antonio, TX, USA).

---

3 Methods

3.1 Operation for Intravital Imaging of the Small Intestine

Preoperative preparations such as fasting and mechanical bowel preparation are not required. Ensure that there is sufficient isoflurane in the vaporizer to perform surgery.

3.1.1 Surgery for the Induction of Intestinal Inflammation

1. Anesthetize a mouse in an induction chamber using 2 % isoflurane inhalation.

2. Place the mouse in the supine position with the nose in a facemask and reduce the isoflurane to 1.5 %. Ensure that the mouse is on a heating pad to maintain the body temperature during surgery.

3. Shave the abdominal hair with a small animal clipper.

4. Disinfect the abdominal area with 70 % ethanol.

5. Put film dressing on the abdominal wall after the disinfectant dries (Fig. 2a).

6. Make a vertical incision (10 mm) in the right side of the abdominal wall (*see* **Note 1**).

7. Pull the small intestine of interest out of the abdominal cavity gently using forceps and wet cotton swabs (Fig. 2b, and *see* **Note 2**).

8. Make small holes in the mesentery and thread two ligatures of 5/0 surgical suture through the holes (*see* **Note 3**).

Fig. 2 Surgical procedures for the induction of intestinal inflammation and intravital imaging. (**a**) The shaved mouse is placed in the supine position and covered with sterile film dressing. (**b**) The small intestine is pulled out of the abdominal cavity. (**c**) Two ligatures of 5/0 surgical suture are placed at the proximal (*white arrowhead*) and distal side (*yellow arrowhead*) of the small intestine. Only the proximal side is ligated. The small intestine is clamped with a bulldog clamp at the caudal side of the distal ligature. (**d**) Intraluminal injection of 1 microgram LPS and/or 100 nM fMLP in 1 ml PBS. Extreme care is required not to damage the mesenteric vessels. (**e**) The distal ligature is ligated and the insulin syringe is removed. The small intestine will be dilated

9. Ligate the proximal side of the small intestine with one of the ligatures of 5/0 surgical suture. Leave the remaining ligature in place, but do not ligate the distal side of the intestine.

10. Clamp the small intestine with a bulldog clamp at the more distal point to the distal ligature (Fig. 2c).

11. Administer 1 μg/ml LPS and/or 100 nM fMLP into the intestinal cavity with an insulin syringe with a permanently attached needle (*see* **Note 4**). Make sure that the point of puncture is between the distal ligature and clamped point. Then, ligate the distal intestine with the distal ligature before pulling the needle from the intestine (Fig. 2d, and *see* **Note 5**).

12. Return the small intestine to the abdominal cavity (*see* **Note 6**).

13. Close the abdominal wound with 6/0 silk suture.

14. Put the mice back into the home cage to recover.

3.1.2 Cannulation of the Jugular Vein

1. Two hours after surgery, anesthetize the mice with 2 % isoflurane again.

2. Place the unconscious mouse in the supine position on a heating pad.

3. Disinfect the skin of the left neck with 70 % ethanol (*see* **Note 7**).

4. Make a longitudinal incision of about 15 mm in the neck of mice, just above the left front leg.

5. Carefully remove the connective tissue surrounding the jugular vein.

6. Place the two ligatures of 5/0 surgical suture at the caudal and rostral side of the vein.

7. Ligate the rostral ligature to prevent bleeding.

8. Make a small incision between the ligatures using micro-scissors and insert the catheter with a syringe (1 ml) filled with

heparin sodium solution in the caudal direction into the vein. Make sure that the blood flows back into the catheter.

9. Ligate the caudal ligature and fix the catheter.

3.1.3 Positioning to the Microscope Stage

1. Re-open the abdominal cavity and pull out the small intestine on a coverslip placed in a heat-stage of TPEM maintained at 37 °C (*see* **Note 8**).

2. Fix a mouse in the right decubitus position on the microscope stage with adhesive tape. Fix the catheter and syringe on the stage with adhesive tape so that it can be handled easily during time-lapse imaging.

3. Place cotton swabs on the ligated points of the small intestine, stretch the small intestine in the longitudinal directions, and fix the cotton swabs on the stage with adhesive tape (*see* **Note 9**).

4. Hang ligatures of 5/0 surgical suture on the stick of cotton swabs and pull the ligatures in the longitudinal directions of the small intestine to fine-tune the tension of the small intestine. In this way, the movement of small intestine can be adjusted mechanically (Fig. 3a, b, and *see* **Note 10**).

5. Through the eyepieces, observe the intestine by epi-fluorescence illumination. Make sure that the view field is minimally affected by peristalsis, heart beat, or breathing (*see* **Note 11**).

6. Cover the intestine with plastic wrap to prevent drying.

3.2 Microscopy Settings

TPEM was performed with an FV1200MPE inverted microscope (Olympus) equipped with a 30×/1.05 NA silicon oil-immersion objective lens (UPLSAPO 30×S) (*see* **Note 12**) and an InSight

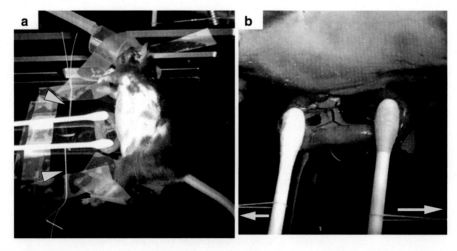

Fig. 3 Fixation of the mouse and small intestine on the stage of a two-photon excitation microscope. (**a**) A mouse is fixed on the heat-stage in the right decubitus position. Strings (*arrowheads*) are hung over the sticks of cotton swabs for the adjustment of the tension of the small intestine to minimize peristalsis. (**b**) The tension of the small intestine is fine-tuned by pulling the ligature in the longitudinal directions (*arrow*)

DeepSee Laser. The laser power used for observation was from 4 to 8 %. Scan speed was set between 12.5 and 20 μs/pixel. The excitation wavelength for CFP and Qtracker 655 was 840 nm. We used an IR-cut filter, BA685RIF-3, two dichroic mirrors, DM505 and DM570, and three emission filters, BA460-500 for CFP, BA520-560 for YFP, and 645/60 for the Qtracker 655.

3.3 Time-Lapse Imaging of Neutrophil Recruitment to the Inflamed Intestines by Two-Photon Excitation Microscopy

1. By epi-fluorescence illumination, observe the lamina propria of the small intestine, find an appropriate postcapillary venule with a diameter of 30–50 μm, and register the XY position of the first field of view (Fig. 4a).

2. Change the observation mode to two-photon excitation and start scanning. Focus the postcapillary venules and set the high-voltage (HV) values of photomultipliers and laser intensity to obtain a sufficient signal. Take a snapshot of the view field as a reference. Typically, the HV values of photomultipliers are 500–550 V and the laser power used for observation is from 4 to 8 %.

3. Start data acquisition, for example, every 20 s (Fig. 4b).

4. If necessary, adjust the Z position manually during the time-lapse imaging by referring to the reference snapshot (*see* **Note 13**).

5. For the time-lapse imaging longer than 2 h, administer 50 μl PBS intravenously every hour.

6. Inject 4 μl Qtracker 655 with inhibitors or other drugs intravenously to visualize the drug delivery (*see* **Note 14**).

7. Sacrifice the mice after time-lapse imaging.

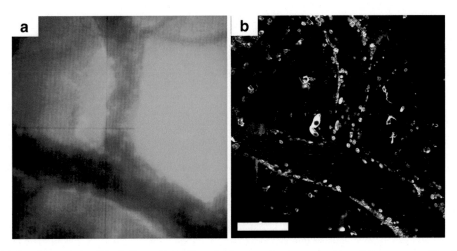

Fig. 4 (**a**) *Lamina propria* of the inflamed intestine of a transgenic mouse under epi-fluorescence illumination. The *dark region* indicates the venule. The *small dots* within the venule are white blood cells. Confirm that peristalsis is minimal and select an appropriate postcapillary venule under epi-fluorescent illumination. (**b**) The same region visualized by two-photon excitation microscopy. Bar: 100 μm

3.4 Data Processing

1. Launch the MetaMorph software.

3.4.1 Generation of FRET/CFP Ratio Images

2. Create FRET/CFP ratio images in the intensity-modulated display (IMD) mode as follows. Find the "Ratio images" function in the "Process" menu of the MetaMorph program. Set each parameter as follows: numerator, the FRET stack file; denominator, the CFP stack file; IMD display and 8 ratios with 32 intensities. Minimum and maximum ratios should be set in a trial-and-error manner so that the image fully uses the eight hues (Fig. 5).

3. If required, make the movie using the "Make Movie" function in the "Stack" menu of the MetaMorph program.

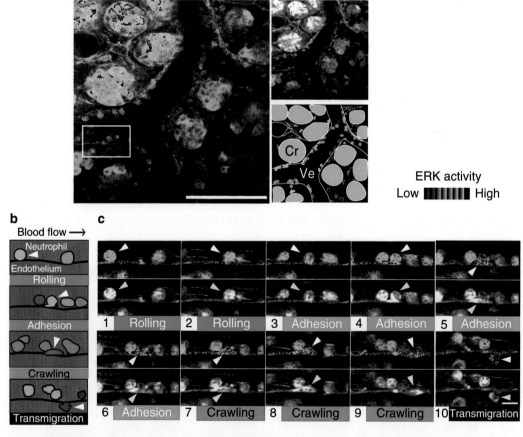

Fig. 5 An example of intravital imaging of molecular activity during neutrophil recruitment in the lamina propria of the intestinal mucosa. (**a**) A representative FRET/CFP image of the lamina propria of the inflamed intestine of an Eisuke mouse, which was subjected to LPS and fMLP 2 h before imaging, is shown in intensity-modulated display mode using 8 ratios with 32 intensities and a CFP image in grayscale, with a schematic view of this region. Cr, crypt; Ve, venule. (**b**) Schematic of the four steps of extravasation. *Arrowheads* indicate the same neutrophil at different time points. (**c**) Time-lapse FRET/CFP and CFP images of neutrophil extravasation. The boxed region (**a**) was magnified and shown in a time series. Bars: (**a**) 100 μm; (**c**) 10 μm. Reproduced from Mizuno R. et al., J Exp Med 2014, 211: 1123–1136

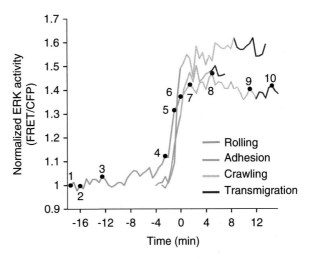

Fig. 6 An example of the quantification of molecular activity in neutrophils during extravasation. ERK activities in three representative neutrophils are plotted against time. The transition point from the adhesion to crawling steps was set as time 0. The rolling, adhesion, crawling, and transmigration steps are indicated by different colors. The numbers in the graph correspond to the numbers in Fig. 5c. Reproduced from Mizuno R. et al., J Exp Med 2014, 211: 1123–1136

3.4.2 Quantification of the FRET/CFP Ratio

1. Open the stack files of the FRET and CFP images. Set ROIs over the cells in the first plane of the FRET stack file. Transfer the ROIs to the CFP stack file (*see* **Note 15**).

2. Measure the average intensity of FRET and CFP by the "Region measurements" function in the "Measure" menu of the MetaMorph program, and export the values in an ASCII or an Excel format.

3. Open the data file in Excel. Plot the FRET/CFP value versus time (Fig. 6).

4   Notes

1. Make an incision in the lateral side, not in the middle, of the abdominal wall in order to pull the shortest possible length of small intestine over the lens. Excessive tension on the mesentery impairs the blood flow of mesenteric vessels. Make an incision on the left side, if the mouse is placed in the left lateral decubitus position during imaging. It is important to make a small incision and pull out as little of the intestine as possible, because an excess of extra-abdominal intestine will cause drifting during imaging.

2. The jejunum is appropriate for the intravital imaging because vessels are more abundant in the mesentery than the ileum. Be careful not to damage the mesenteric vessels.

3. Make sure that the feeding artery is located in the middle of the two small holes used to ligate the intestine. The distance between the two holes is about 1 cm. Extreme care is needed not to damage the marginal artery.

4. Be careful not to dilate the small intestine excessively, because excessive dilation interrupts the blood flow. Make sure that the small vessels on the intestinal surface remain bright red. When the small intestine is dilated excessively, the small vessels will turn a pale color due to the interruption of blood flow.

5. This process is important to prevent the leakage of injected reagents from the puncture hole.

6. This incubation period in situ keeps the intestine in physiological condition and activates the neutrophil recruitment cascade. Keeping the intestine out of the abdominal cavity for a long period of time will perturb neutrophil recruitment.

7. Decide the side of cannulation depending on the position of the mouse during imaging. Use the right jugular vein if the mouse will be placed in the left lateral decubitus position. Retro-orbital injection [17] is an alternative, for those who are not familiar with surgical operations. However, it may be necessary to interrupt time-lapse imaging for a few minutes for the retro-orbital injection.

8. If the small intestine is deflated, add PBS into the intestinal lumen with an insulin syringe.

9. Be careful not to extend the small intestine excessively. Excessive tension stops the blood flow of the small vessels. The stretching strength will be fine-tuned as described in the next step.

10. It has been reported that the small intestine can be fixed with ligatures connected to the small intestine [7]. However, in our experience, it is often difficult to adjust the tension by pulling the small intestine. Scopolamine butylbromide was used in some previous studies to minimize peristalsis; however, it did not suppress peristalsis very effectively in our hands.

11. Fine-tune the tension of the stretched small intestine by checking for small intestine drift under epi-fluorescent illumination. If the small intestine drifts at this time, stretch the small intestine slightly by pulling the 5/0 suture hung over the stick of the cotton swab. Repeat this adjustment until the drift in the XY directions is minimized. It is also important to confirm that the blood flow of the small vessels is not perturbed. If perturbation occurs, loosen the ligature. If the blood flow is not restored after loosening the ligature, the small intestine may have been set in upside-down. Check the direction of the small intestine.

12. To perform intravital imaging by TPEM, an objective lens with a high numerical aperture, such as a 30× NA1.05 silicon oil-immersion lens or 25× NA1.05 water-immersion lens, is desirable. If a water immersion lens is used, the water must be replenished to prevent drying during the imaging. This is one reason why we prefer a silicon oil-immersion lens.

13. Even when the small intestine is fixed to the stage successfully to minimize the drift in the XY directions, the small intestine will still drift slightly in the Z direction over the course of the experiment. Therefore, it is necessary to adjust the Z position to maintain the imaging plane during time-lapse imaging. Although there are some programs to track the same Z position, it may take a few minutes to perform the correction. When the software-based focusing is too slow, the Z position must be adjusted manually.

14. Pay particular attention to the volume and the speed of the intravenous injection. Although intravenous bolus injection achieves high peak concentrations of drugs, it also induces fluid overload and heart strain, resulting in the death of mice. In addition, bolus injection sometimes induces drifting of images in the Z direction. Therefore, it is better to inject the reagents as slowly as possible.

15. Granulocytes can be distinguished from the other cell types by their characteristic segmented nuclei. The cell types may be confirmed by intravenous injection of PE conjugated anti-Gr1 antibody (0.15 mg/kg), which is a marker of granulocytes. However, anti-Gr1 antibody may attenuate the neutrophil invasion into the inflammatory site [18].

Acknowledgements

This work was supported by the Platform for Dynamic Approaches to Living System from the Ministry of Education, Culture, Sports, Science and Technology, Japan. We thank K. Otani for assisting with the jugular vein cannulation.

References

1. Borregaard N (2010) Neutrophils, from marrow to microbes. Immunity 33:657–670

2. Sanz MJ, Kubes P (2012) Neutrophil-active chemokines in in vivo imaging of neutrophil trafficking. Eur J Immunol 42:278–283

3. Megens RT, Kemmerich K, Pyta J, Weber C, Soehnlein O (2011) Intravital imaging of phagocyte recruitment. Thromb Haemost 105:802–810

4. Germain RN, Robey EA, Cahalan MD (2012) A decade of imaging cellular motility and interaction dynamics in the immune system. Science 336:1676–1681

5. Kolaczkowska E, Kubes P (2013) Neutrophil recruitment and function in health and inflammation. Nat Rev Immunol 13:159–175

6. Kamioka Y, Sumiyama K, Mizuno R, Sakai Y, Hirata E et al (2012) Live imaging of protein

kinase activities in transgenic mice expressing FRET biosensors. Cell Struct Funct 37:65–73

7. Mizuno R, Kamioka Y, Kabashima K, Imajo M, Sumiyama K et al (2014) In vivo imaging reveals PKA regulation of ERK activity during neutrophil recruitment to inflamed intestines. J Exp Med 211:1123–1136

8. Swartz DE, Seely AJ, Ferri L, Giannias B, Christou NV (2000) Decreased systemic polymorphonuclear neutrophil (PMN) rolling without increased PMN adhesion in peritonitis at remote sites. Arch Surg 135:959–966

9. Wang MX, Liu YY, Hu BH, Wei XH, Chang X et al (2010) Total salvianolic acid improves ischemia-reperfusion-induced microcirculatory disturbance in rat mesentery. World J Gastroenterol 16:5306–5316

10. Han JY, Horie Y, Fan JY, Sun K, Guo J et al (2009) Potential of 3,4-dihydroxy-phenyl lactic acid for ameliorating ischemia-reperfusion-induced microvascular disturbance in rat mesentery. Am J Physiol Gastrointest Liver Physiol 296:G36–G44

11. Suzuki M, Mori M, Fukumura D, Suzuki H, Miura S et al (1999) Omeprazole attenuates neutrophil-endothelial cell adhesive interaction induced by extracts of Helicobacter pylori. J Gastroenterol Hepatol 14:27–31

12. Ritsma L, Steller EJ, Ellenbroek SI, Kranenburg O, Borel Rinkes IH et al (2013) Surgical implantation of an abdominal imaging window for intravital microscopy. Nat Protoc 8:583–594

13. Klinger A, Orzekowsky-Schroeder R, von Smolinski D, Blessenohl M et al (2012) Complex morphology and functional dynamics of vital murine intestinal mucosa revealed by autofluorescence 2-photon microscopy. Histochem Cell Biol 137:269–278

14. Toiyama Y, Mizoguchi A, Okugawa Y, Koike Y, Morimoto Y et al (2010) Intravital imaging of DSS-induced cecal mucosal damage in GFP-transgenic mice using two-photon microscopy. J Gastroenterol 45:544–553

15. McDole JR, Wheeler LW, McDonald KG, Wang B et al (2012) Goblet cells deliver luminal antigen to CD103+ dendritic cells in the small intestine. Nature 483:345–349

16. Watson AJ, Chu S, Sieck L, Gerasimenko O, Bullen T et al (2005) Epithelial barrier function in vivo is sustained despite gaps in epithelial layers. Gastroenterology 129:902–912

17. Yardeni T, Eckhaus M, Morris HD, Huizing M, Hoogstraten-Miller S (2011) Retro-orbital injections in mice. Lab Anim (NY) 40:155–160

18. Stirling DP, Liu S, Kubes P, Yong VW (2009) Depletion of Ly6G/Gr-1 leukocytes after spinal cord injury in mice alters wound healing and worsens neurological outcome. J Neurosci 29:753–764

Chapter 15

In Vivo Myeloperoxidase Imaging and Flow Cytometry Analysis of Intestinal Myeloid Cells

Jan Hülsdünker and Robert Zeiser

Abstract

Myeloperoxidase (MPO) imaging is a non-invasive method to detect cells that produce the enzyme MPO that is most abundant in neutrophils, macrophages, and inflammatory monocytes. While lacking specificity for any of these three cell types, MPO imaging can provide guidance for further flow cytometry-based analysis of tissues where these cell types reside. Isolation of leukocytes from the intestinal tract is an error-prone procedure. Here, we describe a protocol for intestinal leukocyte isolation that works reliable in our hands and allows for flow cytometry-based analysis, in particular of neutrophils.

Key words Neutrophil granulocytes, Leukocyte isolation, Graft-versus-host disease, Myeloperoxidase, In vivo bioluminescence imaging

1 Introduction

Allogeneic hematopoietic stem cell transplantation (allo-HSCT) is often accompanied by graft-versus-host disease that occurs when donor immune cells react with allo-antigen deriving from the host. Proliferation and activation of the donor cells lead to massive tissue damage in various organs. One important target organ is the gastrointestinal tract that includes the small and large intestine, both containing a robust mucosal immune system. In our previous work [1], we demonstrated that large numbers of neutrophil granulocytes are recruited to the intestine following total body irradiation and chemotherapy. Therefore, we could identify neutrophils as a novel cell type with a major contribution to GvHD. In the described study, we used a non-invasive in vivo method to detect localization of myeloid cells in the gut [2]. Furthermore, we modified a previously developed intestinal tissue digestion protocol involving a combination of dispase, collagenase D, and DNAse I [3] to isolate intestinal leukocytes including neutrophils and analyze them in

Andrei I. Ivanov (ed.), *Gastrointestinal Physiology and Diseases: Methods and Protocols*, Methods in Molecular Biology, vol. 1422, DOI 10.1007/978-1-4939-3603-8_15, © Springer Science+Business Media New York 2016

more detail by flow cytometry. We found that these two methods provide different types of information. MPO imaging shows the activation status of myeloid cells while flow cytometry is the state-of-the-art method for quantification of different subsets of myeloid cells (neutrophils versus macrophages versus inflammatory monocytes and others). The combination of these methods is essential not only to study GvHD, but have general application for various intestinal inflammation models where neutrophils and other myeloid cells are involved.

2   Materials

2.1   Neutrophil Isolation and Analysis by Flow Cytometry

1. Cell Dissociation (CD) Buffer: Hank's buffered salt solution (HBSS without Ca^{2+}, Mg^{2+}), HEPES (10 mM), EDTA (5 mM), pH 7.2 (see **Note 1**).

2. Digestion mix: HBSS (with Ca^{2+}, Mg^{2+}), dispase (0.5 U/ml), collagenase D (0.5 mg/ml), DNAse I (0.5 mg/ml), fetal calf serum (FCS, 2 %), pH 7.2 (see **Note 2**).

3. Stop solution: RPMI 1640 medium, FCS (10 %), Penicillin (100 U/ml), Streptomycin (100 μg/ml).

4. FACS Buffer: Phosphate buffered saline (PBS), FCS (2 %), EDTA (5 mM), NaN_3 (0.02 %), pH 7.2.

5. Antibodies for FACS: Fluorescently labeled anti-mouse CD45, anti-mouse CD11b, anti-mouse Ly-6G diluted in FACS buffer at a previously titrated concentration.

6. Fc-Blocking solution: Fc-Block (anti-mouse CD16/CD32) dissolved at 0.02 mg/ml in FACS Buffer.

7. Dead cell staining solution: Amine reactive fixable dye diluted in PBS in a previously titrated concentration.

8. FACS fixation solution: Paraformaldehyde (1 %) in PBS,

9. V-bottom microtiter well plate (96-well), 5 ml round bottom tubes, 15 ml tubes, 50 ml tubes, cell strainer, 70 μm.

10. Flow Cytometer.

2.2   Myeloperoxidase Imaging

1. Luminol solution: luminol sodium salt dissolved in sterile deionized water at 20 mg/ml. Prepare this solution directly before use.

2. 0.5 ml syringe with a 30G needle.

3. Isoflurane vaporizing system for anesthesia.

4. Bioluminescence imaging system.

5. Software for analysis: Living image.

3   Methods

3.1   Neutrophil Isolation and Analysis by Flow Cytometry

1. This protocol is optimized to analyze intestinal segments up to 5 cm of length.

2. All steps have to be performed fast and on ice.

3. Open the abdominal cavity of euthanized mouse and dissect the entire intestine, distally from the stomach and proximally from the anus.

4. Cut out the segment of interest and remove the surrounding fat and the Peyer patches (*see* **Note 3**).

5. Open the intestinal segment longitudinally and transfer to a petri dish with ice-cold PBS.

6. Wash out the feces using a 5 ml syringe with a 25G needle and ice-cold PBS.

7. Shortly place the intestine on a tissue paper to dry from excess PBS and transfer to a 50 ml Falcon tube containing 10 ml of cold CD-Buffer.

8. Incubate for 15 min at 37 °C with a constant shaking at 200 rpm.

9. Keep CD-Buffer at RT now.

10. Vortex rigorously.

11. Transfer the gut segment into a new 15 ml Falcon tube containing 5 ml CD-Buffer equilibrated at room temperature.

12. Repeat incubation for 10 min at 37 °C with a constant shaking at 200 rpm.

13. Vortex rigorously.

14. Repeat **steps 11–13** one more time.

15. Shorty rinse the segment in PBS and remove the excess of liquid using tissue paper.

16. Cut the tissue segment into very small pieces using a scalpel.

17. Transfer the smashed intestine to a 10 ml Falcon containing 2.5 ml ice-cold Digestion Mix (*see* **Note 4**).

18. Incubate for 15 min at 37 °C with a constant shaking at 200 rpm.

19. Keep the Digestion Mix at RT now.

20. Vortex rigorously.

21. Add 2.5 ml fresh Digestion Mix.

22. Incubate for 15 min at 37 °C with a constant shaking at 200 rpm.

23. Vortex rigorously.

24. Repeat **steps 21–23** one more time.

25. Filter the digestion suspension through a 70 μm cell strainer into a 50 ml Falcon tube (*see* **Note 5**).

26. Add 15 ml stop solution to stop digestion.

27. Spin down at $700 \times g$ for 10 min.

28. Resuspend the pellet in PBS and count the cells if necessary (*see* **Notes 6** and **7**).

29. Resuspend the cells in PBS and spin them down at $700 \times g$ for 10 min (*see* **Note 8**).

30. Resuspend the cells in 150 μl PBS and transfer to a v-bottom microtiter (96-well) plate.

31. Spin down at $450 \times g$ for 5 min at 4 °C.

32. Discard supernatant and add 50 μl of the Dead cell staining solution.

33. Incubate for 20 min at 4 °C in the dark.

34. Add 120 μl of the FACS buffer and spin down at $450 \times g$ for 5 min at 4 °C.

35. Wash using 170 μl of the FACS buffer and spin down at $450 \times g$ for 5 min at 4 °C.

36. Add 25 μl of the Fc-Block solution and incubate for 10 min at 4 °C in the dark.

37. Add the FACS antibody mix.

38. Incubate for 30 min at 4 °C in the dark.

39. Add 120 μl FACS buffer and spin down at $450 \times g$ for 5 min at 4 °C.

40. Wash using 170 μl FACS buffer and spin down at $450 \times g$ for 5 min at 4 °C.

41. Repeat **step 40** one more time.

42. Resuspend the cells in 200 μl of the FACS Fix and transfer into a 5 ml round bottom tube.

43. Perform the analysis using a Flow Cytometer.

44. A representative gating scheme for intestinal leukocytes isolated 48 h after total body irradiation is shown in Fig. 1.

3.2 Myeloperoxidase Imaging

1. Prepare the anesthetics—an isoflurane ventilated chamber and isoflurane ventilation during analysis is recommended.

2. Open the software Living image and initialize the system using the following settings: Exposure time: 300 s; no optical filter, Fstop: 1; Binning small.

3. Inject 200 μl of the luminol salt solution per 25 g mouse intraperitoneally (200 mg/kg of the body weight).

4. Anesthetize the mice and place them into the bioluminescence imaging system with their abdomen facing up to the camera.

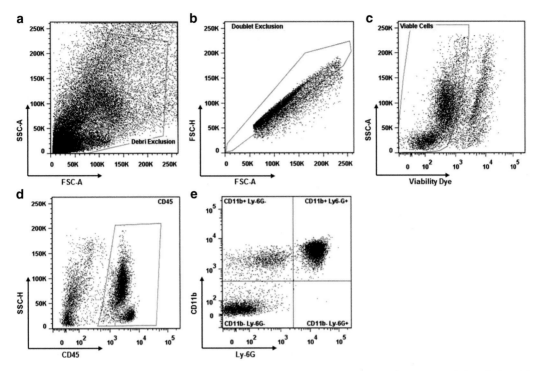

Fig. 1 FACS gating strategy. Cells were isolated from the small intestine 48 h after total body irradiation. (**a**) Debris exclusion, (**b**) doublet exclusion, (**c**) dead cell exclusion, (**d**) CD45+ cells, (**e**) neutrophils defined as CD11b+ Ly-6G+ cells

5. Start the analysis 10 min after injection using the settings mentioned above.

6. After image acquisition draw a region of interest (ROI) and place it to the gut region (*see* **Note 9**).

7. Photons per second in the ROI are calculated by the software (*see* **Note 10**).

8. Figure 2 shows a representative image of an untreated Balb/c mouse (left) and a Balb/c mouse 72 h after total body irradiation (right) after setting the ROIs.

4 Notes

1. EDTA chelates Ca^{2+} which is necessary for strong adhesions between epithelial cells. Losing the cell contacts leads to the dissociation of the epithelial layer from the *lamina propria*. Dithiothreitol (DTT) is sometimes included in various protocols in the first step of epithelial cell dissociation to remove the mucus. In our hands, the epithelial cells dissociate well after three steps of washing without DTT.

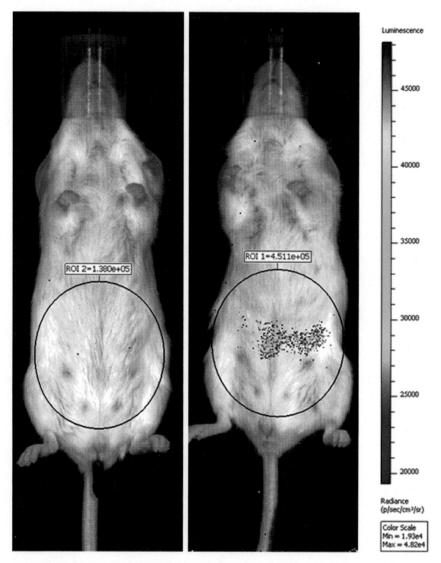

Fig. 2 In vivo MPO bioluminescence imaging of intestinal neutrophils. The *left* image shows an untreated mouse; the *right* image shows a mouse 72 h after total body irradiation

2. The addition of FCS reduces the activity of the digestion enzymes but enhances the viability of the cells.

3. Peyer patches are rich in leukocytes and have to be cut out before digesting the intestine for the *lamina propria* leukocyte/neutrophil analysis.

4. It is necessary to keep the digestion mix cold until all samples are processed. This makes it possible to start the digestion at the same time for all samples when the enzymes are active near 37 °C.

5. Connective tissue remnants remain in the cell strainer when filtering after the third digestion step.

6. Leukocyte yield varies a lot depending on the mouse strain, age, treatment, length and localization of the intestinal segment chosen, and it can be between 2×10^5 and 3×10^6 cells.

7. Performing enrichment of leukocytes by a percoll gradient reduces the cell yield and we eliminated this step. Contamination of isolated leukocytes by epithelial cells is quite low, although a high proportion of dead cells can be seen in the analysis (10–50 %). A viability staining is highly recommended to exclude dead cells by gating.

8. The amine reactive dyes are membrane-impermeable dyes that react with free amines in the cytoplasm of dead cells. It can also react with free amines that are in FCS which makes the washing step with PBS necessary.

9. It is important to draw the same-sized ROI for every mouse. To ensure the same size, duplicate the first ROI and place it over the image of the second mouse.

10. Luminol is oxidized by the myeloperoxidase or its products and emits light that are detected by the camera. The signal is quantified as photons/second and equivalent to MPO activity.

Acknowledgements

This study was supported in part by the Excellence Initiative of the German Research Foundation (GSC-4, Spemann Graduate School).

References

1. Schwab L, Goroncy L, Palaniyandi S, Gautam S et al (2014) Neutrophil granulocytes recruited upon translocation of intestinal bacteria enhance GvHD via tissue damage. Nat Med 20:648–654

2. Gross S, Gammon ST, Moss BL, Rauch D, Harding J, Heinecke JW, Ratner L, Piwnica-Worms D (2009) Bioluminescence imaging of myeloperoxidase activity in vivo. Nat Med 15:455–461

3. Klose CS, Flach M, Möhle L, Rogell L et al (2014) Differentiation of type 1 ILCs from a common progenitor to all helper-like innate lymphoid cell lineages. Cell 157:340–356

Part III

Isolation and Characterization of Intestinal Immune Cells

Chapter 16

Macrophage Isolation from the Mouse Small and Large Intestine

Akihito Harusato, Duke Geem, and Timothy L. Denning

Abstract

Macrophages play important roles in maintaining intestinal homeostasis via their ability to orchestrate responses to the normal microbiota as well as pathogens. One of the most important steps in beginning to understand the functions of these cells is the ability to effectively isolate them from the complex intestinal environment. Here, we detail methodology for the isolation and phenotypic characterization of macrophages from the mouse small and large intestine.

Key words Intestine, Antigen-presenting cell, Macrophage, Dendritic cell

Abbreviations

APC   Antigen-presenting cells
DC   Dendritic cell
LI   Large intestine
LP   Lamina propria
SI   Small intestine

1   Introduction

The mammalian intestine is constantly exposed to a variety of microbes and food antigens [1]. Antigen-presenting cells (APCs), comprised primarily of macrophages and dendritic cells (DCs), are central components of the mucosal immune system that foster homeostasis in the intestine [2–9]. Just beneath the intestinal epithelial layer, the lamina propria (LP) contains a large population of macrophages that are ideally positioned to sample luminal contents and perform surveillance activity [10–12]. The positioning of macrophages in the LP suggests these cells have an important role in modulating innate and adaptive immune responses toward the microbiota [10–19]; however, the mechanisms by which these cells

Andrei I. Ivanov (ed.), *Gastrointestinal Physiology and Diseases: Methods and Protocols,* Methods in Molecular Biology, vol. 1422, DOI 10.1007/978-1-4939-3603-8_16, © Springer Science+Business Media New York 2016

interact with foreign antigens and orchestrate effective immune reactivity remains as area of active investigation [20–25]. Paramount to identifying the functions of intestinal macrophages is the ability to efficiently isolate these cells from a complex cellular environment [26]. While many studies have begun to define the function of intestinal macrophages, continued advancements in the identification and characterization of these cells in the steady state and during inflammatory processes in mouse and humans are desired. Here, we summarize detailed methodology for the isolation and purification of intestinal macrophages that may be employed to investigate these important cell types and the role they play in regulating mucosal immunity.

2   Materials

2.1   Equipment

1. MaxQ 4450 benchtop orbital shaker; any orbital shaker with sufficient capacity should suffice.
2. LS MACS columns and a QuadroMACS separator (Miltenyi Biotec).
3. LSR II benchtop flow cytometer (BD) or other analyzer.
4. FACSAria II benchtop cell sorter (BD) or other sorter.

2.2   Reagents and Solutions

1. 1× PBS, Ca^{2+}- and Mg^{2+}-free (CMF PBS).
2. Hank's balanced salt solution (HBSS) with phenol red, Ca^{2+}- and Mg^{2+}-free (CMF HBSS); HBSS is commonly used for isolation of intestinal immune cells.
3. CMF HBSS with 5 % FBS (CMF HBSS/FBS) and 2 mM EDTA.
4. Sodium bicarbonate.
5. 1 M HEPES in 0.85 % NaCl.
6. Fetal bovine serum, heat-inactivated.
7. 0.5 M EDTA (pH 8.0).
8. Collagenase type VIII.
9. DNase I; Stock solution: 100 mg/mL.
10. Working Collagenase VIII/DNAse I solution: 1.5 mg/mL of collagenase VIII and 40 μg/mL of DNase I in CMF HBSS/FBS.

2.3   Antibodies and Staining Reagents

1. Ice-cold staining buffer: CMF PBS + 5 % FBS.
2. LIVE/DEAD Fixable Aqua Dead Cell Stain Kit for 405 nm excitation; Use at 1:1000 in ice-cold PBS.
3. CD45-PerCP mAb (30 F11); Use at 1:100 dilution in ice-cold staining buffer.

4. CD103-PE mAb (M290); Use at 1:100 dilution in ice-cold staining buffer.

5. FcγRIII/II mAb (2.4G2); Use at 1:200 dilution in ice-cold staining buffer.

6. CD11c-APC mAb (N418); Use at 1:100 dilution in ice-cold staining buffer.

7. MHC-II (I-Ab)-Alexa Fluor 700 mAb (M5/114); Use at 1:100 dilution in ice-cold staining buffer.

8. CD11b-eFluor 450 mAb (M1/70); Use at 1:200 dilution in ice-cold staining buffer.

9. F4/80-PE-Cy7 mAb (BM8); Use at 1:200 dilution in ice-cold staining buffer.

10. CD11b microbeads.

11. CD11c microbeads.

2.4 Disposable Reagents

1. 50 mL conical tubes.

2. Single mesh wire strainer.

3. Small weigh boat.

4. 100 μm cell strainer.

5. 40 μm cell strainer.

6. 5 mL polystyrene round-bottom tubes.

3 Methods

3.1 Isolation of Mouse Small and Large Intestine

1. Prepare the following reagents and equipment:

   - Warm Ca^{2+}/Mg^{2+}-free PBS (CMF PBS) to room temperature.

   - Warm Ca^{2+}/Mg^{2+}-free HBSS with 5 % FBS (CMF HBSS/ FBS) and 2 mM EDTA to room temperature.

   - Warm Orbital shaker to 37 °C.

2. Euthanize mice and spray 70 % ethanol onto the abdomen (*see* **Note 1**).

3. Make a horizontal incision in abdomen with a scissor and peel back the skin and cut open the peritoneum.

4. Cut the intestine at the pyloric sphincter to separate the stomach from the upper small intestine (*see* **Note 2**).

5. Carefully remove the mesentery using forceps and cut at the ileo-cecal valve to separate the small intestine from the large intestine.

6. Continue to tease apart the mesentery from the large intestine and make a cut at the anal verge. Place the large intestine in CMF PBS on ice while first attending to the small intestine.

7. Place the entire small intestine on paper towels pre-wet with CMF PBS. Remove the Peyer's patches along the anti-mesenteric side of the small intestine and cut open longitudinally using scissors and forceps (*see* **Note 3**).

8. Place the intestine in a Petri dish containing room temperature CMF PBS and rapidly move the intestine around using forceps until the PBS becomes cloudy with luminal contents. Move the intestine into a new Petri dish with fresh CMF PBS and repeat this process until the PBS no longer becomes cloudy (usually 3–5 Petri dishes).

9. Cut the small intestine into approximately 1.5 cm pieces and place into a 50 mL conical tube containing 30 mL of pre-warmed CMF HBSS/FBS and 2 mM EDTA (*see* **Note 4**).

10. Cut open the colon longitudinally using scissors and wash off feces and mucus in CMF PBS at room temperature as per **steps 7** and **8** of Subheading 3.1 (*see* **Note 5**).

3.2 Removal of Epithelial Layer

1. Place each 50 mL conical tube horizontally into an orbital shaker and stabilize using tape. Shake at 250 rpm for 20 min at 37 °C in order to begin removing epithelial cells and intraepithelial lymphocytes.

2. After the shaking is done, pour the contents of each 50 mL conical tube through a wire mesh strainer to recover the 1.5 cm pieces of intestine while allowing the epithelial cells and intraepithelial lymphocytes pass through. If analyses of epithelial cells and/or intraepithelial lymphocytes are desirable place a collection tube under the wire mesh strainer during this step.

3. Repeat **steps 1** and **2** of Subheading 3.2 one additional time for a total of two shaking cycles.

    At this stage, the intestine is stripped of most of the epithelial cell layer and is ready for tissue digestion.

3.3 Digestion of Intestinal Tissue

1. Prepare the following reagents and equipment:
   - Pre-warmed working Collagenase VIII/DNAse I solution (see **Note 6**).
   - Pre-warmed CMF HBSS/FBS and 2 mM EDTA.
   - Pre-warmed orbital shaker at 37 °C.
   - Ice-cold CMF HBSS/FBS.

2. Transfer the 1.5 cm pieces of intestine to a small plastic weigh boat after dabbing away excess media using a paper towel (*see* **Note 7**).

3. Rapidly mince the 1.5 cm pieces of intestine using sharp dissection scissors directly in the weigh boat for approximately 10–20 s and then add minced intestine to 20 mL of pre-warmed collagenase solution (*see* **Note 8**).

4. Horizontally place each 50 mL conical tube into an orbital shaker and digest at 200 rpm for 10–20 min at 37 °C (*see* **Notes 9** and **10**).

5. Vortex remaining intestinal tissue for 5–10 s to ensure thorough dissociation and then filter through a 100 μm cell strainer directly into a 50 mL conical tube (*see* **Note 11**).

6. Add CMF HBSS/FBS to top off each 50 mL conical tube and then centrifuge at $300 \times g$ for 5 min at 4 °C (*see* **Note 12**). Repeat this wash step once more.

7. After pouring off the supernatant, resuspend the cell pellets in ice-cold CMF HBSS/FBS and place samples on ice (*see* **Notes 13** and **14**).

3.4 Antibody Staining for Multi-Color Flow Cytometric Analyses

1. Transfer cells into a 5 mL polystyrene round-bottom (FACS) tube or if staining multiple samples use a 96-well V-bottom plate.

2. Wash cells twice using ice-cold CMF PBS.

3. Incubate samples with LIVE/DEAD Fixable Aqua Dead Cell Stain for 15 min on ice in the dark (*see* **Note 15**).

4. Wash cells twice in 200 μL ice-cold CMF PBS.

5. Block non-specific binding of antibodies to cells by using 2.4G2 anti-FcγRIII/II in ice-cold staining buffer for 10 min on ice.

6. Wash cells in ice-cold staining buffer.

7. Incubate samples with specific antibody staining cocktail for 20 min on ice in the dark.

8. Wash cells with ice-cold staining buffer twice and resuspend samples in 400 μL of ice-cold staining buffer and pass through 40 μm filter cap on FACS tubes.

9. Acquire samples on LSR II cytometer (BD) or other appropriate FACS analyzer and analyze as defined by gating strategy in Subheading 3.5 and Fig. 1.

3.5 Antibody Staining for Multi-Color Flow Cytometric Analyses

1. After gating out dead cells, create another dot plot and further gate on CD45$^+$ and I-Ab$^+$ cells, which encompasses both macrophages and DCs (Fig. 1a).

2. Next, using a separate dot plot, gate cells that express either CD11b and/or CD11c. When this population is further analyzed for CD103 and F4/80 expression, two major subsets of cells (R1 and R2; Fig. 1a) will be evident. Cells in the R1 gate are CD103$^+$ F4/80$^{dull/-}$ cells, which are mostly DCs, while cells in region R2 are CD103$^-$F4/80$^+$ cells, which are mostly macrophages.

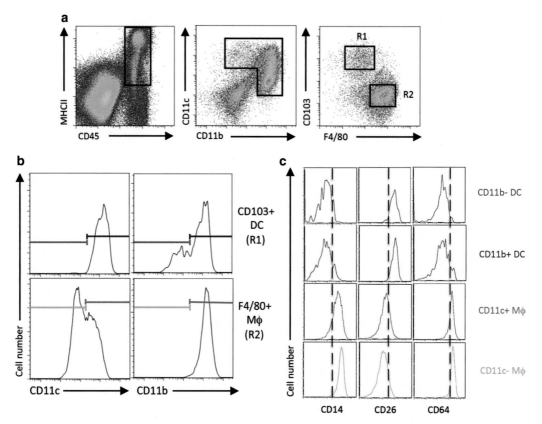

Fig. 1 Representative analysis for macrophages and DCs in the intestine by multi-color flow cytometry. (**a**) Small intestinal lamina propria living CD45 + MHCII+ cells were analyzed for CD11c and CD11b expression and consequently the resulting populations were separated for CD103+ DCs (R1) and F4/80+ macrophages (R2). (**b**) CD103+ DCs (R1) and F4/80+ macrophages (R2) were divided by CD11c and CD11b expression. (**c**) Expression of CD14, CD26, and CD64 was analyzed for CD11b- DC, CD11b + DC, CD11c + macrophages, and CD11c- macrophages with colors corresponding to the populations defined in panel **b**

3. Analyze regions R1 and R2 further for CD11c and CD11b expression to further differentiate macrophage and DC subsets, respectively [27]. CD103+F4/80$^{dull/-}$ cells in R1 universally express high levels of CD11c and are either CD11b + or CD11b$^{dull/-}$. Alternatively, CD103 F4/80+ cells in R2 universally express high levels of CD11b and are either CD11c^{hi} or CD11c^{int} (Fig. 1b).

4. Recently several additional markers including CD14, CD26, and CD64 have been used to discriminate intestinal macrophages and DCs [28, 29]. As shown in Fig. 1c, these markers can be used to further confirm the DC and macrophage subsets (*see* **Note 16**).

3.6 Magnetic Bead-Based Enrichment of Intestinal Macrophages

(Optional—only if more highly-purified cells are desired for functional studies).

Prepare the following reagents and equipment:

- Ice-cold staining buffer (CMF PBS + 5 % FBS).

1. Incubate the cell suspension obtained from **step 7** of Subheading 3.3 with CD11b and/or CD11c MACS beads according to the manufacturer's instructions (*see* **Note 17**).

2. Wash cells with ice-cold staining buffer followed by centrifugation.

3. Discard supernatant and resuspend the cell pellet in 1 mL ice-cold staining buffer and pass through a 100 μm cell strainer followed by a 40 μm cell strainer (*see* **Note 18**).

4. Enrich for magnetic bead-attached cells by positive selection using MACS LS magnetic column.

5. Repeat step 3.2 and discard supernatant.

6. Incubate cells with surface marker antibodies as described in Subheadings 3.4 and 3.5.

7. Wash magnetic bead-enriched cells twice with ice-cold staining buffer. Resuspend cell pellets in 500 μL ice-cold staining buffer without sodium azide, and pass through 40 μm cell strainer into a FACS tube (*see* **Note 19**).

8. Proceed to FACS-sorting on the BD ARIA II Cell Sorter to sort intestinal macrophage (or DC) subsets of interest.

4 Notes

1. **Steps 1–9** of Subheading 3.1 must be performed as quickly as possible to minimize the extent of cell death and to achieve maximum cell yield.

2. Be careful when removing the mesentery from the gut wall. Hold the intestine with one pair of forceps while gently pulling the mesentery with another pair.

3. To easily visualize Peyer's patches for removal, begin removing them from the ileum first and push darker luminal contents toward the jejunum and duodenum in order to contrast the light Peyer's patches from the lumen.

4. Each small intestine after being cut into 1.5 cm pieces is placed into a single 50 mL conical tube. When isolating more than 1 small intestine, it is important to not place more than one small intestine per single 50 mL conical tube. This prevents increased cell death.

5. When isolating more than one colon, up to three colons can be combined per single 50 mL conical tube.

6. We have had success with both type VIII and type IV collagenase from Sigma-Aldrich. While optimized lots of type VIII and type IV collagenase can both provide excellent digestion of the mouse small intestine, in our experience, collagenase type IV provides superior digestion of the mouse large intestine.

7. Make sure that as much excess media is dabbed away to ensure the most uniform cutting of tissue.

8. It is important to use sharp scissor that efficiently and thoroughly mince intestinal tissue. Also, it is important that the minced intestine is added into the collagenase solution immediately before placing in the orbital shaker. If more than one intestine is being processed, minced intestines can be left in the plastic weigh boats until they are all ready to be placed into their respective collagenase tubes.

9. Optimizing the concentration of collagenase and the duration of tissue digestion is important to obtain optimal cell yield without compromising viability and surface antigen expression. Under-digestion yields a low total cell number while over-digestion dramatically increases the number of dead cells and the quality of the FACS staining is compromised. Overall cell yield, viability, and surface antigen expression are affected by several factors including the source, type, concentration and enzymatic activity of collagenase, the duration of tissue digestion, the degree of mincing, the temperature of media, and the status of inflammation in the intestine [26]. Based on our experience, the concentration of collagenase is usually in the range of 1–1.5 mg/mL, and optimization of the different factors mentioned above for tissue digestion is important to ensure the highest quality and reproducibility of data.

10. One of the most important parameters for consideration in this protocol is the manufacturer, type, and specific lot of collagenase. Extreme variability in collagenase activity exists between different manufacturers, types of collagenase, and production lots, thus the potency of digestion may vary greatly and requires optimization.

11. If intestinal tissue is already dissociated after this step there is no need to perform vortexing.

12. If a solid pellet is not observed for colon samples after centrifugation, invert the sample several times and centrifuge again for 5 min.

13. Pouring off the supernatant is an important checkpoint for the quality of the cell digestion. The goal is a tightly packed cell pellet with a small ring of RBCs.

14. In our experience, use of a ~45/70 % Percoll gradient to further enrich for macrophages and dendritic cells leads to reduced

cell yield, likely as a result of a fraction of these cells residing on top of the upper 45 % layer.

15. Make sure to perform live/dead staining in PBS without serum. For a positive control for dead cell staining, we place a separate aliquot of cells into a 100 °C heat block for 1 min and then place these "dead" cells on ice and add in an equal amount of cells that were not heat-killed to ensure nice positive and negative peaks for live/dead staining. For all FACS staining, unstained intestinal cells may be utilized as a negative control to assist in the proper placement of the gates to separate positive and negative populations.

16. DC subsets express high levels of CD26, while macrophage subsets express high levels of CD14 and CD64. Other markers such as CD68, CX3CR1, and CD272 may also be used to identify DCs and macrophages [3, 4, 30].

17. If simultaneous enrichment of macrophages and DCs is desired, magnetic beads targeting CD11b and CD11c can be added at the same time. We have had success using 50 μL of beads + 450 μL buffer per each intestine.

18. In order to prevent clogging of the MACS LS columns, it is incredibly helpful to pass cells first through a 100 μm cell strainer followed by a 40 μm cell strainer before adding cells onto the column.

19. For functional studies where healthy, live cells are required, it is imperative to use ice-cold staining buffer *without sodium azide*.

Acknowledgements

This work was supported by National Institutes of Health Grants 1R01DK097256 (to T.L.D.) and 1F30DK097904-03 (to D.G.).

References

1. Backhed F, Ley RE, Sonnenburg JL, Peterson DA, Gordon JI (2005) Host-bacterial mutualism in the human intestine. Science 307:1915–1920

2. Bain CC, Mowat AM (2011) Intestinal macrophages—specialised adaptation to a unique environment. Eur J Immunol 41: 2494–2498

3. Bain CC, Mowat AM (2014) Macrophages in intestinal homeostasis and inflammation. Immunol Rev 260:102–117

4. Mowat AM, Bain CC (2011) Mucosal macrophages in intestinal homeostasis and inflammation. J Innate Immun 3:550–564

5. Platt AM, Mowat AM (2008) Mucosal macrophages and the regulation of immune responses in the intestine. Immunol Lett 119:22–31

6. Coombes JL, Powrie F (2008) Dendritic cells in intestinal immune regulation. Nat Rev Immunol 8:435–446

7. Maloy KJ, Powrie F (2011) Intestinal homeostasis and its breakdown in inflammatory bowel disease. Nature 474:298–306

8. Bogunovic MF, Ginhoux J, Helft L, Shang D et al (2009) Origin of the lamina propria dendritic cell network. Immunity 31:513–525

9. Varol C, Vallon-Eberhard A, Elinav E, Aychek T, Shapira Y et al (2009) Intestinal lamina pro-

pria dendritic cell subsets have different origin and functions. Immunity 31:502–512

10. Viney JL, Mowat AM, O'Malley JM, Williamson E, Fanger NA (1998) Expanding dendritic cells in vivo enhances the induction of oral tolerance. J Immunol 160:5815–5825

11. Hume DA, Robinson AP, MacPherson GG, Gordon S (1983) The mononuclear phagocyte system of the mouse defined by immunohistochemical localization of antigen F4/80. Relationship between macrophages, Langerhans cells, reticular cells, and dendritic cells in lymphoid and hematopoietic organs. J Exp Med 158:1522–1536

12. Liu LM, MacPherson GG (1993) Antigen acquisition by dendritic cells: intestinal dendritic cells acquire antigen administered orally and can prime naive T cells in vivo. J Exp Med 177:1299–1307

13. Mucida D, Park Y, Kim G, Turovskaya O, Scott I, Kronenberg M, Cheroutre H (2007) Reciprocal TH17 and regulatory T cell differentiation mediated by retinoic acid. Science 317:256–260

14. Sun CM, Hall JA, Blank RB, Bouladoux N, Oukka M, Mora JR, Belkaid Y (2007) Small intestine lamina propria dendritic cells promote de novo generation of Foxp3 T reg cells via retinoic acid. J Exp Med 204:1775–1785

15. Coombes JL, Siddiqui KR, Arancibia-Carcamo CV, Hall J, Sun CM, Belkaid Y, Powrie F (2007) A functionally specialized population of mucosal CD103+ DCs induces Foxp3+ regulatory T cells via a TGF-beta and retinoic acid-dependent mechanism. J Exp Med 204:1757–1764

16. Denning TL, Wang YC, Patel SR, Williams IR, Pulendran B (2007) Lamina propria macrophages and dendritic cells differentially induce regulatory and interleukin 17-producing T cell responses. Nat Immunol 8:1086–1094

17. Macpherson AJ, Lamarre A (2002) BLySsful interactions between DCs and B cells. Nat Immunol 3:798–800

18. Macpherson AJ, Uhr T (2004) Induction of protective IgA by intestinal dendritic cells carrying commensal bacteria. Science 303:1662–1665

19. Macpherson AJ, Geuking MB, McCoy KD (2005) Immune responses that adapt the intestinal mucosa to commensal intestinal bacteria. Immunology 115:153–162

20. Johansson-Lindbom B, Svensson M, Pabst O, Palmqvist C, Marquez G, Forster R, Agace WW (2005) Functional specialization of gut CD103+ dendritic cells in the regulation of tissue-selective T cell homing. J Exp Med 202:1063–1073

21. Pabst O, Bernhardt G (2010) The puzzle of intestinal lamina propria dendritic cells and macrophages. Eur J Immunol 40:2107–2111

22. Schulz O, Jaensson E, Persson EK, Liu X, Worbs T, Agace WW, Pabst O (2009) Intestinal CD103+, but not CX3CR1+, antigen sampling cells migrate in lymph and serve classical dendritic cell functions. J Exp Med 206:3101–3114

23. Persson EK, Uronen-Hansson H, Semmrich M, Rivollier A et al (2013) IRF4 transcription-factor-dependent CD103(+)CD11b(+) dendritic cells drive mucosal T helper 17 cell differentiation. Immunity 38:958–969

24. Atarashi K, Tanoue T, Shima T, Imaoka A (2011) Induction of colonic regulatory T cells by indigenous Clostridium species. Science 331:337–341

25. Ivanov II, Atarashi K, Manel N, Brodie EL et al (2009) Induction of intestinal Th17 cells by segmented filamentous bacteria. Cell 139:485–498

26. Geem D, Medina-Contreras O, Kim W, Huang CS, Denning TL (2012) Isolation and characterization of dendritic cells and macrophages from the mouse intestine. J Vis Exp. e4040

27. Denning TL, Norris BA, Medina-Contreras O, Manicassamy S et al (2011) Functional specializations of intestinal dendritic cell and macrophage subsets that control Th17 and regulatory T cell responses are dependent on the T Cell/APC ratio, source of mouse strain, and regional localization. J Immunol 187:733–742

28. Tamoutounour S, Henri S, Lelouard H, de Bovis B et al (2012) CD64 distinguishes macrophages from dendritic cells in the gut and reveals the Th1-inducing role of mesenteric lymph node macrophages during colitis. Eur J Immunol 42:3150–3166

29. Scott CL, Bain CC, Wright PB, Sichien D et al (2015) CCR2CD103 intestinal dendritic cells develop from DC-committed precursors and induce interleukin-17 production by T cells. Mucosal Immunol 8:327–339

30. Cerovic V, Bain CC, Mowat AM, Milling SW (2014) Intestinal macrophages and dendritic cells: what's the difference? Trends Immunol 35:270–277

Chapter 17

Isolation and Functional Analysis of Lamina Propria Dendritic Cells from the Mouse Small Intestine

Naoki Takemura and Satoshi Uematsu

Abstract

Dendritic cells (DCs) are the most professional antigen-presenting cells that are indispensable for the initiation of adaptive immune responses. DCs are heterogeneous in terms of their origin, anatomical location, cell-surface markers, and functions. Previous studies have demonstrated that there exist several groups of DCs in the lamina propria (LPDC) of gastrointestinal tract, which collectively contribute to the maintenance of gut homeostasis through the regulation of the balance between active immunity and tolerance. However, although intestinal LPDCs are attractive research target for understanding the immunological mechanisms in the gut, isolation of the LPDCs is complicated and technically difficult for unskilled people. Therefore, establishment of the method to isolate intestinal LPDCs is a major obstacle in this research. Here, we describe the methods that we have established for the isolation of primary DCs from the LP of mouse small intestine. Our isolation method provides high yield of viable LP leukocytes (LPLs) including DCs. Combination with FACS sorting allows for the selective isolation of $CD103^+CD8\alpha^+$ DCs and $CD103^+CD8\alpha^-$ DCs from the LPLs. Furthermore, isolated LPDCs can be subjected to immunological assays, such as measurement of cytokine productions following stimulation of Toll-like receptors. Thus, our methods would be useful for studying the functions of LPDCs of mouse small intestine.

Key words Mouse small intestine, Lamina propria, Dendritic cell, Flow cytometry, $CD8\alpha$, CD11b, CD11c, CD103, Toll-like receptor

1 Introduction

Dendritic cells (DCs) are specialized antigen-presenting cells and are highly heterogeneous population with the various subsets defined by their distinct surface marker profiles, localization, and functions [1, 2]. DCs exist as immature cells in the resident tissues, where they sample foreign and self-antigens. Following recognition of antigens, DCs mature and migrate to T cell area in the draining lymph nodes, where they interact with naïve T cells. Through antigen presentation and cytokine production, DCs regulate differentiation of naïve T cells into effector and regulatory T cells. Thus, DCs are essential mediators of active immunity and tolerance.

Andrei I. Ivanov (ed.), *Gastrointestinal Physiology and Diseases: Methods and Protocols*, Methods in Molecular Biology, vol. 1422, DOI 10.1007/978-1-4939-3603-8_17, © Springer Science+Business Media New York 2016

The luminal surface of gastrointestinal tract is constantly exposed to foods and commensal microorganisms. Although the gut immune system has evolved mechanisms to maintain immunological tolerance to innocuous exogenous antigens derived from foods and commensal microorganisms, it also senses the invasion of pathogens and properly induces protective immune responses against them. DCs play critical roles in distinguishing between commensal microorganisms and potentially harmful pathogens and in regulating the balance between immune tolerance and active immunity in the gut. In the intestine, DCs are present not only in gut-associated lymphoid tissues, such as the isolated lymphoid follicles and Peyer's patches, but also in the lamina propria (LP) [3, 4]. CD103[+] DCs are representative DCs in the intestinal LP (LPDCs). CD103[+] DCs constitutively migrate from the LP to the mesenteric lymph nodes (MLNs) in a CCR7-dependent manner, where they interact with and induce differentiation of recirculating naïve T cells [5–8]. Interestingly, CD103[+] DCs specifically produce retinoic acid and thereby induce regulatory T cells [7, 8]. Retinoic acid-producing CD103[+] DCs in MLNs confer gut tropism to naïve lymphocytes through the induction of $\alpha 4\beta 7$ integrin and CCR9 expression [9]. Thus, CD103[+] DCs have critical functions in regulating acquired immune responses in gut, and their roles in the maintenance of gut homeostasis are still being actively investigated.

Although intestinal DCs are attractive research target for understanding the immunological mechanisms in gastrointestinal tract, isolation of the DCs from intestinal LP is complicated and technically difficult. Therefore, establishment of the methods to isolate intestinal LPDCs is a major obstacle in this research. In this protocol, we introduce reproducible and robust methods that we established for the isolation of LPDCs from mouse small intestine [10]. Using the isolation methods, we have previously reported that CD103[+] DCs in the LP of small intestine are divided into two subsets distinguished by the presence or absence of CD8α[+] expression, and also showed the various functions of these two LPDC subsets for the induction of immune responses, such as expression of Toll-like receptors (TLR) [10–12]. Our methods would be helpful for investigators who will study on the functions of intestinal LPDCs.

2   Materials

2.1   Isolation of LPLs from Mouse Small Intestine

1. Phosphate-buffered saline without Ca^{2+} and Mg^{2+} (PBS).

2. 500 mM EDTA in PBS.

3. Isolation buffer: PBS supplemented with 10 % (vol/vol) fetal calf serum, 20 mM HEPES, 100 U/ml penicillin, 100 μg/ml streptomycin, 1 mM sodium pyruvate, 10 mM EDTA, and 10 μg/ml polymyxin B.

4. 90 % (vol/vol) Percoll solution: a 9:1 mixture of Percoll (GE Healthcare Life Sciences) and ×10 Hank's Balanced Salt Solution (HBSS).

5. 75 % (vol/vol) Percoll solution: a 3:1 mixture of 90 % Percoll solution and Isolation buffer (approximately density, 1.093 g/ml).

6. 40 % (vol/vol) Percoll solution: a 2:3 mixture of 90 % Percoll solution and Isolation buffer (approximately density, 1.058 g/ml).

7. DNase I solution: 1 mg/ml DNase I (Roche) in PBS, sterile with 0.45 μm filter (*see* **Note 1**).

8. Collagenase solution: 4000 Mandl units/ml collagenase D (Roche) in PBS, sterilized by passage through a 0.45 μm membrane filter (*see* **Note 1**).

9. Tissue culture medium: RPMI 1640 supplemented with 10 % (vol/vol) fetal calf serum (FCS).

10. Digestion medium: a 1:1:8 mixture of DNase I solution, collagenase solution, and 10 % FCS-supplemented RPMI 1640 medium.

11. 100 μm cell strainer.

2.2 Flow Cytometry of LPLs

1. Isolation buffer.

2. Antibodies: fluorescein isothiocyanate-conjugated anti-mouse CD8α antibody (clone: 53-6.7), purified anti-mouse CD16/32 (clone: 2.4G2), phycoerythrin–cyanine 7-conjugated anti-mouse CD11b (clone: M1/70), allophycocyanin-conjugated anti-mouse CD11c (clone: HL3), and phycoerythrin-conjugated anti-mouse CD103 antibody (clone: M290) (all from BD Bioscience).

3. Flow cytometry tube with cell strainer cap.

2.3 Culture of LPDCs

1. Culture medium: RPMI 1640 medium supplemented with 10 % FCS, 100 U/ml penicillin, 100 μg/ml streptomycin.

2. TLR ligands: poly I:C (InvivoGen), flagellin (InvivoGen), R-848 (InvivoGen), CpG oligodeoxynucleotide (ODN1668), purified as previously described [10].

3 Methods

3.1 Isolation of LPLs from Mouse Small Intestine

Our experimental procedures are specialized to isolate the LPLs that maximally contain DCs from mouse small intestine. Although other leukocytes including macrophages, eosinophils, and lymphocytes are also isolated, it would be necessary to modify the experimental conditions in order to more efficiently collect cell populations other than DCs. Briefly, we treat small intestinal segments with isolation buffer containing EDTA to remove epithelial cells and then wash extensively with PBS. We digest small intestinal segments with collagenase D and DNase I. After removing tissue residues with cell strainer, we collect single cells by centrifugation and subject them to density-gradient centrifugation using Percoll. Cells collected from the interface are used as LPLs. We usually obtain $3–5 \times 10^6$ cells of viable LPLs from a small intestine.

1. Excise small intestines from mice (*see* **Note 2**).

2. Remove mesenchymal adipose tissue, pancreas, MLNs, and Peyer's patches from small intestines using tweezer and surgical scissors (*see* **Notes 3–5**).

3. Cut open small intestines longitudinally.

4. Remove luminal contents from small intestines by rinsing in PBS twice (*see* **Note 2**).

5. Wash small intestines in 30–35 ml of PBS in a 50 ml tube by vortex (*see* **Note 6**).

6. Cut small intestines into 1–2 cm of length pieces.

7. Incubate small intestinal segments in 50 ml of isolation buffer with continuous stirring in a conical flask at 37 °C for 20 min (*see* **Notes 7–9**).

8. Filter with strainer to collect small intestinal segments (*see* **Notes 8** and **10**).

9. Put small intestinal segments into 30–35 ml of PBS in a 50 ml tube.

10. Wash extensively by vortex.

11. Repeat **steps 8–10** five times.

12. Put small intestinal segments into the tissue culture medium.

13. Transfer small intestinal segments to a glass vial and add 0.5–1 ml of digestion medium (*see* **Notes 8** and **9**).

14. Mince small intestinal segments into small pieces (~1 mm).

15. Add 5–10 ml of digestion medium and resuspend the small intestinal segments.

16. Leave for 1–2 min and remove the floating adipose tissues by aspiration.

17. Transfer the suspension to a conical flask and add rest of digestion medium (*see* **Note 11**).

18. Incubate the suspension with continuous stirring at 37 °C for 45–90 min (*see* **Notes 12** and **13**).

19. Add 500 mM EDTA/PBS (10 mM final concentration) and incubate the suspension for an additional 5 min at 37 °C.

20. Remove tissue residues with 100 μm cell strainer and collect the flow-through.

21. Remove medium after centrifugation at $440 \times g$ for 5 min.

22. Resuspend the cell pellet with 5 ml of 75 % Percoll solution and layer equal volume of 40 % Percoll solution on top of the cell suspension, being careful to minimize mixing of the solutions.

23. Centrifuge at $800 \times g$ for 20 min at room temperature with brake off.

24. Remove all liquid and debris 2 cm above the 40–75 % Percoll interface.

25. Collect cells at the 40–75 % Percoll interface.

26. Wash cells with isolation buffer twice.

27. Resuspend cells with isolation buffer (or other preferred medium).

3.2 Flow Cytometry of LPLs

LPDC subsets can be sorted using an FACSAria system (BD Biosciences). As we have previously reported [10], LPLs are divided into four subsets on the basis of their expressions of CD11b and CD11c: CD11c^{hi}CD11b^{low} DCs (R1), CD11c^{hi}CD11b^{hi} DCs (R2), CD11c^{int}CD11b^{int} macrophages (R3), and CD11c^{int}CD11b^{hi} eosinophils (R4) (Fig. 1a). After drawing a gate

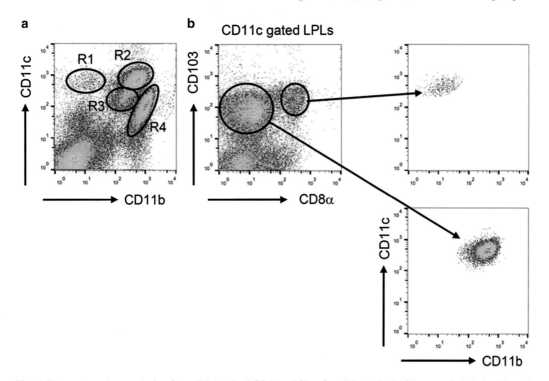

Fig. 1 Flow cytometry analysis of small intestinal CD11c$^+$ LPLs. Small intestinal LPLs were labeled with antibodies for CD8α, CD11b, CD11c, and CD103. (**a**) Flow cytometry analysis of small intestinal LPLs stained for CD11b and CD11c. (**b**) Separation of CD103$^+$CD8α$^+$ and CD103$^+$CD8α$^-$ DCs. CD11c$^+$ LPLs were gated by CD103 versus CD8α expression (*left* panel). CD11c$^+$CD103$^+$CD8α$^+$ and CD11c$^+$CD103$^+$CD8α$^-$ cells were further analyzed for the expression of CD11b (*right* panels)

that include CD11c⁺ populations, CD11c⁺ populations are further separated on the basis of CD103 and CD8α expression (Fig. 1b). CD103⁺CD8α⁺ and CD103⁺CD8α⁻ populations are equivalent to CD11c^hiCD11b^low and CD11c^hiCD11b^hi DC subsets, respectively [12]. The purity of the sorted DCs is routinely >95 %.

1. Resuspend LPLs in the isolation buffer.

2. Incubate LPLs with anti-mouse CD16/32 antibody (0.5 µg/ml) on ice for 15 min for blocking of non-specific Fc receptor-mediated antibody binding.

3. Incubate LPLs with fluorescence-labeled anti-CD8α (0.5 µg/ml), anti-CD11b (0.2 mg/m), anti-CD11c (0.2 µg/ml), and anti-CD103 antibodies (0.2 µg/ml) in the dark on ice for 20–30 min.

4. Centrifuge at $440 \times g$ for 5 min.

5. Remove the supernatant and wash LPLs with isolation buffer.

6. Repeat **steps 4–5** again.

7. Resuspend LPLs in the isolation buffer and transfer to flow cytometry tube.

8. Sort DCs on the bases of the expressions of their specific markers by FACS and collect in isolation buffer or 10 % FCS-supplemented RPMI1640 (Fig. 1b).

3.3   Culture of LPDCs

If CD103⁺CD8α⁺ LPDCs and CD103⁺CD8α⁻ LPDCs are properly collected, these LPDCs produce inflammatory cytokines following TLR ligand stimulation based on their expression patterns of TLRs. CD103⁺CD8α⁺ LPDCs produce IL-6 and IL-12p40 but not TNF-α, IL-10, or IL-23 upon stimulation with ligands for TLR3, TLR7, and TLR9 [12]. CD103⁺CD8α⁻ LPDCs express TLR5 and TLR9 and similarly produce inflammatory cytokines following their respective ligand stimulation [11, 12]. Isolated CD103⁺CD8α⁺ LPDCs and CD103⁺CD8α⁻ LPDCs are also applicable to other immunological assays as we previously described [11, 12].

1. Resuspend LPDCs into the culture medium (*see* **Note 14**).

2. Seed the LPDCs at 5×10^4 cells/100 µl/well in a 96-well plate.

3. Add 100 µl of the culture medium containing TLR ligands. Stimulate CD103⁺CD8α⁺ LPDCs medium alone, poly I:C (50 µg/ml), R-848 (100 nM), or CpG ODN (1 µM) and CD103⁺CD8α⁻ DCs with medium alone, flagellin (1 µg/ml), or CpG ODN (1 µM), respectively.

4. Incubate at 37 °C for 24 h in a 5 % CO_2 incubator.

5. Resuspend the LPDCs in the cell culture medium and transfer the suspension to 1.5 ml tube.

6. Centrifuge at $440 \times g$ for 5 min.

7. Collect the supernatant and store at −80 °C until use.

8. Measure the levels of cytokines by Bio-plex system or ELISA.

4 Notes

1. We normally store the reconstituted collagenase and DNase stock solutions at −20 °C until use.

2. We normally keep resected and washed small intestines in ice-clod PBS during the collection and washing treatment of another small intestine.

3. We normally place a small intestine on paper towel moistened with PBS to prevent the drying during this process.

4. Adipose tissues should be thoroughly removed, because adipocytes easily die and form cell aggregation involving other cells, which causes decrease of cell yield.

5. There are several populations of DCs in MLNs and Peyer's patches differing from LPDCs of small intestine. To prevent the contamination of other DCs from MLNs and Peyer's patches, these tissues should be completely removed.

6. We normally wash up to six small intestines in a single tube.

7. We normally use 50 ml of isolation buffer for up to six small intestines.

8. All equipment, including flasks, glass vials, and steel strainers should be sterilized by autoclaving.

9. We normally pre-warm isolation buffer and digestion medium to 37 °C before use.

10. We normally use a commercial tea strainer.

11. We normally use 40 ml of digestion medium for up to three small intestines.

12. Collagenase D can be substituted by other products. In our experience, cell yields when Liberase (0.425 mg/ml, Roche) or collagenase (0.6 mg/ml, crude type, Wako) is used are almost equivalent to that in the case of using collagenase D.

13. Be careful of excessive digestion of small intestinal segments owing to high concentrations of digestive enzymes and/or long-time incubation. It would result in substantial decrease in cell viability and yield.

14. If necessary, addition of GM-CSF (granulocyte macrophage colony-stimulating factor) to the culture medium (10 ng/ml final concentration) would improve viability of LPDCs during incubation.

Acknowledgement

We thank for Myoung Ho Jang at Pohang University of Science and Technology for his great support in establishing the isolation methods of LPDCs of mouse small intestine.

References

1. Mildner A, Jung S (2014) Development and function of dendritic cell subsets. Immunity 40:642–656

2. Pearce EJ, Everts B (2015) Dendritic cell metabolism. Nat Rev Immunol 15:18–29

3. Iwasaki A (2007) Mucosal dendritic cells. Annu Rev Immunol 25:381–418

4. Uematsu S, Akira S (2009) Immune responses of TLR5(+) lamina propria dendritic cells in enterobacterial infection. J Gastroenterol 44: 803–811

5. Annacker O, Coombes JL, Malmstrom V et al (2005) Essential role for CD103 in the T cell-mediated regulation of experimental colitis. J Exp Med 202:1051–1061

6. Jang MH, Sougawa N, Tanaka T et al (2006) CCR7 is critically important for migration of dendritic cells in intestinal lamina propria to mesenteric lymph nodes. J Immunol 176: 803–810

7. Sun CM, Hall JA, Blank RB et al (2007) Small intestine lamina propria dendritic cells promote de novo generation of Foxp3 T reg cells via retinoic acid. J Exp Med 204:1775–1785

8. Coombes JL, Siddiqui KR, Arancibia-Cárcamo CV et al (2007) A functionally specialized population of mucosal CD103+ DCs induces Foxp3+ regulatory T cells via a TGF-beta and retinoic acid-dependent mechanism. J Exp Med 204:1757–1764

9. Iwata M, Hirakiyama A, Eshima Y et al (2004) Retinoic acid imprints gut-homing specificity on T cells. Immunity 21:527–538

10. Uematsu S, Jang MH, Chevrier N et al (2006) Detection of pathogenic intestinal bacteria by Toll-like receptor 5 on intestinal CD11c+ lamina propria cells. Nat Immunol 7:868–874

11. Uematsu S, Fujimoto K, Jang MH et al (2008) Regulation of humoral and cellular gut immunity by lamina propria dendritic cells expressing Toll-like receptor 5. Nat Immunol 9:769–776

12. Fujimoto K, Karuppuchamy T, Takemura N et al (2011) A new subset of CD103+ CD8alpha+ dendritic cells in the small intestine expresses TLR3, TLR7, and TLR9 and induces Th1 response and CTL activity. J Immunol 186:6287–6295

Chapter 18

Purification and Adoptive Transfer of Group 3 Gut Innate Lymphoid Cells

Xiaohuan Guo, Kevin Muite, Joanna Wroblewska, and Yang-Xin Fu

Abstract

Recent studies have identified several related but distinct innate lymphoid cells (ILCs) populations that are relatively enriched in the intestinal mucosal and protect the host from various infections. Among ILCs, group 3 ILCs (ILC3s) produce lymphotoxin and IL-22, and play important roles in the development of the immune system and host–bacteria interactions. Here, we describe methods for the isolation and purification of ILC3s from the mouse intestine, and the adoptive transfer of purified ILC3s into recipient mice.

Key words Group 3 Innate lymphoid cells, Intestine, Purification, *Rag1*, Adoptive transfer

1 Introduction

Innate lymphoid cells (ILCs) represent a novel family of immune cells that resemble adaptive lymphocytes in effector function, yet lack rearranged antigen receptors. Within the past few years, based on cytokine production and transcription factors associated with T helper 1, 2, and 17 cell, three different ILCs population have been defined, including group 1 ILCs, group 2 ILCs, and group 3 ILCs (ILC3s) [1]. Although they are a tiny of lymphoid population, ILCs are enriched in mucosal areas and play essential roles in mucosal homeostasis, initiation of immune responses against pathogens, and tissue repair [2, 3].

ILC3s are Lineage$^-$, RORγt$^+$, IL-7Rα^+, and LTα1β2$^+$, with variable expression of NK receptors (NCR), and upon IL-1β and IL-23 stimulation, ILC3s have the ability to produce an array of effector cytokines, including but not limited to IL-17 and IL-22, that correspond tightly with their T helper cell counterpart, Th17 cells. To date, three ILC3 subsets have been described, including lymphoid tissue inducer (LTi) cells, NCR$^-$ ILC3s and NCR$^+$ ILC3s [2, 4]. LTi cells play a critical role in the development of secondary lymphoid organs, such as lymph nodes, Peyer's patches, and isolated lymphoid follicles in the intestine. ILC3s are located in the

Andrei I. Ivanov (ed.), *Gastrointestinal Physiology and Diseases: Methods and Protocols*, Methods in Molecular Biology, vol. 1422, DOI 10.1007/978-1-4939-3603-8_18, © Springer Science+Business Media New York 2016

lamina propria of the intestine, have an important role in intestinal homeostasis and host defense against gut pathogens. Defects in the development or function of ILC3s result in increased susceptibility to infection, inflammatory bowel disease, tumor development, and allergies [5–9]. Our and others' recent studies have shown that adoptive transfer of ILC3s could result in homeostatic regulation of the gut microbiota and rescue susceptible mice from the gut pathogen *Citrobacter rodentium* [6, 7, 10]. Therefore, we will describe our detailed protocols for the isolation, purification, and adoptive transfer of ILC3s.

2   Materials

2.1   Animals

C57BL/6 and *Rag1$^{-/-}$* mice were purchased from Harland Teklad and kept under specific pathogen-free conditions. Animal care and use should be in accordance with protocols approved by the appropriate Institutional Animal Care and Use Committee.

2.2   Reagents
(See Note 1*)*

1. Phosphate buffered saline (PBS), pH 7.2–7.4.

2. Washing Buffer I: 1× Hank's buffered salt solution (HBSS) without Ca^{2+} and Mg^{2+}, 10 mM HEPES, 5 mM EDTA, 1 mM Dithiothreitol (DTT) (*see* **Note 2**), 3 % FBS (*see* **Note 3**), pH 7.2–7.4.

3. Washing Buffer II: 1× HBSS without Ca^{2+} and Mg^{2+}, 10 mM HEPES, pH 7.2–7.4.

4. Digestion Buffer: RPMI-1640, 10 mM HEPES, 100 U/ml Penicillin-Streptomycin, 0.1 mg/ml Liberase TL (Roche) (*see* **Note 4**), 0.05 % DNase I (*see* **Note 5**), 3 % FBS, pH 7.2–7.4.

5. RPMI-1640 medium: RPMI-1640, 10 mM HEPES, 100 U/ml Penicillin-Streptomycin, 3 % FBS.

6. DMEM medium: DMEM, 10 mM HEPES, 100 U/ml Penicillin-Streptomycin, 3 % FBS.

7. Complete RPMI-1640 culture medium: RPMI-1640, 10 mM HEPES, 100 U/ml Penicillin-Streptomycin, 20 % FBS.

8. 100 % Percoll solution: Percoll, 10 mM HEPES, 100 U/ml Penicillin-Streptomycin, 10 % 10× HBSS with Ca^{2+} and Mg^{2+}.

9. 80 % Percoll solution: 80 % of 100 % Percoll solution and 20 % of PBS.

10. FACS buffer: 1× PBS, 2 % FBS, 0.02 % Sodium azide (NaN$_3$) (*see* **Note 6**).

11. Anesthetic Agent: ketamine (60 mg/kg), xylazine (2.5 mg/kg) (*see* **Note 7**).

12. Antibodies: anti-mouse CD16/32 purified (clone number: 2.4G2), anti-mouse CD90.2 (clone number: 30-H12), and

anti-mouse CD45 (clone number: 30-F11). Dilute the antibodies in FACS buffer with one million cells to final concentrations 0.2 μg/100 μl for anti-CD16/32 and CD45 antibodies and 0.1 μg/100 μl for CD90.2 antibody.

13. 7-Aminoactinomycin D (7-AAD) stock solution: 50 μg/ml in PBS with 0.09 % sodium azide, pH 7.2.

2.3 Equipment

1. Dissecting scissors.
2. Forceps.
3. Metal mesh cell strainer or kitchen strainer.
4. 70 μM cell strainer.
5. Constant-temperature incubator/shaker.
6. 6-well plate (for manual dissociation).
7. Centrifuge.
8. Microscope.
9. Hemocytometer.
10. FACS Aria (BD Biosciences) for cell sorting.
11. (Optional) GentleMACS dissociator and gentleMACS C tubes (Miltenyi Biotec).

3   Methods

3.1 Isolation of ILC3s from the Lamina Propria Compartment of the Intestine

Since there are more ILC3s in the small intestine than the colon, this procedure mainly describes the isolation of ILC3s from small intestine. Compared to wild type mice, $Rag1^{-/-}$ mice lack adaptive lymphocytes and the total number of ILC3s is greatly increased in the gut. Thus, if adoptive transfer of ILC3s is desired, isolating ILC3s from $Rag1^{-/-}$ mice is a good choice (*see* **Note 8**).

1. Make the Washing Buffer II freshly and pre-warm the Washing Buffers I and II in a 37 °C incubator.

2. Euthanize the mouse with CO_2 and surface-sterilize the skin with 70 % ethanol. Open the peritoneal cavity and extract the small intestine. Carefully remove all the mesenteric material and fat. Then remove the Peyer's Patches with dissecting scissors while pushing out fecal material from the intestine. The white Peyer's Patches could be easily seen in contrast to the brown-colored luminal contents within intestine. The total number of Peyer's Patches in wild type C57BL/6 mice is between 6 and 10 and occur in increasing frequencies toward the terminal small intestine. For the preparation of small intestine from $Rag1^{-/-}$ mice or colon, there is no need to remove the Peyer's Patches or Colonic Patches respectively (*see* **Note 9**).

3. Cut the intestine open longitudinally and cut into 1 cm pieces.

4. Transfer all the pieces into 50 ml conical tubes with 15 ml 1× PBS buffer in room temperature. Then shake vigorously with a vortex for 30 s to remove the remaining feces and mucus (*see* **Note 10**).

5. Strain the sample through a kitchen strainer or metal mesh strainer, retaining the tissue fragments. Transfer the tissue to 50 ml conical tubes with 15 ml of pre-warmed Washing Buffer I. Then Shake the tubes for 20 min on a platform shaker at 200 rpm at 37 °C to remove the intestinal intraepithelial fraction.

6. Shake vigorously for 10 s using a vortex and strain the sample through a kitchen strainer, retaining the tissue fragments.

7. Repeat **steps 5** and **6**.

8. Transfer the tissue to 50 ml conical tubes with 15 ml Washing Buffer II. Shake for another 20 min on platform shaker at 200 rpm and 37 °C. Shake vigorously for 10 s using a vortex and apply the sample onto a kitchen strainer (*see* **Note 11**).
   *If using gentleMACS C tubes and a gentleMACS Dissociator proceed to **step 10**. If using manual dissociation proceed to **step 9** and skip **step 10**.

9. Transfer the intestinal tissues into a 6 wells plate and mince into 1–2 mm pieces with surgical scissors (about 2 min). Then add 2.5 ml of pre-warmed Digestion Buffer. Shake for 20 min on a platform shaker at 150 rpm, 37 °C. Shake for 30 min for colon ILCs isolation. Very few tissue pieces should remain after digestion and the solution should be a cloudy cell suspension.

10. (Alternatively, if available) Transfer the intestine tissues from **step 8** into the gentleMACS C Tube containing 2.5 ml Digestion Buffer. Shake for 20 min on a platform shaker at 150 rpm, 37 °C. Then tightly close C Tube and attach it upside down onto the sleeve of the gentleMACS Dissociator. Run the gentleMACS Program m_intestine_01. No tissue pieces should remain and the solution should be a cloudy cell suspension.

11. Add 5 ml of RPMI-1640 medium, and apply the sample suspension to a 70 μm cell strainer placed on a 50 ml conical tube to remove cellular debris. Manually homogenize the tissue pieces in the strainer with the flat end of the plunger from a 1-ml syringe. Wash the strainer once with 5 ml of RPMI-1640 medium. Discard the strainer and centrifuge cell suspension at $300 \times g$ for 7 min at room temperature.

12. Aspirate the supernatant completely. Resuspend the cells in 4 ml DMEM medium and add 4 ml of the 80 % percoll solution (for a final percoll concentration of 40 %). Mix well by pipetting and transfer into a new 15 ml conical tubes. Carefully underlay with 4 ml 80 % percoll solution. A clear separation of layers should be observed.

13. Centrifuge at $1200 \times g$ for 20 min at room temperature with no break (this is essential). At the end of the spin the ILC3s will appear as an opaque ring between the 40 and 80 % percoll fraction. A top layer of epithelial cells will rest on top of the 40 % fraction. Aspirate and discard top layer of epithelial cells. Carefully collect the lymphoid cells at the interphase using 1 ml pipet and transfer to a fresh 15 ml tube. Wash the cells once with cold FACS buffer and spin down at $400 \times g$ for 5 min at 4 °C.

14. Resuspend the cells in FACS buffer and place it on ice. Count the cells with a hemocytometer (*see* **Note 12**).

3.2 Sorting ILC3s from the Lamina Propria Cells by Flow Cytometry

In the intestine, the ILCs are identified as Lineage⁻, CD45⁺, and IL-7Rα⁺. Compared to other ILCs, all ILC3s are positive for RORγt. Because there is not a distinct panel of surface marker to identify ILC3s vs. other ILCs, the sorting of ILC3s mainly relies on $Rorc^{gfp/+}$ mice, in which the RORγt⁺ ILC3s were GFP-positive cells. However, because of extremely low number of ILC3s in wild type mice and the dim GFP of $Rorc^{gfp/+}$ mice, it was difficult to obtain enough pure ILC3s for adoptive transfer or other purposes. Our recent study has found that ILC3s from the intestines could be strictly identified by their differential surface expression of bright CD45 and CD90 in both C57BL/6 wild type mice and $Rag1^{-/-}$ mice. As shown before and in Fig. 1, the intestinal RORγt⁺ ILC3s are all CD90^high CD45^low cells [7].

1. Resuspend the cells from **step 14** in FACS buffer (without azide) containing anti-mouse CD16/CD32 (2.4G2) at a concentration of 1×10^7 cells/ml to block Fc receptor internalization.

2. Stain the ILC3s with fluorescence conjugated antibodies against CD45 and CD90. Incubate the cells for 20 min at 4 °C in the dark. Wash the cells with FACS buffer by centrifugation at $400\ g$ for 5 min at 4 °C.

3. Resuspend the cells with FACS buffer and filter the cell suspension using a 70 μm cell strainer to remove cell debris. Add 5 μl of 7-AAD stock solution per million cells in 0.5 ml of FACS buffer to gate out dead cells, and then keep the cells on ice until the cell sorter is ready for separation.

4. To sort ILC3s, first select cells within the lymphocyte gate on the basis of forward-scatter (FSC) and side-scatter (SSC) properties. Then gate out dead cells: 7-AAD-positive cells. ILC3s can be identified as CD90^high CD45^low lymphoid cells. Place 5 ml collection tubes with cold complete RPMI-1640 culture medium in the appropriate position of the cell sorter and sort ILC3s (*see* **Note 13**).

5. Wash the sorted ILC3s twice with 1× PBS buffer by centrifugation at $400\ g$ for 5 min at 4 °C. Resuspend the cells in the

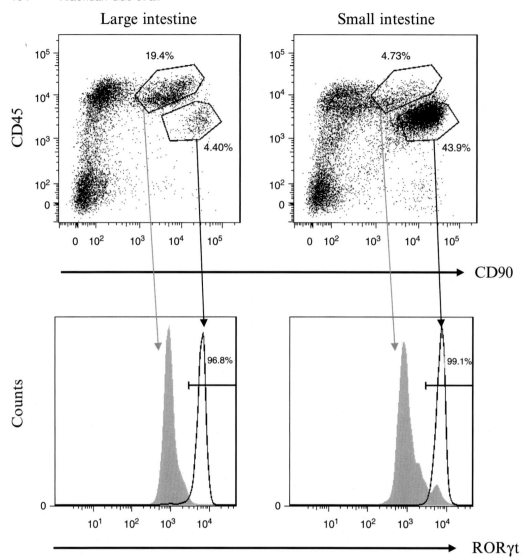

Fig. 1 Intestinal CD90^{hi}CD45^{lo} LPLs are RORγt$^+$ ILC3s. LPLs were isolated from both the large intestine and small intestine of *Rag1$^{-/-}$* mice and gated by CD45 and CD90 expression. Two separate populations, CD90^{high}CD45^{low} and CD90^{low}CD45^{high} were further analyzed with the expression of RORγt

PBS buffer at a concentration of 2×10^6 cells/ml. It might be important to have confirmation of purity or identity of the cells while keeping the cells on ice until adoptive transfer.

3.3 Adoptive Transfer of ILC3s into Recipient Mice

1. Anaesthetize the recipient mouse with ketamine (60 mg/kg) and xylazine (2.5 mg/kg) solution via intraperitoneal injection.

2. Wait until the mouse is fully anaesthetized and then adoptive transfer 2×10^5 ILC3s (100 μl per mouse) via retro-orbital injections.

3. After the injection is complete, place the mouse back into its cage for observing recovery.

4. Confirm and compare the efficacy of adoptive transfer by flow cytometry analysis or functional study.

4   Notes

1. If sorting and adoptive transfer of cells is desired, prepare all the solutions under sterilize conditions and perform all the experiments after **step 11** in a sterile space such as biosafety cabinet.

2. Make a stock solution of DTT and add it freshly into the washing buffer immediately before use.

3. Heat-inactivate FBS in a water bath at 56 °C for 30 min before use.

4. If Librase is not available, it can be replaced by Collagenase D (0.5 mg/ml) and Dispase (0.5 mg/ml).

5. DNase I could remove the DNA from digested samples to prevent the clumping of cells.

6. NaN3 is toxic. Avoid contact with skin, eyes, and mucous membranes.

7. Access to anesthetic is regulated by US state and federal law. The anesthetics should be locked up and the using of anesthetics should be recorded.

8. This method describes isolation of ILC3s from the small intestine of one mouse. If you are using multiple mice, do not pool the intestines until you get the lamina propria leukocytes at **step 13**.

9. Peyer's Patches are enriched in adaptive lymphocytes, thus it is necessary to remove Peyer's Patches in order to obtain an enriched population of ILCs from the lamina propria.

10. It is important to remove luminal contents and mucus, which may influence the effect of the digestion buffer.

11. After this step, the small intestine pieces should be pink in color. If the intestinal tissue contains white segments, which suggest dead tissues, you should discard the white tissues.

12. Usually, total $3–10 \times 10^6$ lamina propria leukocytes could be isolated from the small intestine of one naïve C57BL/6 mouse, and about 2×10^6 cells from $Rag1^{-/-}$ mouse.

13. Usually, around 1×10^4 ILC3 can be sorted from the small intestine of one naïve C57BL/6 mouse, and about 5×10^5 cells from one $Rag1^{-/-}$ mouse.

Acknowledgement

This work was supported by the National Natural Science Foundation of China (31570923 to X.G.) and National Institutes of Health grant (DK100427-01A1 to Y.X.F.).

References

1. Spits H, Artis D, Colonna M, Diefenbach A et al (2013) Innate lymphoid cells—a proposal for uniform nomenclature. Nat Rev Immunol 2:145–149

2. Artis D, Spits H (2015) The biology of innate lymphoid cells. Nature 7534:293–301

3. McKenzie AN, Spits H, Eberl G (2014) Innate lymphoid cells in inflammation and immunity. Immunity 3:366–374

4. Sonnenberg GF, Artis D (2015) Innate lymphoid cells in the initiation, regulation and resolution of inflammation. Nat Med 21:698–708

5. Eisenring M, Vom Berg J, Kristiansen G, Saller E, Becher B (2010) IL-12 initiates tumor rejection via lymphoid tissue-inducer cells bearing the natural cytotoxicity receptor NKp46. Nat Immunol 11:1030–1038

6. Guo X, Liang Y, Zhang Y, Lasorella A, Kee BL, Fu YX (2015) Innate lymphoid cells control early colonization resistance against intestinal pathogens through ID2-dependent regulation of the microbiota. Immunity 4:731–743

7. Guo X, Qiu J, Tu T, Yang X, Deng L, Anders RA, Zhou L, Fu YX (2014) Induction of innate lymphoid cell-derived interleukin-22 by the transcription factor STAT3 mediates protection against intestinal infection. Immunity 1:25–39

8. Qiu J, Guo X, Chen ZM, He L, Sonnenberg GF, Artis D, Fu YX, Zhou L (2013) Group 3 innate lymphoid cells inhibit T-cell-mediated intestinal inflammation through aryl hydrocarbon receptor signaling and regulation of microflora. Immunity 2:386–399

9. Stefka AT, Feehley T, Tripathi P, Qiu J et al (2014) Commensal bacteria protect against food allergen sensitization. Proc Natl Acad Sci U S A 36:13145–13150

10. Sonnenberg GF, Monticelli LA, Elloso MM, Fouser LA, Artis D (2011) CD4(+) lymphoid tissue-inducer cells promote innate immunity in the gut. Immunity 1:122–134

Chapter 19

Immunotherapy with iTreg and nTreg Cells in a Murine Model of Inflammatory Bowel Disease

Dipica Haribhai, Talal A. Chatila, and Calvin B. Williams

Abstract

Regulatory T (Treg) cells that express the transcription factor Foxp3 are essential for maintaining tolerance at mucosal interfaces, where they act by controlling inflammation and promoting epithelial cell homeostasis. There are two major regulatory T-cell subsets, "natural" CD4+ Treg (nTreg) cells that develop in the thymus and "induced" Treg (iTreg) cells that develop from conventional CD4+ T (Tconv) cells in the periphery. Dysregulated Treg cell responses are associated with autoimmune diseases, including inflammatory bowel disease (IBD) and arthritis. Adoptive transfer of Treg cells can modulate innate and adaptive immune responses and cure disease in animal models, which has generated considerable interest in using Treg cells to treat human autoimmune disease, prevent rejection of transplanted organs, and to control graft-versus-host disease following hematopoietic stem cell transplantation. Herein, we describe our modifications of a treatment model of T-cell transfer colitis designed to allow mechanistic investigation of the two major Treg cell subsets and to compare their specific roles in mucosal tolerance.

Key words Natural regulatory T cells, Induced regulatory T cells, Foxp3, Adoptive T-cell transfer, Inflammatory bowel disease, Immunotherapy

1 Introduction

Natural CD4+ regulatory T (nTreg) cells that express the transcription factor Foxp3 arise during T-cell development in the thymus [1]. Foxp3 expression depends on a strong TCR activation signal (agonist ligand) delivered by self-antigens [2, 3]. Following completion of the Treg developmental program, nTreg cells migrate to the periphery where they function by keeping immune responses in check. In contrast to this central pathway, induced regulatory T (iTreg) cells differentiate from conventional CD4+ T (Tconv) cells in peripheral lymphoid organs and tissues. These iTreg cells also express Foxp3 and act to control immune responses, particularly those that are directed toward foreign antigens [4, 5]. Both Treg cell subsets are necessary to maintain mucosal tolerance [6–8]. In humans and mice, loss-of-function mutation in *Foxp3* results in a multi-organ autoimmune lymphoproliferative syndrome [9–12].

Andrei I. Ivanov (ed.), *Gastrointestinal Physiology and Diseases: Methods and Protocols*, Methods in Molecular Biology, vol. 1422, DOI 10.1007/978-1-4939-3603-8_19, © Springer Science+Business Media New York 2016

Importantly, dysregulation of Treg function is associated with many autoimmune diseases [13].

A variety of experimental conditions have been described that induce Tconv cells to express Foxp3 [4]. T-cell receptor (TCR) and TGFβR signaling appear to be common requirements. TCR transgenic iTreg cells develop in the gastrointestinal tract following oral administration of their cognate antigen [14]. Similarly, chronic administration of intravenous antigen drives polyclonal and TCR transgenic CD4+ Tconv cells to express Foxp3 [15, 16]. Environmental factors such as agonists of the aryl hydrocarbon receptor, all-trans retinoic acid, and polysaccharides derived from commensal bacteria have been shown to increase the generation of iTreg cells [17–19]. In treatment models of Foxp3 deficiency and colitis, the iTreg cells that develop in vivo from Tconv cells are essential for reestablishing mucosal tolerance [6, 7]. In vitro-derived iTreg cells can substitute for those iTreg cells that arise in vivo [7]. In these treatment studies, both nTreg and iTreg cells share similar gene expression profiles. These gene expression data suggest that two major Treg subsets share suppressive mechanisms. Consistent with this hypothesis, IL-10 produced by either Treg subset is sufficient to suppress immune responses in a model of colitis [20]. In contrast to this functional reciprocity linked to suppressive mechanisms, the TCR repertoires of iTreg and nTreg cells are distinct [7, 21]. Thus any "division of labor" between iTreg and nTreg cells may well be based more on differences in the pools of antigens that they recognize rather than on unique suppressive mechanisms [7, 22, 23]. These data support the notion of large-scale in vitro production of iTreg cells for use in human immunotherapy as a supplement to the endogenous Treg pool [24–26].

Heritable expression of Foxp3 is linked to demethylation of CpG motifs in the conserved non-coding region 2 (CNS2) of the Foxp3 promoter [27]. The CNS2 region of the Foxp3 promoter in nTreg cells is highly demethylated, consistent with their stability as a cell lineage. In contrast, demethylation of the CNS2 region in iTreg cells is variable, which indicates comparatively unstable Foxp3 expression [20, 28, 29]. Loss of Foxp3 expression by iTreg cells over time can result in the development of a population of ex-iTreg cells that produce pro-inflammatory cytokines such as IL-17 and IFN-γ [20].

Although many aspects of Treg biology remain to be explored, immunotherapy of established autoimmune diseases using Treg cell transfers holds great promise. Here, we describe a method to re-establish mucosal immune tolerance by treating established colitis with nTreg and iTreg cells in a preclinical model of inflammatory bowel disease [6, 20]. Powrie and colleagues first developed this model of colitis [30, 31], which we have modified to allow tracking, isolation, and analysis of both iTreg and nTreg populations from treated mice. We describe two different options that utilize different sources of iTreg cells. In the first option, *Foxp3*[EGFP] mice [32] are used as the donor of naïve T cells. Most transferred CD4[+] EGFP[−] CD45RB[hi] cells become proinflammatory effector T cells that drive disease. However, some of the transferred cells will express Foxp3 and become iTreg cells in vivo. Successful treatment of mice with colitis requires these in vivo-derived iTreg cells as well as the subsequent adoptive transfer of nTreg cells [6]. In the second option, we use *Foxp3*[ΔEGFP] mice as donors of naïve T cells [33]. Here the CD4[+] EGFP[−] CD45RB[hi] cells carry a non-functional *Foxp3* allele (*Foxp3*[ΔEGFP]), where EGFP replaces the nuclear translocation signal and forkhead domain of Foxp3. The Foxp3–EGFP fusion protein remains in the cytoplasm and these EGFP[+] "wanna be" Treg cells lack suppressive function. In this second option, both iTreg and nTreg are supplied through adoptive transfer immunotherapy [20]. This strategy allows the iTreg cells to be produced in vitro under a variety of conditions, permitting direct manipulation of the quantity and quality of the iTreg compartment. As in the first option, both iTreg and nTreg cells are required for complete recovery from disease.

2   Materials

Prepare all solutions using ultrapure water (deionized water purified to 18 MΩ cm at 25 °C). Use analytical grade reagents prepared and stored at room temperature, and filter-sterilized as noted.

2.1   Mice

1. Donor mice: 6- to 12-week-old Thy1.2[+] *Foxp3*[EGFP] BALB/c (**options 1 and 2**, *see* **Note 1**), Thy1.1[+] *Foxp3*[EGFP] BALB/c (options 1 and 2), and rescued Thy1.1[+] *Foxp3*[ΔEGFP] BALB/c (**option 2**, *see* **Note 2**).

2. Recipient mice: 7-week-old *Rag1*[−/−] BALB/c mice.

2.2   Buffers, Media, and Other Reagents

1. Heat-inactivated fetal bovine serum (FBS): Thaw FBS under cold running tap water. Place FBS in 56 °C water bath for 45 min. Cool on ice for 1 h and aliquot into desired volumes. Freeze at −20 °C.

2. PBS (Mg^{2+}/Ca^{2+} free).

3. Red blood cell lysis buffer: 8.25 g NH_4Cl, 1 g $KHCO_3$, 100 nM EDTA, 900 ml sterile milliQ water, pH to 7.2 and add up to 1 L H_2O. Filter sterilize.

4. Complete medium: RPMI 1640 (with l-glutamine) supplemented with 10 % heat-inactivated FBS, 20 nM Glutamax, 50 μM 2-mercaptoethanol, 50 μg/ml gentamicin, 50 U/ml penicillin, and 50 μg/ml streptomycin. Filter sterilize.

5. FACS buffer: PBS, 0.1 % sodium azide, and 0.5 % bovine serum albumin (BSA); pH 7.2–7.4.

6. Modified FACS buffer: PBS, 0.1 % sodium azide, 0.5 % BSA, and 10 μg/ml Brefeldin A; pH 7.2–7.4.

7. Human IL-2: 1000 U/ml IL-2 in complete medium.

8. Triton X-100 buffer: FACS buffer, and 0.1 % Triton X-100.

2.3 Cell Preparations

1. Sterilized glass slides with frosted ends.

2. 5 cm sterile cell culture dishes.

3. 14 and 50 ml sterile polypropylene tubes.

4. 5 ml sterile polystyrene tubes.

5. Cell strainers: 70 and 30 μM pore size.

2.4 Antibodies

1. For cell sorting: anti-CD4 (clone RM4), anti-CD45RB (clone C363.16A), and anti-Thy1.2 (clone OX-7).

2. For FACS analysis: anti-CD4 (clone RM4), anti-TCRβ (clone H57-597), anti-CD25 (clone PC61 5.3), anti-CD103 (clone 2E7), anti-KLRG1 (clone 2 F1), anti-CD62L (clone MEL-14), anti-CD44 (clone IM7), anti-Thy1.2 (clone OX-7), anti-IFN-γ (clone XMG1.2), and anti-IL-17A (clone TC11-18 H10.1).

2.5 Cell Sorting

1. 14 ml polypropylene tubes.

2. Cell strainer: 30 μM pore size.

3. FACSAria (BD Biosciences).

2.6 Cell Transfer

1. Syringes: 1 ml.

2. Needles: 27 G ½.

3. Ethanol swabs.

2.7 Generation of Induced Regulatory T Cell In Vitro

1. Cell culture plates: 6 wells.

2. Anti-CD3 (clone 145-2C11) and anti-mouse CD28 (clone 37.51).

3. Human TGF-β1.

4. Incubator at 37 °C, 5 % CO_2.

5. 14 ml polypropylene tubes.

2.8 Serum Collection

1. Micro-hematocrit capillary tubes-heparinized.

2. Serum separator tubes.

3. Syringe: 1 ml.

4. Needle: 27 G ½.

2.9 Histology	1. Zinc formalin.
	2. Histology cassettes.
	3. Storage containers for histology cassettes.
2.10 Lamina Propria Lymphocyte Isolation	1. Media 1: PBS (Mg^{2+}/Ca^{2+} free).
	2. Media 2: RPMI 1640.
	3. Media 3: RPMI 1640, 10 % FBS, 5 mM EDTA, 1 mM DTT, 50 µg/ml gentamicin, 50 U/ml penicillin, and 50 µg/ml streptomycin; pH 7.2–7.4.
	4. Media 4: RPMI 1640, 20 % FBS, 1 mg/ml Collagenase D, 50 U/ml DNase I, 50 µg/ml gentamicin, 50 U/ml penicillin, and 50 µg/ml streptomycin; pH 7.2–7.4.
	5. Media 5: DMEM high glucose, 10 % FBS, 50 µg/ml gentamicin, 50 U/ml penicillin, and 50 µg/ml streptomycin; pH 7.2–7.4.
	6. Percoll 90 %: Percoll, and 10 % 1.5 M NaCl.
	7. Percoll 67 %: 90 % Percoll diluted in DMEM high glucose.
	8. Percoll 44 %: 90 % Percoll diluted in DMEM high glucose.
2.11 Intracellular Cytokine Staining	1. Phorbol Myristate Acetate (PMA).
	2. Ionomycin.
	3. GolgiPlug: aliquot into 50 µl vials and freeze at –20 °C.
	4. Brefeldin A.

3   Methods

1. We present two methods for inducing colitis and treating mice with a combination of iTreg and nTreg cells. The major difference between the two methods is the source of the iTreg cells. In **option 1**, the iTreg cells are derived in vivo through normal endogenous mechanisms, while in **option 2** the iTreg cells are derived in vitro using exogenously added reagents. In both methods, serum and tissue samples are collected at investigator-defined time points. FACS is performed on lymphocytes to determine their expression of proinflammatory molecules, and these multi-parameter data sets are correlated with histological grading of colitis severity.

Option 1. Naïve CD4+ CD45RBhi conventional T (Tconv) cells sorted from *Foxp3*EGFP mice are used to induce colitis [6, 32]. Some CD4+ Tconv cells will express Foxp3 and become EGFP+ iTreg cells in vivo. In vivo-derived iTreg cells delay colitis but do not prevent colitis from occurring, as both iTreg and nTreg cells are needed to treat disease and reestablish tolerance. When the mice begin to lose weight, nTreg cells are provided through adoptive transfer immunotherapy. Congenic markers and EGFP expression are used to identify the two Treg populations.

Option 2. Naïve CD4+ CD45RB^hi Tconv cells sorted from rescued *Foxp3*^ΔEGFP mice are used to induce colitis [6, 7, 20, 33]. Transferred CD4+ Tconv cells cannot express a functional Foxp3 protein and iTreg cells do not develop in vivo. The adoptive transfer of iTreg and nTreg cells is used to treat the mice when they begin to lose weight, and the iTreg cells can be conveniently prepared from Tconv cells in vitro. As in **option 1**, both iTreg and nTreg cells are needed to treat colitis and reestablish tolerance. Congenic markers and EGFP expression are used to identify the two Treg populations.

3.1 Naïve T-Cell Sample Preparation

1. All mice are euthanized according to institutional guidelines.

2. Dissect and pool peripheral lymph nodes (inguinal, axillary, brachial, mesenteric) and spleen from Thy1.2+ *Foxp3* EGFP BALB/c (**option 1**) or rescued Thy1.1+ *Foxp3* ΔEGFP (**option 2**).

3. Prepare single-cell suspensions in a sterile tissue culture hood by disrupting tissues between sterilized frosted slides in 5 cm culture dishes (*see* **Note 3**).

4. Filter cells with a 70 μm cell strainer into a newly labeled 50 ml conical tube. Fill tube up to 50 ml and spin at $400 \times g$, 5 min at 4 °C.

5. Discard supernatant and resuspend cells in appropriate volume of red blood cell lysis buffer (5 ml/spleen). Incubate for 30 s and add PBS up to 50 ml (*see* **Note 4**).

6. Spin cells down as above, discard supernatant and resuspend in 10 ml PBS. Add this cell suspension to lymph node cell suspension (*see* **Note 5**).

7. Add up to 50 ml PBS and spin as above. Discard supernatant and add complete medium: 1 ml/mouse (if two mice were taken, add 2 ml complete medium). Add antibodies: anti-CD4, anti-CD45RB for Thy1.2 *Foxp3*^EGFP BALB/c mice (**option 1**), or anti-CD4, anti-CD45RB and anti-Thy1.2 for rescued Thy1.1 *Foxp3*^ΔEGFP BALB/c. The concentration of the antibodies needs to be titrated depending on the dye to which the antibody is conjugated. We frequently find that 2 μg/ml of each antibody is an appropriate concentration for staining cells prior to cell sorting due to the large number of cells.

8. Incubate cells on ice and protected from light for 10 min. Add 10 ml of complete medium and spin as above. Discard the supernatant, resuspend in 10 ml complete medium, and filter the cells with a 30 μm cell strainer. For **option 1**, purify CD4+ EGFP- CD45RB^hi naïve Tconv cells by cell sorting. For **option 2**,

purify CD4$^+$ EGFP$^-$ Thy1.2$^-$ CD45RB^{hi} naïve Tconv cells by cell sorting (Fig. 1). Note that in **option 2**, Thy1.2$^+$ cells are the donor cells used to rescue newborn *Foxp3*$^{\Delta EGFP}$ mice that are not needed in the colitis experiments (*see* **Note 6**).

3.2 Adoptive Transfer of Naïve T Cells

1. Resuspend the sorted naïve T cells in PBS at 4×10^5 cells/ml.

2. Induce colitis in *Rag1*$^{-/-}$ BALB/c mice by the intra-peritoneal injection of 4×10^5 sorted naïve T cells: CD4$^+$ EGFP$^-$ CD45RB^{hi} from Thy1.2$^+$ *Foxp3*^{EGFP} mice (**option 1**) or CD4$^+$ EGFP$^-$ Thy1.2$^-$ CD45RB^{hi} from rescued Thy1.1$^+$ *Foxp3*$^{\Delta EGFP}$ mice (**option 2**).

3.3 Measure Change in Weight

1. Weigh mice twice-weekly beginning on the day that naïve T cells are transferred (day zero).

2. Euthanize the mice when they lose 20 % of their initial weight or become moribund. Weight loss of 20 % is considered evidence of severe colitis and is used as the end point of the experiment.

3. Calculate weight change: ((current weight – initial weight)/ initial weight) $\times$ 100.

3.4 In Vitro Generation of iTreg Cells

1. Sort CD4$^+$ EGFP$^-$ Thy1.2$^+$ naïve Tconv cells from Thy 1.2$^+$ *Foxp3*^{EGFP} BALB/c mice. Culture 1×10^6/ml purified Tconv cells in anti-CD3 mAb (clone 14-2C11 at 2.5 µg/ml)-coated dishes in the presence of soluble anti-CD28 mAb (clone 37.51 at 1 µg/ml), 100 U/ml IL-2 and 5 ng/ml TGF-β1.

2. After 72 h in culture, purify the newly generated Foxp3$^+$ iTreg cells by cell sorting on the basis of CD4 and EGFP expression (CD4$^+$ EGFP$^+$).

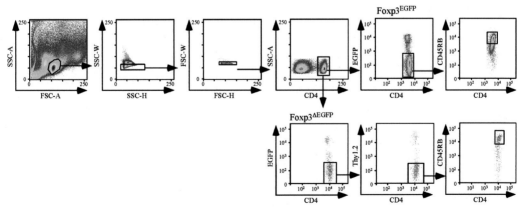

Fig. 1 Isolation of naïve T cells. Flow cytometry plots showing the gating strategy for acquiring naïve T lymphocytes (CD4$^+$ EGFP$^-$ CD45RB^{hi}) for adoptive transfer into *Rag1*$^{-/-}$ mice to initiate colitis. T cells are isolated from Thy1.1$^+$ *Foxp3*^{EGFP} BALB/c mice (**option 1**) or rescued Thy1.1$^+$ *Foxp3* ΔEGFP BALB/c mice (**option 2**). Single-cell suspensions from spleen and lymph nodes were stained with anti-CD4 and anti-CD45RB, and with anti-Thy1.2 antibodies. Cells were sorted on the basis of CD4, EGFP, and CD45RB expression. A FACSAria (BD Biosciences) was used for all cell sorting

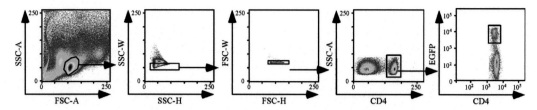

Fig. 2 Isolation of nTreg cells. Flow cytometry plots showing the gating strategy for isolating nTreg cells from *Foxp3*EGFP mice by cell sorting. Single-cell suspensions from spleen and lymph nodes were stained with anti-CD4. The nTreg cells were sorted based on CD4 and EGFP expression

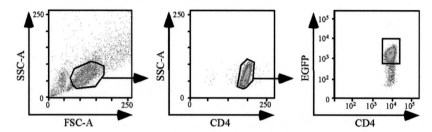

Fig. 3 Isolation of iTreg cells. Flow cytometry plots showing the gating strategy for isolating iTreg cells by cell sorting. The iTreg cells were generated in vitro from CD4+ Tconv cells that were sorted from *Foxp3*EGFP mice. After 72 h of TCR cross-linking in the presence of TGF-β1, the cultures were re-stained with anti-CD4, and the newly generated iTreg cells were isolated by cell sorting based on CD4 and EGFP expression

3.5 Adoptive Transfer of Treg Cells for Immunotherapy

1. In **option 1**, when the $Rag1^{-/-}$ host mice have lost 5–7 % of their initial weight (indicative of active colitis), transfer 1×10^6 nTreg cells by IP injection to treat the disease. The nTreg cells are purified by cell sorting from Thy1.1+ *Foxp3*EGFP mice (Fig. 2). In this version of the experiment, iTreg cells are EGFP+ Thy1.2+, nTreg cells are EGFP+ Thy1.1+, and ex-nTreg cells that have lost Foxp3 expression are EGFP− Thy1.1+. Note that ex-iTreg cells cannot be distinguished from the Thy1.1+ effector T cells that drive bowel inflammation using this congenic marker arrangement.

2. In **option 2**, $Rag1^{-/-}$ host mice received naïve T cells from rescued Thy1.1+ *Foxp3*ΔEGFP mice to induce colitis. When the $Rag1^{-/-}$ recipient mice have lost 5–7 % of their initial weight (indicative of active colitis), treat the mice with Thy1.1+ nTreg cells purified by cell sorting from Thy1.1+ *Foxp3*EGFP mice, together with Thy 1.2+ iTreg cells derived in vitro (*see* Subheading 3.4) and purified by cell sorting (Fig. 3). The Treg subsets are mixed at 1:1 ratio, and a total dose of 1×10^6 Treg cells is sufficient to cure colitis (*see* **Notes 7** and **8**).

3. Weigh mice twice weekly for up to 125 days.

3.6 Collection of Blood for Serum Cytokines

1. Anesthetize mice according to institutional guidelines.

2. Collect blood in a serum collection tube through retro-orbital bleed using a capillary tube. Cardiac puncture is also an option for collecting blood if the mice will be sacrificed.

3. Spin down the blood.

4. Collect serum and freeze aliquots at –20 °C.

3.7 Histology and Colitis Scores

1. Collect colons in a zinc formalin fixative (*see* **Note 9**).

2. Process and stain the tissue with hematoxylin and eosin (H&E).

3. Use a colitis scoring system to grade disease (e.g. ref. 34). We use a pathologist blinded to the experimental conditions and a scoring system based on a 4-point semi-quantitative scale (Fig. 4). The following features are considered: severity, depth and chronic nature of the inflammatory infiltrate, crypt abscess formation, granulomatous inflammation, epithelial cell hyperplasia, mucin-producing goblet cell depletion, ulceration, and crypt loss.

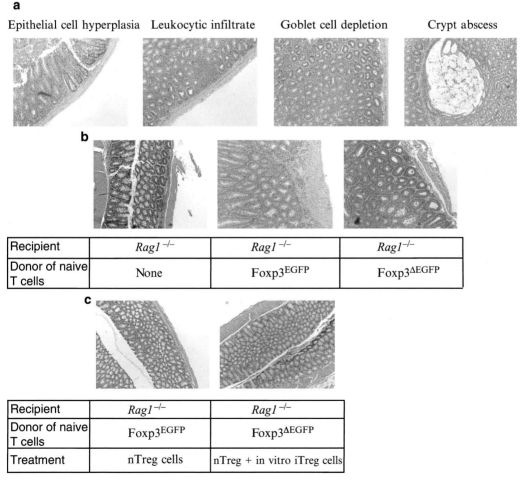

Fig. 4 Scoring of colon histopathology. (**a**) Representative H&E stained slides of colons from colitic mice showing features of disease that are scored including epithelial cell hyperplasia with villus elongation, leukocyte infiltrate, goblet cell depletion, and crypt abscess formation. (**b**) Representative H&E stained slides of colons from controls and mice with colitis, where colitis was induced with naïve cells from *Foxp3*EGFP or rescued *Foxp3* ΔEGFP mice. (**c**) Representative H&E stained slides of colons from mice treated with nTreg cells (plus those iTreg cells that were derived in situ), or with combination of nTreg and iTreg cells generated in vitro

3.8 Isolation and Analysis of Lymphocytes from the Spleen and Mesenteric Lymph Nodes

1. Prepare single-cell suspensions in the tissue culture hood by disrupting spleens and lymph nodes between sterilized frosted slides in 5 cm culture dishes. The lymph node cell suspension is ready to be used.

2. Spin splenocytes at $400 \times g$ for 5 min at 4 °C.

3. Discard supernatant and resuspend cells in 5 ml of red blood cell lysis buffer.

4. Incubate for 30 s and add PBS up to 15 ml.

5. Spin cells down as above, discard supernatant and resuspend in 10 ml PBS.

6. Count cells from spleen and lymph nodes.

7. Stain up to 5×10^6 cells per sample for cell surface antigens.

8. For intracellular cytokine staining, take 2–4 ml of the cell suspension, spin down and add media containing PMA, Ionomycin, and Golgi Plug. Plate cells in 12-well plates (one tissue sample per well), and follow the protocol in Subheading 3.10 (*see* **Note 10**).

3.9 Isolation and Analysis of Lymphocytes from the Lamina Propria of the Colon and Small Intestine

1. Take colon and small intestine (SI—last 15 cm) and open by cutting lengthwise.

2. Place organs in ice-cold Media 1 (Mg^{2+}/Ca^{2+} free PBS) and shake vigorously.

3. Place tubes on ice.

4. Transfer organs into new tube containing ice-cold Media 1 (10 ml/ tube).

5. Shake and transfer into another tube containing ice-cold Media 1. Repeat this procedure three more times.

6. Shake and transfer into another tube containing ice-cold Media 2 (10 ml/ tube).

7. Colon and SI should now be clean.

8. Cut colon and SI into 0.5 cm pieces and place them into tubes containing 40 ml Media 3 per sample.

9. Place tubes in shaker at 300 rpm, 37 °C for 30 min.

10. Take samples and filter into new conical tube using a 100 μM filter.

11. Keep this for intra-epithelial lymphocytes (IEL) isolation.

12. Follow **steps 21–35** to further isolate IEL.

13. Take the filter and the remaining gut tissue and wash with Media 5.

14. Chop tissue finely with razor blade.

15. Add tissue into tubes containing 24 ml Media 4.

16. Place tubes in shaker at 300 rpm, 37 °C for 60 min.

17. Filter samples into new conical tube using a 100 μM filter.

18. Use syringe plunger for macerating left over tissue.

19. Wash filter with Media 5 and fill tube with Media 5.

20. Spin down cells at 400×*g*, 4 °C for 5 min.

21. Discard supernatant.

22. Add Media 5 and resuspend the cell pellet.

23. Spin down cells at 400×*g*, 4 °C for 5 min.

24. Discard supernatant.

25. Add 8 ml of 44 % Percoll to each sample to resuspend cell pellet.

26. Slowly layer cell suspension onto 67 % Percoll placed in 15 ml conical tube (5 ml of 67 % Percoll per tube).

27. Spin at 308×*g*, 35 min at room temperature, with *NO* centrifuge brake.

28. Collect lymphocytes (between top and bottom layers) with a transfer pipette and place in 15 ml conical tube.

29. Add 10 ml Media 5.

30. Spin down cells at 400×*g*, 4 °C for 5 min.

31. Discard supernatant.

32. Resuspend cells in 2–3 ml Media 5.

33. Count cells.

34. Cells are ready for cell surface staining or ex-vivo stimulation followed by intracellular cytokine staining for effector cytokines.

3.10 Intracellular Staining and Cytokine Analysis

1. Perform intracellular cytokine staining after 5-h of re-stimulation with 5 ng/ml PMA and 0.5 μM ionomycin in the presence of 1 μl/ml Golgi Plug.

2. Perform surface staining of cells using a modified FACS buffer containing 10 μg/ml brefeldin A.

3. Stain cells on ice for 20 min with primary antibodies: 0.1–0.3 μg/ml anti-CD4 and anti-TCRβ. Wash cells with the modified FACS buffer and fix with 1 % paraformaldehyde overnight at 4 °C.

4. Wash cells with 1 ml PBS and permeabilize with 1 ml 0.1 % Triton-X buffer by centrifugation.

5. Perform intracellular staining for 30 min at room temperature with 0.1–0.4 μg/ml anti-IL-17A and anti-IFN-γ.

6. Add 4 ml FACS buffer and spin down cells. Discard supernatant.

7. Re-fix cells with 1 % paraformaldehyde and refrigerate until analyzing by FACS (Fig. 5).

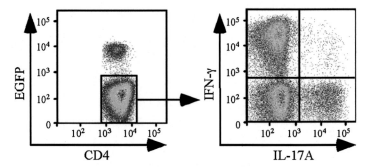

Fig. 5 Analysis of T-cell cytokine production. Flow cytometry analysis of ex-vivo stimulated MLN lymphocytes by intracellular staining for IFN-γ and IL-17A. Production of these cytokines by Tconv cells isolated from mice with colitis is shown

4   Notes

1. *Foxp3* is X-linked. Keep *Foxp3*^{EGFP} female mice homozygous for the EGFP-tagged allele to increase the number of Treg cells in female mice that express EGFP. Also note that all experiments are performed in the BALB/c background. There are differences between strains in the kinetics of colitis development, and in both the frequency and number of iTreg cells that develop in vivo. Strain differences in iTreg generation can affect the results of adoptive transfer immunotherapy with nTreg cells. *Foxp3*^{EGFP} mice on the BALB/c background are commercially available (The Jackson Laboratory—*C.Cg-Foxp3tm2Tch/J*).

2. *Foxp3*$^{\Delta EGFP}$ mice lack functional Treg cells and develop uncontrolled T-cell activation resulting in autoimmune lymphoproliferative disease [33]. To create a source of naïve T cells that carry the non-functional *Foxp3*$^{\Delta EGFP}$ allele (**option 2**), newborn Thy1.1$^+$ *Foxp3*$^{\Delta EGFP}$ BALB/c mice are given 4–6×10^6 unfractionated splenocytes isolated from Thy1.2$^+$ BALB/c mice by IP injection within 48 h of birth. The splenocytes are washed and resuspended in PBS, then placed in a 1 cc syringe with a 29G needle. The volume delivered should not exceed 200 μl per pup. It is important to include bedding material from the cage when holding the pup, and to angle the needle down toward the bladder and away from the liver for the injection. The splenocytes contain a sufficient number of Treg cells and Tconv cells with a functional Foxp3 allele (source of iTreg cells) to treat the mice. Note that in Fig. 1, Thy1.2$^-$ cells are the host T cells that carry the *Foxp3*$^{\Delta EGFP}$ allele. *Foxp3*$^{\Delta EGFP}$ mice are commercially available (The Jackson Laboratory—*C.129X1-Foxp3tm3Tch/J*).

3. Work with cells using sterile conditions. Cells are being used for transfer into mice and for in vitro stimulation in cell culture.

4. When processing spleen and lymph nodes, keep splenocytes in separate tube until red blood cell lysis has been performed. Otherwise, yields will decrease dramatically.

5. Once RBC lysis buffer has been washed away, pool splenocytes and lymph node cell suspension for cell sorting. For phenotyping by flow cytometry, do not pool splenocytes and lymphocytes from lymph nodes.

6. Keep cells on ice until ready to sort, otherwise yields decrease. Sort Treg cells into IL-2 media (1000 U/ml). By the time all cells are sorted the final IL-2 concentration will be 100 U/ml. Expect up to 1×10^6 nTreg cells per mouse. Note that such "peripheral" Treg cells may contain ~5–10 % iTreg cells that were generated in vivo and cannot be readily distinguished from the nTreg cells derived in the thymus.

7. In this configuration, effector T cells are EGFP⁻ Thy1.1⁺, nTreg cells are EGFP⁺ Thy1.1⁺, iTreg cells are EGFP⁺ Thy1.2⁺, and iTreg cells that lose Foxp3 expression (Thy1.2⁺ EGFP⁻) can be clearly distinguished from the effector T cells that drive colitis. This enables mechanistic study of iTreg stability, as well as the fate of iTreg cells that lose Foxp3 expression.

8. Pre-enrichment of CD4⁺ T cells by magnetic bead isolation using negative selection will help reduce sorting time on the flow cytometer. This strategy will yield untouched CD4⁺ T cells that can be used for downstream applications.

9. When collecting the colon for histology, also check small intestine, lungs, and liver. In some mice the small intestine is enlarged due to inflammatory infiltrates. Infiltrates can also occur at other mucosal sites and in the liver.

10. EGFP staining for flow cytometry analysis of Treg cells is not necessary. We routinely acquire 50,000 CD4⁺ T-cell events in order to characterize the different subsets.

Acknowledgement

This work was supported by Senior Research Award #296598 from the Crohn's and Colitis Foundation of America (to C.B.W.), NIH R01 AI073731 and R01 AI085090 (to C.B.W. and T.A.C.), the D.B. and Marjorie Reinhart Family Foundation (to C.B.W.), and the Children's Hospital of Wisconsin (to C.B.W.).

References

1. Josefowicz SZ, Lu LF, Rudensky AY (2012) Regulatory T cells: mechanisms of differentiation and function. Annu Rev Immunol 30:531–564

2. Jordan MS, Boesteanu A, Reed AJ, Petrone AL, Holenbeck AE, Lerman MA, Naji A, Caton AJ (2001) Thymic selection of CD4+ CD25+ regulatory T cells induced by an agonist self-peptide. Nat Immunol 2:301–306

3. Relland LM, Mishra MK, Haribhai D, Edwards B, Ziegelbauer J, Williams CB (2009) Affinity-based selection of regulatory T cells occurs independent of agonist-mediated induction of Foxp3 expression. J Immunol 182:1341–1350

4. Schmitt EG, Williams CB (2013) Generation and function of induced regulatory T cells. Front Immunol 4:152

5. Samstein RM, Josefowicz SZ, Arvey A, Treuting PM, Rudensky AY (2012) Extrathymic generation of regulatory T cells in placental mammals mitigates maternal-fetal conflict. Cell 150:29–38

6. Haribhai D, Lin W, Edwards B, Ziegelbauer J et al (2009) A central role for induced regulatory T cells in tolerance induction in experimental colitis. J Immunol 182:3461–3468

7. Haribhai D, Williams JB, Jia S, Nickerson D et al (2011) A requisite role for induced regulatory T cells in tolerance based on expanding antigen receptor diversity. Immunity 35:109–122

8. Josefowicz SZ, Niec RE, Kim HY, Treuting P, Chinen T, Zheng Y, Umetsu DT, Rudensky AY (2012) Extrathymically generated regulatory T cells control mucosal TH2 inflammation. Nature 482:395–399

9. Chatila TA, Blaeser F, Ho N, Lederman HM, Voulgaropoulos C, Helms C, Bowcock AM (2000) JM2, encoding a fork head-related protein, is mutated in X-linked autoimmunity-allergic disregulation syndrome. J Clin Invest 106:R75–R81

10. Brunkow ME, Jeffery EW, Hjerrild KA, Paeper B et al (2001) Disruption of a new forkhead/winged-helix protein, scurfin, results in the fatal lymphoproliferative disorder of the scurfy mouse. Nat Genet 27:68–73

11. Bennett CL, Christie J, Ramsdell F, Brunkow ME et al (2001) The immune dysregulation, polyendocrinopathy, enteropathy, X-linked syndrome (IPEX) is caused by mutations of FOXP3. Nat Genet 27:20–21

12. Wildin R, Ramsdell S, Peake FJ, Faravelli F et al (2001) X-linked neonatal diabetes mellitus, enteropathy and endocrinopathy syndrome is the human equivalent of mouse scurfy. Nat Genet 27:18–20

13. Chatila TA (2005) Role of regulatory T cells in human diseases. J Allergy Clin Immunol 116:949–959

14. Mucida D, Kutchukhidze N, Erazo A, Russo M, Lafaille JJ, Curotto de Lafaille MA (2005) Oral tolerance in the absence of naturally occurring Tregs. J Clin Invest 115:1923–1933

15. Thorstenson KM, Khoruts A (2001) Generation of anergic and potentially immunoregulatory CD25+ CD4 T cells in vivo after induction of peripheral tolerance with intravenous or oral antigen. J Immunol 167:188–195

16. Apostolou I, von Boehmer H (2004) In vivo instruction of suppressor commitment in naive T cells. J Exp Med 199:1401–1408

17. Mucida D, Pino-Lagos K, Kim G, Nowak E, Benson MJ, Kronenberg M, Noelle RJ, Cheroutre H (2009) Retinoic acid can directly promote TGF-beta-mediated Foxp3(+) Treg cell conversion of naive T cells. Immunity 30:471–472

18. Quintana FJ, Basso AS, Iglesias AH, Korn T et al (2008) Control of T(reg) and T(H)17 cell differentiation by the aryl hydrocarbon receptor. Nature 453:65–71

19. Round JL, Mazmanian SK (2010) Inducible Foxp3+ regulatory T-cell development by a commensal bacterium of the intestinal microbiota. Proc Natl Acad Sci U S A 107:12204–12209

20. Schmitt E, Haribhai GD, Williams JB, Aggarwal P et al (2012) IL-10 Produced by Induced Regulatory T Cells (iTregs) Controls Colitis and Pathogenic Ex-iTregs during Immunotherapy. J Immunol 189:5638–5648

21. Relland LM, Williams JB, Relland GN, Haribhai D, Ziegelbauer J, Yassai M, Gorski J, Williams CB (2012) The TCR repertoires of regulatory and conventional T cells specific for the same foreign antigen are distinct. J Immunol 189:3566–3574

22. Curotto de Lafaille MA, Kutchukhidze N, Shen S, Ding Y, Yee H, Lafaille JJ (2008) Adaptive Foxp3+ regulatory T cell-dependent and -independent control of allergic inflammation. Immunity 29:114–126

23. Curotto de Lafaille MA, Lafaille JJ (2009) Natural and adaptive foxp3+ regulatory T cells: more of the same or a division of labor? Immunity 30:626–635

24. Trzonkowski P, Bieniaszewska M, Juscinska J, Dobyszuk A, Krzystyniak A, Marek N, Mysliwska J, Hellmann A (2009) First-in-man

clinical results of the treatment of patients with graft versus host disease with human ex vivo expanded CD4+ CD25+ CD127– T regulatory cells. Clin Immunol 133:22–26

25. Trzonkowski P, Szarynska M, Mysliwska J, Mysliwski A (2009) Ex vivo expansion of CD4(+) CD25(+) T regulatory cells for immunosuppressive therapy. Cytometry A 75:175–188

26. Brunstein CG, Miller JS, Cao Q, McKenna DH et al (2011) Infusion of ex vivo expanded T regulatory cells in adults transplanted with umbilical cord blood: safety profile and detection kinetics. Blood 117:1061–1070

27. Zheng Y, Josefowicz S, Chaudhry A, Peng XP, Forbush K, Rudensky AY (2010) Role of conserved non-coding DNA elements in the Foxp3 gene in regulatory T-cell fate. Nature 463:808–812

28. Floess S, Freyer J, Siewert C, Baron U et al (2007) Epigenetic control of the foxp3 locus in regulatory T cells. PLoS Biol 5:e38

29. Polansky JK, Kretschmer K, Freyer J, Floess S (2008) DNA methylation controls Foxp3 gene expression. Eur J Immunol 38:1654–1663

30. Asseman C, Mauze S, Leach MW, Coffman RL, Powrie F (1999) An essential role for interleukin 10 in the function of regulatory T cells that inhibit intestinal inflammation. J Exp Med 190:995–1004

31. Uhlig HH, Coombes J, Mottet C, Izcue A et al (2006) Characterization of Foxp3+ CD4+ CD25+ and IL-10-secreting CD4+ CD25+ T cells during cure of colitis. J Immunol 177: 5852–5860

32. Haribhai D, Lin W, Relland LM, Truong N, Williams CB, Chatila TA (2007) Regulatory T cells dynamically control the primary immune response to foreign antigen. J Immunol 178:2961–2972

33. Lin W, Haribhai D, Relland LM, Truong N, Carlson MR, Williams CB, Chatila TA (2007) Regulatory T cell development in the absence of functional Foxp3. Nat Immunol 8:359–368

34. Leach MW, Bean AG, Mauze S, Coffman RL, Powrie F (1996) Inflammatory bowel disease in C.B-17 scid mice reconstituted with the CD45RBhigh subset of CD4+ T cells. Am J Pathol 148:1503–1515

Chapter 20

Isolation of Eosinophils from the Lamina Propria of the Murine Small Intestine

Claudia Berek, Alexander Beller, and Van Trung Chu

Abstract

Only recently has it become apparent that eosinophils play a crucial role in mucosal immune homeostasis. Although eosinophils are the main cellular component of the lamina propria of the gastrointestinal tract, they have often been overlooked because they express numerous markers, which are normally used to characterize macrophages and/or dendritic cells. To study their function in mucosal immunity, it is important to isolate them with high purity and viability. Here, we describe a protocol to purify eosinophils from the lamina propria of the murine small intestine. The method involves preparation of the small intestine, removal of epithelial cells and digestion of the lamina propria to release eosinophils. A protocol to sort eosinophils is included.

Key words Eosinophils, Lamina propria, Small intestine, Epithelial cells, Peyer's patches, Digestion

1 Introduction

The mucosal lamina propria (LP) is a loose connective tissue underlying the layer of epithelial cells lining the gut [1]. Within the LP one finds many cell types of the immune system, including B and plasma cells, T cells, macrophages, dendritic cells, and also eosinophils. Of central importance are plasma cells producing IgA antibody, which is secreted into the mucosal tissues and transported into the gut lumen, where it has an important function in shaping the repertoire of intestinal bacteria [2–4].

Recently, it was demonstrated that eosinophils are required for the maintenance of IgA-producing plasma cells in the LP and hence for the establishment of normal levels of IgA antibody in mucosal tissues and in serum [5, 6]. Furthermore it was shown that eosinophils are essential for the efficient generation of IgA$^+$ B cells in the germinal centers of Peyer's patches (PP), a surprising result since eosinophils are practically absent from PP [5]. In eosinophil-deficient mice germinal center B cells in PP switch preferentially to IgG [5]. As a consequence, a strong reduction of IgA$^+$

Andrei I. Ivanov (ed.), *Gastrointestinal Physiology and Diseases: Methods and Protocols*, Methods in Molecular Biology, vol. 1422, DOI 10.1007/978-1-4939-3603-8_20, © Springer Science+Business Media New York 2016

B and plasma cells is observed in the LP of eosinophil-deficient mice [5]. These findings suggest a crucial role of eosinophils for immune homeostasis in the gastrointestinal tract [5, 6].

Only recently has the use of antibodies highly specific for eosinophils, such as anti-Siglec F antibodies [7], demonstrated the abundance of eosinophils in the gastrointestinal tract [8, 9]. Indeed, in contrast to lymphocytes, eosinophils home to the LP before birth and before the gut is colonized by the microbiota and thus their presence in the LP of the gastrointestinal tract is independent of inflammatory processes [10, 11]. When tissue sections are stained with antibodies specific for the major basic protein, which is expressed only in eosinophils (rat monoclonal antibody developed by Lee NA and Lee JJ, Mayo Clinic, Scottsdale, Arizona, USA), the distribution of eosinophils within the small intestine is seen to vary. Whereas in the duodenum they are found associated with the entire basal surface of the villi, in the jejunum and the ileum they are mainly localized to the crypt region at the base of the villi [10]. While a specific subset of T cells is present as intraepithelial cells, eosinophils are not.

Here, we describe a protocol to isolate eosinophils from the murine LP. The protocol focuses on the isolation from the LP of the small intestine [5, 6, 8, 12], however, using the same methods eosinophils may also be isolated from the LP of the large intestine [13]. In the LP about 6 % of total cells are eosinophils (Fig. 1), after Percoll density centrifugation they are enriched to about 20 % and after sorting eosinophils, to more than 95 %.

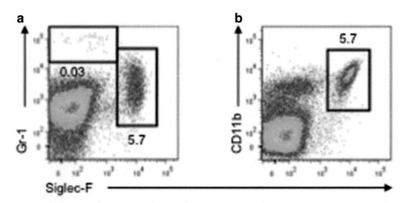

Fig. 1 Abundance of eosinophils in the LP of the small intestine of BALB/c mice. The small intestine was dissected, epithelial cells removed by predigestion and a cell suspension prepared by enzymatic digestion. Cell suspension was stained with GR-1 and Siglec-F (**a**) or CD11b and Siglec-F (**b**) specific antibodies. The percentages of eosinophils (5.7 % GR-1^{lo}, CD11b$^+$ Siglec-F$^+$ cells) and of neutrophils (0.03 % GR-1^{hi}, CD11b$^-$ Siglec-F$^-$ cells) are indicated (reprinted from ref. 3 with kind permission of Elsevier)

2 Materials

2.1 Stock Solutions

1. EDTA, 0.5 M: Add 186.1 g of EDTA to 800 ml of distilled water in a 1 l graduated glass beaker. Stir and adjust the pH to 8.0 by slowly adding 5 N NaOH. Add deionized water to a final volume of 1 l.

2. 10× phosphate buffered saline (PBS).

3. 1× PBS/BSA: 1× PBS containing 0.5 % bovine serum albumin.

4. Fetal calf serum (FCS).

5. Antibiotics: 10,000 IU/ml penicillin, 10 mg/ml streptomycin (*see* **Note 1**).

6. 1 M HEPES.

7. HBSS: Hanks' Balanced Salt Solution, Calcium and Magnesium free.

8. RPMI culture medium: RPMI containing 10 % FCS, 0.01 M HEPES, 100 IU/ml penicillin, and 100 μg/ml streptomycin c (*see* **Note 1**).

9. Isotonic Percoll separation solution: Percoll 1.130 g/ml density (GE Healthcare). Add one part (v/v) of 10× PBS to nine parts (v/v) of Percoll.

2.2 Equipment

1. Ice bucket.

2. Vortex.

3. Cell centrifuge (e.g. Megafuge, Heraeus).

4. 15 and 50 ml Falcon tubes.

5. 1 ml Eppendorf tubes.

6. Shaking water bath or incubator (rotation 120–130 rpm).

2.3 Preparation of the LP

1. Petri dishes 10 cm in diameter.

2. Forceps, normal and with curved tips, scissors, scalpel.

3. 5 ml syringes and needles (18 G × 1¹/₂).

2.4 Digestion of the LP

1. Ice-cold 1× PBS for washing the small intestine.

2. Pre-digestion buffer (HBSS containing 10 % FCS, 0.01 M HEPES, 5 mM EDTA, 100 IU/ml penicillin, and 100 μg/ml streptomycin c): Add 10 ml of FCS, 1 ml 0.5 M EDTA (pH 8.0), 1 ml 1 M HEPES, 1 ml Pen/Strep to 100 ml HBSS. Mix the reagents shortly before use and warm to 37 °C.

3. Digestion buffer: 100 ml RPMI culture medium containing 1 mg/ml Collagenase D (Roche), 1 mg/ml Dispase II (Sigma) and 0.1 mg/ml DNase I (Sigma) (*see* **Note 2**).

4. Metal sieve.

5. 70 μm cell strainer (Falcon, BD).

6. 10 ml syringes and needles ($18 \, G \times 1^1/_2$).

**2.5 Density
Centrifugation**

1. Percoll 40 % (vol/vol): Use 44.5 ml isotonic Percoll separation solution and dilute with 55.5 ml of RPMI culture medium.

2. Percoll 75 % (vol/vol): Use 83 ml of isotonic Percoll separation solution and dilute with 17 ml of 1× PBS (*see* **Note 3**).

**2.6 Sorting
of Eosinophils**

1. Resuspending and antibody dilution buffer: PBS/BSA.

2. Blocking reagent: anti-FcγR (clone 2.4G2, BD Life Science) antibody and rat IgG serum (Sigma).

3. Staining antibodies: antibodies specific for CD45 (anti-CD45-PE-Cy7, clone 30-F11), Ly6G (anti GR-1-PE, clone RB6-8C5), MHCII (anti MHCII-APC-Cy7, clone M5/114.15), CD11b (anti-CD11b-FITC, clone M1/70).

4. Control antibody specific for Siglec-F (clone E50-2440, BD Pharmingen).

5. DAPI: DAPI nuclei staining solution (1 μg/ml) diluted in PBS/BSA.

6. Fluorescence-activated cell sorter.

3 Methods

Isolation of eosinophils from the LP of the small intestine is time-consuming, because each of the small intestine preparations should be handled separately. If large numbers of eosinophils are required, it is therefore advisable that a second person helps with the preparation of the LP cell suspension, as the yield of eosinophils is higher when cells are rapidly isolated. One person can handle 4–6 animals.

From a single intestine 3×10^5 to 8×10^5 eosinophils can be isolated. The number is variable with each preparation. In addition, the yield of eosinophils depends on the age of the animals.

**3.1 Preparation
of the Small Intestine**

1. Open the abdomen of the mouse, slowly pull out the intestines and dissect the small intestine (Fig. 2). Move each sample of small intestine tissue separately into a petri dish containing 10 ml ice-cold 1× PBS. Remove carefully and discard all mesenteric fat. Lift PP from the surface of the small intestine using forceps with curved tips and collect separately (*see* **Note 4**).

2. Rinse the gut lumen by flushing using a syringe filled with ice-cold 1× PBS.

3. Open the intestine longitudinally and cut into 1–2 cm pieces.

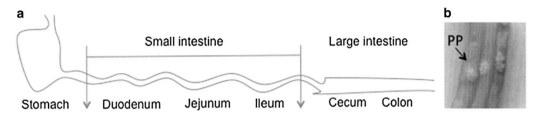

Fig. 2 Preparation of small intestine. (**a**) Schematic presentation indicates the different parts of the gut. (**b**) PPs on the outer surface of the small intestine. Prepare the small intestine by cutting the gut about 1 cm downstream from the stomach and 1 cm upstream from the cecum, as indicated. Using forceps with curved tips, carefully remove PP and attached mesenteric fat

4. Wash the pieces in ice-cold HBSS.

5. Repeat washing several times by transferring the pieces into new petri dishes filled with fresh ice-cold HBSS.

3.2 Removal of Epithelial Cells

1. Move the pieces recovered from one intestine into a 50 ml Falcon tube containing 5 ml pre-warmed, pre-digestion solution.

2. Incubate for 15 min at 37 ° C in an incubator or water bath under rotation (120–130 rpm/min). Take out of the incubator and put immediately on ice.

3. Briefly vortex (10 s) each tube and pour the pre-digestion mixture into a metal sieve (*see* **Note 5**). Rinse the sieve three times with ice-cold HBSS buffer and move the remaining pieces from the sieve into fresh pre-digestion solution. Repeat **steps 2** and **3**.

4. After pre-digestion (**steps 1–3**) collect the pieces from one intestine into a new 50 ml tube and wash twice by adding 20 ml ice-cold HBSS (*see* **Note 6**). After each wash the tissue pieces are collected by centrifugation at $400 \times g$ at 4 °C.

3.3 Preparation of Cell Suspension

1. Move the washed pieces of intestine into a petri dish containing 1 ml of ice-cold RPMI culture medium. Macerate the tissue using scissors and/or a scalpel and transfer the finely divided tissue into a fresh 50 ml Falcon tube on ice. Use a separate tube for each individual small intestine.

2. Add 5 ml pre-warmed digestion buffer to each LP homogenate and incubate at 37 °C for 20 min with gentle shaking.

3. After incubation, vortex the cell solution at full speed for 20 s (*see* **Note 7**) and pass through a 70 μm cell strainer (Falcon, BD). Collect the LP cell suspension in a 50 ml Falcon tube in an ice bucket.

4. Undigested tissue pieces held back on the cell strainer are taken up in 5 ml fresh pre-warmed digestion buffer and incubated for another 20 min at 37 ° C.

5. Disaggregate tissue pieces remaining after **step 5** by passing several times through an 18-gauge needle attached to a 10 ml syringe and filter the cell suspension through a 70 μm cell strainer (Falcon, BD) into a 50 ml Falcon tube in an ice bucket (*see* **Note 8**).

6. Combine cells from all digestion steps in a fresh 50 ml tube and centrifuge for 7 min at 400×*g* at 4 ° C. Discard the supernatant.

7. Resuspend the cell pellet in 20 ml of culture medium. Pass the suspension through a 70 μm cell strainer and collect the cells in a 50 ml Falcon tube in an ice bucket. Recover the cells by centrifuging for 7 min at 400×*g* at 4 ° C. Discard the supernatant.

8. Discard the supernatant and repeat this washing step (**step 8**) one more time (*see* **Note 9**).

3.4   Isolation of Mononuclear Cells by Percoll Discontinuous Density Gradient Centrifugation

1. Resuspend the cell pellet of each LP cell preparation in 10 ml isotonic 40 % Percoll.

2. Add 5 ml isotonic 70 % Percoll in a 15 ml Falcon tube. Using a Pasteur pipette carefully overlay the cell suspension in 40 % Percoll.

3. Centrifuge the discontinuous Percoll gradient for 20 min at 800×*g* at room temperature. Make sure the brake on the centrifuge is switched off.

4. Aspirate debris, which has collected on the top of the gradient and carefully collect the cells from the interphase between the 40 and 75 % Percoll layers using a Pasteur pipette. Transfer the cells into a fresh 15 ml Falcon tube (*see* **Note 10**). Add ice-cold RPMI medium to fill the tube to 15 ml. Recover the cells by centrifuging for 10 min at 400×*g* at 4 °C.

5. Resuspend cell pellet and wash cells for a second time with ice-cold RPMI culture media.

6. Final yield is around 2–5×10^6 LP mononuclear cells per small intestine. Approximately 20 % of the mononuclear cells are eosinophils (Fig. 3).

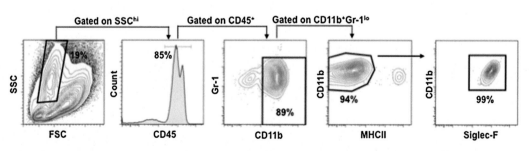

Fig. 3 Gating strategy to isolate eosinophils. LP cell suspension was stained with CD45, GR-1 CD11b, and MHCII-specific antibodies. Before sorting, approximately 20 % of leucocytes are eosinophils (SSC^{hi}). CD45$^+$, GR-1^{lo} CD11b$^+$, and MHCII$^-$ cells were sorted. Control staining of sorted cells with Siglec-F specific antibodies showed that after sorting 99 % of cells are eosinophils. DAPI was used to exclude dead cells

**3.5 Isolation
of Eosinophils**

1. Resuspend the cell pellet in 1 ml ice-cold PBS/BSA.

2. Count cells and adjust the concentration to 2×10^6 cells per 100 µl PBS/BSA.

3. Add 10 µl blocking reagent for 1 ml of cell suspension. Mix gently.

4. Incubate for 10 min on ice.

5. Add 10 µl diluted antibody combination. Mix gently and incubate for 20 min on ice (*see* **Note 11**).

6. Wash cells twice with ice-cold PBS/BSA. After each wash the cells are collected by centrifugation for 10 min at $400 \times g$.

7. Resuspend cell pellet in 1 ml ice-cold PBS/BSA and add directly before sorting 3 µl DAPI solution.

8. Sort eosinophils as SSC^{hi}, $CD45^+$, $MHCII^-$, $Gr-1^{lo}$, and $CD11b^+$ cells (*see* **Note 12**). For the gating strategy *see* Fig. 3. Collect the cells into 1 ml RPMI culture medium. Pre-rinse the collecting tube with RPMI culture medium.

9. Each small intestine yields approximately $3–8 \times 10^5$ eosinophils, which have a purity of >95 % (*see* Fig. 3). The numbers vary with each preparation.

4   Notes

1. Antibiotics should be added to control bacterial contamination, in particular, when eosinophils are isolated for in vitro cultures.

2. Prepare digestion buffer immediately before use.

3. Changes in the Percoll gradient concentrations between 35 and 45 % for the upper layer and 65–75 % for the lower layer produced similar results, indicating that small differences in the density centrifugation step do not substantially influence the yield of cells.

4. PP are practically free of eosinophils. PP may be used for isolation of lymphocytes.

5. The flow-through contains epithelial cells and remaining intestinal content. It may be decanted or stored for the isolation of intraepithelial lymphocytes (IEL) (using Percoll density centrifugation) or for the analysis of epithelial cells.

6. Carefully washing with EDTA-free HBSS is important, as proteases require divalent ions for efficient digestion.

7. Short, vigorous vortexing is required to release leukocytes from the connective tissue.

8. If necessary repeat the digestion step for a third time. Ideally, all of the intestine pieces should be digested to a single cell suspension.

9. Cells can be resuspended in PBS/BSA to directly sort eosinophils. If eosinophils are needed for tissue culture analysis, it is better to purify cell suspension by performing a Percoll density gradient. The final yield of eosinophils will be lower, however the viability of eosinophils is higher, when enriched by Percoll density gradient centrifugation.

10. Viable LP cells are visible in a white ring at the interphase between the two different Percoll concentrations. Erythrocytes should be visible in a red ring below the interphase.

11. Siglec-F antibodies are highly specific for eosinophils [7]. Injection of mice with Siglec-F specific antibodies results in depletion of eosinophils by the induction of apoptosis [5, 14]. Nevertheless, a recent paper described an isolation protocol of peritoneal murine eosinophils by magnetic cell sorting (MACS, Miltenyi) using Siglec-F specific antibodies. Although this antibody induces apoptosis, its brief application for magnetic sorting yielded viable eosinophils [15].

12. To maximize the yield of eosinophils and to improve the viability of sorted cells, it is important to decrease the flow rate and to use a nozzle with a wide opening.

Acknowledgement

The work was supported by the DFG grant (BE 1171/2). The DRFZ, an Institute of the Leibniz Association, is supported by the Berlin Senate of Research and Education.

References

1. Brandtzaeg P (2012) Gate-keeper function of the intestinal epithelium. Benef Microbes 4:1–16

2. Tsuji M, Suzuki K, Kinoshita K et al (2008) Dynamic interactions between bacteria and immune cells leading to intestinal IgA synthesis. Semin Immunol 20:59–66

3. Cerutti A, Rescigno M (2008) The biology of intestinal immunoglobulin A responses. Immunity 28:740–750

4. Geuking MB, McCoy KD, Macpherson AJ (2012) The function of secretory IgA in the context of the intestinal continuum of adaptive immune responses in host-microbial mutualism. Semin Immunol 24:36–42

5. Chu VT, Beller A, Rausch S et al (2014) Eosinophils promote generation and maintenance of immunoglobulin-A-expressing plasma cells and contribute to gut immune homeostasis. Immunity 40:582–593

6. Jung Y, Wen T, Mingler MK et al (2015) IL-1beta in eosinophil-mediated small intestinal homeostasis and IgA production. Mucosal Immunol 8:930–942

7. Crocker PR, Paulson JC, Varki A (2007) Siglecs and their roles in the immune system. Nat Rev Immunol 7:255–266

8. Carlens J, Wahl B, Ballmaier M et al (2009) Common gamma-chain-dependent signals confer selective survival of eosinophils in the murine small intestine. J Immunol 183: 5600–5607

9. Mowat AM, Bain CC (2010) News and highlights: the curious case of the intestinal eosinophil. Mucosal Immunol 3:420–421

10. Mishra A, Hogan SP, Lee JJ et al (1999) Fundamental signals that regulate eosinophil homing to the gastrointestinal tract. J Clin Invest 103:1719–1727

11. Rothenberg ME, Mishra A, Brandt EB et al (2001) Gastrointestinal eosinophils. Immunol Rev 179:139–155

12. Verjan Garcia N, Umemoto E, Saito Y et al (2011) SIRPalpha/CD172a regulates eosinophil homeostasis. J Immunol 187:2268–2277

13. Weigmann B, Tubbe I, Seidel D et al (2007) Isolation and subsequent analysis of murine lamina propria mononuclear cells from colonic tissue. Nat Protoc 2:2307–2311

14. Zimmermann N, McBride ML, Yamada Y et al (2008) Siglec-F antibody administration to mice selectively reduces blood and tissue eosinophils. Allergy 63:1156–1163

15. Carretero R, Sektioglu IM, Garbi N et al (2015) Eosinophils orchestrate cancer rejection by normalizing tumor vessels and enhancing infiltration of CD8(+) T cells. Nat Immunol 16:609–617

Part IV

Animal Models of Gastrointestinal Inflammation and Injury

Chapter 21

Investigation of Host and Pathogen Contributions to Infectious Colitis Using the Citrobacter rodentium Mouse Model of Infection

Else S. Bosman, Justin M. Chan, Kirandeep Bhullar, and Bruce A. Vallance

Abstract

Citrobacter rodentium is used as a model organism to study enteric bacterial infections in mice. Infection occurs via the oral–fecal route and results in the pathogen forming attaching and effacing lesions on infected epithelial cells. Moreover, infection leads to a subsequent host-mediated form of colitis. *C. rodentium* infection is thus an excellent model to study infectious colitis in vivo, while the ability to genetically manipulate *C. rodentium* virulence genes provides the opportunity to develop clear insights into the pathogenesis of this and related infectious microbes. This chapter outlines the basic techniques involved in setting up a *C. rodentium* infection in mice and several different methodologies to assess the severity of the infection.

Key words *Citrobacter rodentium*, Mouse infection model, Intestinal pathology, Infectious colitis, Epithelial barrier, Permeability

1 Introduction

The gram-negative bacterial pathogens enteropathogenic *Escherichia coli* (EPEC) and enterohemorrhagic *E. coli* (EHEC) belong to the *Enterobacteriaceae* family and are leading causes of infantile diarrhea and hemorrhagic colitis respectively. Although EPEC/EHEC are highly virulent in humans, they are poorly infectious in mice, resulting in significant difficulty studying infection dynamics and serving as a poor in vivo model of infectious colitis [1]. Therefore, the natural murine pathogen, *Citrobacter rodentium*, which expresses similar virulence factors as EPEC/EHEC and shares 67 % genetic homology, has been widely adopted by the research community and has proven to be an excellent model to study EHEC/EPEC virulence mechanisms in mice [2, 3].

The pathogenicity of these microbes is dependent on their formation of attaching and effacing (A/E) lesions on the surface of infected intestinal epithelial cells. This ability is encoded within the

Andrei I. Ivanov (ed.), *Gastrointestinal Physiology and Diseases: Methods and Protocols*, Methods in Molecular Biology, vol. 1422, DOI 10.1007/978-1-4939-3603-8_21, © Springer Science+Business Media New York 2016

Locus of Enterocyte Effacement (LEE), a pathogenicity island carrying the genes responsible for the structural components of a type three secretion system (T3SS) along with several effector proteins [4, 5]. Upon T3SS assembly, and its penetration into host cells, effector proteins are manufactured and translocated into the host cell, ultimately resulting in actin rearrangement, formation of the attaching/effacing lesion [5], as well as a number of other actions that subvert the normal functions of the cell. The functions of these different effector proteins have been extensively investigated by mutagenesis of the different virulence genes to study their role during each stage of infection.

C. rodentium is transmitted via the fecal–oral route resulting in initial colonization of the cecal patch (specialized lymphatic tissue at the tip of the cecum) and later, of the distal colon [6]. Bacterial load peaks at 7–10 days post-infection at roughly 10^9 colony forming units $(CFU) g^{-1}$ and subsequently diminishes, beginning in the cecum until finally C. rodentium is no longer detectable in the colon at approximately 21–28 days post-infection [6]. The bacteria that are excreted in the stool of the infected mice are representative of the colonizing bacteria, this is also called bacterial shedding [6] (see below: Subheading 3.1). The bacterial load (measured by CFUs) can be assessed in the stool and tissue homogenates after oral infection with C. rodentium. The bacterial load represents the susceptibility of the mouse to the bacterial strain and should be compared to the bacterial load following infection with a wild-type strain of C. rodentium [7].

Various tools and techniques are available for studying the colonization dynamics of C. rodentium. For example, C. rodentium deletion mutants can be made using homologous recombination. This system promotes recombination between the bacterial chromosome and double-stranded DNA molecules introduced in the bacterial cell via electroporation (see below: Subheading 3.2). In addition, bioluminescent imaging (BLI) has emerged as an important tool for studying biological processes in vivo [8, 9]. C. rodentium can be engineered to express the luminescence phenotype through mini-Tn5 transposon mutagenesis [10] and imaged throughout infection using in vivo imaging systems (IVIS) (reviewed by Prescher and Contag [8]) (see below: Subheading 3.3). As well, the cecal-loop model, adapted from previously described rabbit ileal loop models [11], creates a temporarily ligated cecum, forming a controlled in vivo environment to study early colonization dynamics of C. rodentium infection (see below: Subheading 3.4). This model allows researchers to study infection dynamics at specific time points post-infection in the cecal environment.

Infection with C. rodentium causes colitis, characterized by colonic hyperplasia, histologically visible between 5 and 14 days post-infection [12]. Transmissible murine colonic hyperplasia

(TMCH) causes thickening of the colonic tissue through epithelial cell proliferation and crypt elongation, while immune cell infiltration, goblet cell depletion, and edema are also evident. Damage to the infected tissues triggers intestinal barrier dysfunction, allowing for increased passage of luminal bacterial products across the epithelial barrier [13]. A FITC-dextran assay quantifies the translocation of FITC-labeled sugar from the intestine into the peripheral blood, providing an assessment of intestinal barrier permeability (*see* below: Subheading 3.5) [7]. Histological assessment of tissues to identify and assess the severity of disease can be performed through histological scoring on hematoxylin and eosin (H&E) stained tissue sections. Immunofluorescence staining can also be used to characterize the inflammatory/immune processes associated with *C. rodentium*-induced colitis (*see* below: Subheading 3.6).

Environmental factors such as commensal microbial composition play an important role in the *C. rodentium* infection outcome. For example, treatment with antibiotics, such as streptomycin, 24 h prior to infection skews the microbial composition and results in a 10- to 50-fold increase in intestinal *C. rodentium* burdens [14]. Different mouse knock out strains could harbor different microbiota and therefore could account for variation in susceptibility to infection with *C. rodentium* [15]. Changes in the number of commensals or the relative proportions of *C. rodentium* shed in the stool can be assessed using DAPI staining of the stool (*see* below: Subheading 3.7). It can also be used for monitoring commensal depletion, a common characteristic of *C. rodentium* infection [16]. Finally, another technique to localize the bacteria in the infected tissues is fluorescence in situ hybridization (FISH). The high sensitivity and specificity (DNA specific probes) of FISH allow for spatial localization and visualization of *C. rodentium* and commensals in this model [17] (*see* below: Subheading 3.8). Taken together, the *C. rodentium* infection mouse model is a very useful tool to study intestinal colitis and pathology as well as the infection dynamics of attaching and effacing bacterial pathogens.

2 Materials

2.1 Oral Gavage and Measuring Bacterial Colonization

1. *Citrobacter rodentium*, DBS100; a gift from Dr. Brett Finlay.
2. Luria Bertani (LB) broth: weigh 10 g tryptone, 5 g yeast extract, 10 g NaCl and dissolve in 1 l dH$_2$O.
3. LB agar plate: supplement LB with 1.5 % agar.
4. 37 °C incubator.
5. Sterile culture tube.
6. Spectrophotometer.
7. Oral gavage needle with bulbous tip.

8. 1 ml syringe.

9. Metal beads.

10. Mixer mill MM 400 (Retsch).

2.2 Making C. rodentium Deletion Mutants Using Homologous Recombination

1. PCR Thermocycler.

2. Agarose gel apparatus.

3. *C. rodentium* electrocompetent cells.

4. Electroporator (e.g. Bio-Rad MicroPulser).

5. 0.1 cm Electrocuvette (Bio-Rad).

6. Gel imaging system.

7. PCR Clean up kit (e.g. Qiagen PCR Purification kit).

8. LB agar plates and antibiotics.

2.3 Luciferase-Based Reporter Assays for C. rodentium

1. *C. rodentium* DBS100 expressing luxCDABE (for bioluminescence imaging).

2. Preclinical In vivo Imaging System (IVIS).

3. Isoflurane.

2.4 Cecal-Loop Model for Studying Early Colonization Dynamics

1. Dulbecco's Modified Eagle Medium.

2. Ketamine and Xylazine.

3. 1 ml syringes.

4. 25-gauge needle.

5. Electrical fur trimmer.

6. Surgical forceps.

7. Hemostat.

8. Surgical thread.

9. Surgical scissors.

2.5 FITC-Dextran Assay for Measuring Epithelial Barrier Permeability

1. 4 kDa FITC-dextran.

2. Acid-citrate dextrose: 38 mM citric acid, 107 mM sodium citrate, 136 mM dextrose.

3. Black 1.5 ml microcentrifuge tubes.

4. Gavage needle.

5. 22-gauge needle.

6. Black 96-well flat-bottom plate.

7. Spectrofluorometer.

2.6 H&E and Immunofluorescence Staining of Formalin-Fixed Tissues

1. Hematoxylin and Eosin (H&E) for histological staining.

2. Sealed coplin jar.

3. Xylene and ethanol.

4. Sodium citrate buffer: 10 mM Sodium citrate, 0.05 % Tween 20, pH 6.0.

5. Blocking buffer: 2 % goat serum, 1 % bovine serum albumin (BSA), 0.2 % Triton X-100 (TX-100), 0.05 % Tween-20 in 1× phosphate-buffered saline (PBS).

6. Primary antibody dilution buffer: 1 % BSA, 0.2 % TX-100, 0.05 % Tween-20 in 1× PBS.

7. Secondary antibody dilution buffer: 0.2 % TX-100, 0.05 % Tween-20 in 1× PBS.

8. Primary antibodies: Rat antiserum against *C. rodentium* Tir (1:500, gift from W. Deng) and Rabbit anti-Muc2 (H-300, 1:100, Santa Cruz).

9. Secondary antibodies: AlexaFluor 488-conjugated goat anti-rat IgG (1:5000, Life Technologies) and AlexaFluor 568-conjugated goat anti-rabbit IgG (1:5000, Life Technologies).

10. PBS azide: 0.02 % sodium azide in 1× PBS.

11. DAPI Prolong Gold mounting medium (Life technologies).

12. Liquid blocker PAP pen.

13. Slides and coverslips.

14. Nail polish.

15. Steamer for antigen retrieval protocol.

2.7 DAPI Commensal Staining

1. Stool samples.

2. Mixer mill (Retsch MM 400).

3. Vacuum filtration system.

4. Whatman filter circles (size, 185 mm).

5. Nucleopore membrane filter (size, 25 mm).

2.8 Fluorescence In Situ Hybridization (FISH)

1. FISH hybridization buffer: 0.1 M Tris, 0.9 M NaCl, 0.1 % SDS, pH 7.2.

2. FISH washing buffer: 0.1 M Tris, 0.9 M NaCl, pH 7.2.

3. Microscope slides and coverslips.

4. DNA probes.

5. Glass staining dish.

3 Methods

3.1 Oral Gavage and Measuring Bacterial Colonization

1. Use a sterile inoculation loop to recover viable *C. rodentium* from −80 °C frozen 10 % glycerol stock (*see* **Note 1**) by plating bacteria onto a LB agar plate. Incubate the plate overnight at 37 °C. For all steps related to bacterial culture, use aseptic technique to prevent contamination (Fig. 1).

Fig. 1 LB agar plate with *C. rodentium* culture. Representative image of LB agar plate streaked with *C. rodentium* colonies grown overnight at 37 °C. Ensure that all colonies have uniform morphology and color. Contamination can be circumvented by using selective antibiotics in the agarose media if the *C. rodentium* strain contains antibiotic resistance genes

2. The next day, pick 3–5 colonies with a sterile inoculation loop and dilute in minimum 3 ml of sterile LB broth and grow at 37 °C overnight for stationary phase bacteria in a laboratory incubator shaker at 200 rpm (*see* **Notes 2** and **3**). The LB broth should look clear at inoculation and turbid after the incubation period (*see* **Note 4**).

3. Measure the optical density of the culture with a spectrophotometer at 600 nm. The OD_{600} should be approximately 1.7, equivalent to $1.3–1.4 \times 10^9$ CFU/ml (*see* **Note 5**).

4. Prepare a 1 ml syringe for gavage. Swirl the bacterial culture to ensure a homogenous mixture before loading the syringe. Firmly attach the bulbous oral gavage needle onto the syringe and remove air bubbles by flicking the syringe and push liquid through until it exits the tip.

5. Restrain the mouse with thumb and middle finger by gently pulling the skin of the neck upwards and back. Use index finger to immobilize the head and to straighten the esophagus of the mouse. Use your pinky (little) finger to press the tail toward your palm for a more secure grip. Hold the mouse in an upright position.

6. Slowly insert the gavage needle along the side of the oral cavity and circumvent the tongue. Gently progress the tip of the gavage needle into the esophagus. If you notice any resistance, withdraw the needle and check for secure restraint before reattempting.

7. Slowly inject 100 µl of the culture into the esophagus and remove the gavage needle at the same angle (*see* **Notes 6** and **7**).

8. Monitor mice during the course of infection, typically 6–10 days.

9. On the final day post-infection, prepare 2 ml tubes, each with 1 ml sterile PBS and an autoclaved metal bead. Label and weigh all tubes individually (one tube for each tissue from each mouse that will be analyzed).

10. After euthanasia of the mice, open the abdominal cavity and dissect the cecum and colon and separate them. Tissues for histology can be removed and fixed in 10 % buffered formalin or other fixative of choice (*see* **Note 8**).

11. Remove luminal contents by cutting the tissues open longitudinally and gently scrape off the stool with forceps and collect this in the designated 2 ml tube.

12. Wash the tissues in a separate dish with PBS and collect tissues in the labeled 2 ml tube (*see* **Note 9**).

13. Weigh the individual tubes again to determine the weight of tissue collected for analysis and homogenize the tissues in a mixer mill for 6 min at 30 Hz.

14. Prepare 6 serial dilutions in a round-bottomed 96-well plate with PBS using an aseptic technique. Start the dilutions with a minimum of 20 µl from the homogenized tissues (*see* **Note 10**). Plate 10 µl of each well in triplicate on a square LB agar plate containing selective antibiotics. Incubate the plates O/N in a 37 °C incubator.

15. Count the number of individual colonies from the dilution that shows between 10 and 50 colony forming units (CFU). Calculate the average CFU per ml (average $CFU \times$ dilution factor $\times 100$) and calculate the CFU per gram of tissue (CFU ml^{-1}/g tissue).

3.2 Making C. rodentium Deletion Mutants Using Homologous Recombination

1. Design primers flanking your target gene (*see* **Note 12**). Decide which drug-resistant cassette to use for replacing your target gene.

2. Streak out the bacteria to generate single colonies and inoculate in LB (as previously described).

3. Isolate genomic DNA (template DNA) using a genomic DNA Isolation Kit (e.g. QIAamp DNA Mini Kit) (*see* **Note 13**).

4. Set up a PCR reaction with primers for target gene and template DNA. Examine the PCR product through agarose gel electrophoresis and gel-purify the desired/amplified band (*see* **Note 14**). Purify the PCR product through ethanol precipitation (*see* **Note 15**).

5. Prepare electrocompetent *C. rodentium* DBS100 cells carrying pSIM9 recombineering plasmid with recombinase (*see* **Note 16**) (as previously described [18]) (*see* **Note 17**).

6. Transform the resulting PCR fragment (from **step 4**) into *C. rodentium* DBS100 electrocompetent cells carrying pSIM9 recombineering plasmid (through electroporation) (*see* **Note 18**).

7. In labeled cuvettes on ice, add 50 µl of recombinant cells. Mix 1 µl of clean PCR fragment to the electroporation cuvettes and mix up and down several times (*see* **Notes 19** and **20**).

8. Transfer the DNA into the cells through electroporation (voltage—1.8 kV).

9. Immediately after transformation, add 1 ml of pre-warmed LB media to cuvette and transfer the 1 ml mix into a sterile culture tube.

10. Incubate the culture tubes while shaking at 37 °C for 1 h. This facilitates the recombination process and transcription of drug resistance genes. Following recombination, the recombinant genes carried by the recombineering plasmid exchange the target gene with the resistant marker flanked by FRT (Flippase Recognition Target) sites, hence resulting in a deletion mutant.

11. Antibiotic-resistant transformants are selected on LB agar plates supplemented with the desired antibiotic. The resistance cassette can be eliminated using a plasmid that expresses Flp recombinase gene (to remove DNA that is flanked by two FRT sites) [19].

12. Confirm the deletion mutant through PCR. Design a set of primers—one primer in the DNA region flanking the target gene and one primer in the drug-resistant cassette. Also design a set of flanking primers to confirm the loss of the deleted gene.

13. Grow the antibiotic-resistant colonies in 2 ml LB broth and isolate genomic DNA. Screen ~20 colonies for deletion mutants by PCR analysis (using primers from the **step 12**).

14. Run 5–10 µl of the PCR products on an agarose gel along with molecular weight size markers. If the products are of the expected size, the target gene is successfully deleted and the knockout strain is ready for use. Purify the positive colonies and make a freezer stock in 20 % glycerol for the deletion mutant.

15. Infect mice through oral gavage with overnight culture of luciferase expressing *C. rodentium* construct (as previously described).

3.3 Luciferase-Based Reporter Assays for C. rodentium

1. On day 2 post-infection, take the mice from the cage and gently place them in an anesthesia chamber connected to an anesthesia vaporizer linked to isoflurane and oxygen cylinders. Adjust the isoflurane concentration between 2 and 5 % and O_2 concentration between 0.5 and 1 %.

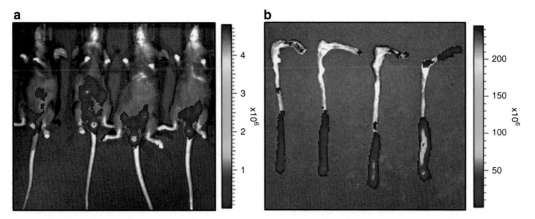

Fig. 2 Bioluminescent imaging of *C. rodentium* expressing luciferase. Bioluminescent image of luciferase expressing *C. rodentium* at 7 days post-infection in (**a**) mice (1 min exposure) and (**b**) exteriorized colons (1 s exposure). The *colour bars* displayed on the *right* indicate high signal intensity (*red*) and low signal intensity (*blue*) with corresponding units of light measurement (photons s^{-1} cm^{-2} steradian^{-1})

2. Mice inactivity indicates successful anesthesia. Place mice in the light-tight chamber of the CCD camera system (IVIS machine) (*see* **Notes 21** and **22**).

3. Grayscale (reference) image is taken at low illumination. After switching off the light source, photons emitted from luciferase expressing *C. rodentium* are quantified using LIVING IMAGE software during the set exposure time (Xenogen) (*see* **Notes 23** and **24**).

4. Colour bar is based on the signal intensity—red corresponds to higher signal intensity (heavy *C. rodentium* colonization) whereas blue corresponds to lower signal intensity (less *C. rodentium* colonization). *C. rodentium* colonization is predominantly seen in cecum and distal colon (Fig. 2).

5. Return the live mice to their cages and monitor for any abnormal activity/breathing. Mice should recover and be moving within 1–2 min. Repeat the same procedure at desired time points (*see* **Notes 25** and **26**).

3.4 Cecal-Loop Model for Studying Early Colonization Dynamics

1. Activate the bacterial T3SS for accelerated infection by transferring 50 µl of prepared overnight *C. rodentium* inoculum into 3 ml of DMEM and incubate at 37 °C and 5 % CO_2 for 3 h without shaking (*see* **Note 27**).

2. Anesthetize the mouse using an intraperitoneal injection of ketamine and xylazine cocktail and wait for mice to exhibit surgical plane (*see* **Note 28**).

3. Shave the mouse abdomen and the proposed area of incision using an electrical trimmer and sterilize the area with 70 % ethanol.

4. Make a 1 cm incision along the midline of the abdomen and gently exteriorize the cecum beginning from the tip until 1 cm of the proximal colon is exteriorized.

5. Using a pair of surgical forceps, a hemostat and surgical thread, tie a surgical knot at the cecal-colonic junction (*see* **Note 29**). It is not necessary to tie the cecum off proximally, as the cecal-ileal valve should be sufficient to prevent leakage.

6. Take a 1 ml syringe and inject 300 μl of activated *C. rodentium* inoculum directly into the cecum, 1 cm from the dorsal tip.

7. The cecum and colon can be replaced into the abdominal cavity and the cavity closed with discontinuous sutures.

8. Mice can be euthanized at various time points following recovery and tissues collected for histological processing and assessment (*see* **Notes 30–32**).

3.5 FITC-Dextran Assay for Measuring Epithelial Barrier Permeability

1. Infect the mice as previously described until you reach the desired time point to assess intestinal barrier integrity (*see* **Notes 33** and **34**).

2. Dissolve 4 kDa FITC-dextran to 80 mg/ml and prepare enough for 150 μl per mouse and enough for creation of the standard curve (10 μl)(*see* **Note 35**).

3. Prepare black 1.5 ml tubes by adding 20 μl acid-citrate dextrose, one for each mouse, for blood collection.

4. Orally gavage the mouse (as previously described) with 150 μl of FITC-dextran and remove food from cage.

5. 4 h post-gavage, euthanize the mice and collect approximately 500 μl of circulating blood via cardiac puncture into the prepared black 1.5 ml tubes containing acid-citrate-dextrose.

6. Centrifuge blood at $1000 \times g$ for 12 min at 4 °C and transfer serum into a black 96-well plate for dilution into 1/10 and 1/100 samples with PBS in triplicates. Ensure a final volume of 100 μl per sample.

7. Create a standard curve using the 80 mg/ml FITC-dextran at 0, 6.25, 12.5, 25, 50, 100, 200, 400, and 800 μg/ml concentrations and add 100 μl of each to the 96-well plate.

8. Read the plate using a spectrofluorometer at excitation wavelength of 485 nm and an emission wavelength of 535 nm.

9. Analyze data by plotting the standard curve and apply to samples. Use either the 1/10 or 1/100 dilution for analysis of samples.

3.6 H&E and Immunofluorescence Staining of Formalin-Fixed Tissues

1. Collect tissues as described in Subheading 3.1 and store in 10 % formalin overnight followed by storage in 70 % ethanol (up to 2 weeks at 4 °C). Embed tissues in paraffin, arranging them to be cut in cross-section at 5 μm thickness. For histological scoring, perform H&E staining (Fig. 3a, b).

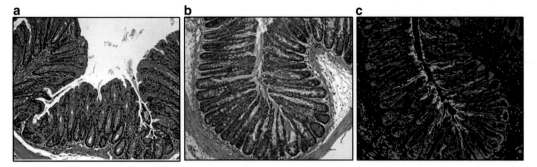

Fig. 3 H&E and immunofluorescence staining of uninfected and *C. rodentium*-infected mice. 200× magnification images of (**a**) H&E stained uninfected C57BL/6 mouse distal colon, (**b**) H&E stained *C. rodentium*-infected C57BL/6 mouse distal colon, and (**c**) immunofluorescently stained *C. rodentium*-translocated effector Tir (*green*), Muc2 (*red*), and DNA (*blue*) 12 days post-infection. H&E staining enables pathological assessment of tissue while immunofluorescence visualizes *C. rodentium* and specific proteins found in the tissue

2. For immunofluorescence staining, deparaffinize tissues by placing slides in a sealed coplin jar which is then placed in a water bath at 65 °C for 20 min.

3. Wash slides in a series of xylene and ethanol baths to complete deparaffinization and rehydration:

   (a) 4 washes of xylene, each for 3 min.

   (b) 2 washes of 100 % ethanol, each for 5 min.

   (c) 1 wash of 95 % ethanol for 5 min.

   (d) 1 wash of 75 % ethanol for 5 min.

   (e) 1 wash of distilled water for 5 min.

4. Place slides into the sealed coplin jar and cover with preheated antigen retrieval buffer (sodium citrate buffer) and place into a steamer for 30 min to maintain sample at 95 °C without boiling. Allow samples to cool to room temperature (*see* **Note 36**).

5. Wash slides three times with PBS azide, each for 5 min (*see* **Note 37**).

6. Draw a hydrophobic border around tissue using a PAP pen to minimize required reagents.

7. Cover tissues with blocking buffer for 1 h at room temperature (*see* **Note 38**).

8. Dilute primary antibody in dilution buffer. Remove blocking buffer and cover tissues with primary antibody and incubate at 4 °C overnight or 2 h at room temperature.

9. Wash three times with PBS azide, each for 3 min.

10. Add diluted secondary antibody solution to tissues and incubate in the dark for 1 h at room temperature.

11. Wash two times with PBS azide followed by 1 wash with distilled water, each for 5 min. Allow slides to dry.

12. Remove the hydrophobic border with a laboratory wipe.

13. Add a single drop of DAPI prolong gold mounting media and seal with coverslip.

14. Seal coverslip with clear nail polish or other sealant along the edges and view slides with a fluorescence microscope (Fig. 3c).

3.7 DAPI Commensal Staining

1. Add 1 ml sterile PBS and an autoclaved metal bead to a round-bottomed 2 ml centrifuge tube using an aseptic technique. Collect stool samples from each mouse and place them in separate PBS tubes (*see* **Note 39**). Weigh the tubes prior to and after the addition of stool samples. This is considered as the baseline stool sample/control sample (day 0, uninfected mice). Collect stool samples in a similar manner on day 2, day 4, and day 6 (etc.) post-infection with *C. rodentium*.

2. Homogenize the stool samples using a mixer mill for 6 min at 30 Hz. Dilute 1:10 in sterile PBS.

3. Add 450 μl of diluted stool sample in 50 μl 37 % Formalin (final Formalin concentration is 3.7 %). Dilute 1:100 in PBS.

4. Place Whatman filter (185 mm) disc on the washer connected with a vacuum pump and wet the filter paper with 1 ml PBS. Place nucleopore membrane filter (25 mm) on top of the Whatman filter. Add a glass tube and assemble the entire apparatus with a clip.

5. Vortex diluted sample and pipet the sample on the top of the nucleopore filter and turn on the vacuum pump. Remove the nucleopore membrane filter and let it dry (*see* **Note 40**). Use a new nucleopore membrane for each stool sample.

6. Place the dry nucleopore membrane filter on a microscope slide and stain with DAPI Prolong Gold mounting medium. Apply a coverslip and let the slides dry.

7. Count the DAPI-positive bacteria and enumerate total bacteria/gram in the stool using a fluorescence microscope (*see* **Notes 41** and **42**).

3.8 Fluorescence In Situ Hybridization (FISH)

1. Deparaffinize the slides (as previously described in Subheading 3.6) (*see* **Note 43**).

2. Dry slides and draw a hydrophobic border around the tissue with a PAP pen.

3. Use 100 μl FISH hybridization buffer (*see* **Note 44**) and add 1:100 dilution of probes in the hybridization buffer. Mix it well. If you want to largely differentiate *C. rodentium* from non-*C. rodentium*, use EUB (recognizes all bacteria with 16S rDNA) and GAM42 (recognizes gamma-proteobacteria that includes *C. rodentium*) (*see* **Notes 45** and **46**).

4. Incubate slides overnight at 37 °C (make sure the tray is wet to prevent the buffer from evaporating overnight). Cover the tray with aluminum foil.

5. Next day, wash slides with FISH hybridization buffer for 2×5 min.

6. Place the slides in the glass staining dish with 50 ml FISH washing buffer. Cover the container with aluminum foil and let it shake slowly for 15 min. Wash with dH_2O for 1 min.

7. Add a drop of DAPI Prolong Gold mounting medium, apply a coverslip and let it dry.

8. Visualize staining under fluorescence microscope.

4 Notes

1. A backup stock of the bacterial strains should be stored in liquid nitrogen.

2. Before inoculating colonies from the agar plate, check if the morphology of the bacteria is uniform to exclude contamination.

3. While starting the bacterial culture in LB, approximately two-thirds of the tube should be empty in order to grow the bacteria aerobically. Ensure that the cap of the culture tube is loose, letting oxygen enter.

4. To ensure the stock of LB broth is not contaminated, include a tube with only LB media in the incubator as a control (i.e. growth in control tube indicates contamination).

5. Optional: plate 10 µl of serial dilutions on a LB agar plate in triplicate to determine the bacterial density of the inoculate more precisely.

6. After the oral gavage, return mouse to the cage and monitor for any abnormal behavior (heavy breathing, etc.). Monitor the mice regularly throughout the infection to ensure that the mice do not experience any unnecessary morbidity.

7. Besides the commonly used oral gavage technique, a natural infection route can be used. In this model, one mouse per cage receives an oral dose of bacteria. At specific time point post-infection (often 2 days), additional mice are added to the cage, and they will become infected through coprophagia of the infected mouse's feces. Unfortunately the researcher is unable to measure the exact number of bacteria that will be ingested by the exposed mice, which is a drawback to this model. On the other hand, this approach is most similar to the natural entry route of bacteria in mice [20].

8. Other commonly used fixatives besides formalin are 4 % paraformaldehyde or Carnoy's solution (60 % ethanol, 30 % chloroform, 10 % acetic acid).

9. It is useful to cut the colon into smaller pieces before collecting it in the 2 ml tube to enhance the homogenization.

10. Cut the pipette tips before starting the serial dilution, as the homogenate is often too dense for accurate pipetting.

11. The majority of laboratory mouse strains are able to clear *C. rodentium* infection (C57BL/6, NIH Swiss and Balb/c mice) however infection of C3H/HeJ mice (as well as other select strains) results in accelerated colitis and by day 10 post-infection, frequent mortality is seen in these susceptible mice [21].

12. Polarity effect—knocking out a gene may affect the polarity of downstream genes which will alter their expression. Ensure that primers are designed carefully. Primer sets should have similar annealing temperatures.

13. Instead of isolating genomic DNA template, a well-isolated single colony may be used as a template for PCR. Inoculate a single colony directly into PCR mixture and mix it well.

14. Avoid overexposing the DNA gel during gel extraction. Too much exposure to UV light may cause DNA mutations.

15. Check the purity of DNA samples using a micro-volume UV–Vis spectrophotometer (a 260/280 ratio of ~1.8 is generally accepted as pure for DNA).

16. While making recombinant cells, keep everything cold and sterile. Use freshly made competent *C. rodentium* cells because the efficiency will be 10- to 100-fold higher than pre-frozen competent cells.

17. Heat-shock may be required to induce recombination functions before making electrocompetent cells. After the O/N culture reaches desired OD_{600} for preparation of electrocompetent cells, transfer half of the culture into Erlenmeyer flask and place this culture in the 42 °C water bath (for inducting recombination activity). Leave the remaining culture at 37 °C as an un-induced control. Make electrocompetent cells using both induced and un-induced cultures.

18. Different kinds of recombineering plasmids may be used for homologous recombination. pSIM9 recombineering plasmid is used as an example.

19. During electroporation, it is important to mix the template and recombinant cells.

20. Salt contamination may interfere with the electroporation reaction. Lowering the amount of DNA used for electroporation may help in solving this issue. Setting up five electroporation reactions with varying amounts of plasmid DNA is advised.

21. This imaging procedure can be done on the dorsal or ventral side of the mouse.

22. Ensure that the plastic nose cones are placed properly to allow for continued anesthesia of mice during imaging.

23. Expression of the luciferase operon can be either plasmid-based or chromosomal. If a plasmid-based construct is used, the plasmid may be not be stable or may be lost over the course of infection (especially true for longer infections). In that case, chromosomal constructs are advised. However, the advantage of plasmid-based luciferase construct is that the luciferase will be expressed to a much higher level than chromosomally expressed luciferase.

24. Different mouse strains (with different fur color) may require different exposure times for optimal signal detection. This may require some optimization. It is important to use a consistent exposure time for data analysis. For the first experimental run, it is necessary to determine the optimal exposure time for peak distribution/peak signal.

25. It can be difficult to differentiate between luminal bacteria, loosely attached bacteria, and adherent bacteria. Euthanizing the mice and separating the luminal contents from tissue and washing the tissue with PBS may help determine bacterial localization.

26. This is a low-throughput method as only five mice can be imaged at a time. It can thus be a time-consuming procedure. However, BLI imaging is a powerful tool as it allows for non-invasive live imaging without requiring of euthanizing the animals.

27. Pre-grown bacteria are exposed to DMEM for 3 h in order to activate the expression of the T3SS for accelerated bacterial effector delivery.

28. Surgical plane of anesthesia is exhibited when the mouse is unconscious and does not respond to any external stimuli (i.e. pinching the hind leg paw).

29. For the highest level of consistency among experiments, it is recommended to fast the mice (with-hold food) for 12–24 h prior to the procedure.

30. Because of the invasiveness of the procedure, the cecal-loop model is only suitable for partial recovery. Though the procedure can be modified for complete recovery, the listed protocol should only be used for up to 12 h infections.

31. Soft food and easily accessible water should be provided for the mice following the cecal-loop procedure. It is recommended to house mice in separate cages during recovery.

32. Monitor mice every hour post-infection and record physiological and behavioral observations.

33. The optimal timeframe for measuring epithelial barrier integrity varies based on the mouse strain and the strain of *C. rodentium*. Typically, 6–7 days post-infection yields representative results, as infection is widespread but it is before peak tissue damage.

34. All experiments should include control uninfected mice in the study design.

35. FITC is light-sensitive so ensure that all tubes and plates are black and/or covered in aluminum foil to prevent sample bleaching.

36. Antigen retrieval is optional but highly recommended as it removes formalin-associated protein linkages that may interfere with antigen sites.

37. It is recommended to use a laboratory wipe to draw off extra liquid following washes to prevent dilution of antibodies and mounting media.

38. Add 2 % goat serum to the blocking buffer if the secondary antibody is retrieved from this animal. Adjust if the secondary antibody is retrieved from another animal.

39. Collect 1–2 pellets/mouse for uniform stool homogenization.

40. Be careful when handling the small filter. Avoid touching the surface as it may result in loss of filtered bacteria.

41. For *C. rodentium*, day 6 post-infection is optimal as infection is well established by this time. Significant changes in commensal depletion are noted by this time-point.

42. This method can be used to determine relative proportions of *C. rodentium* as well as different commensal populations. After **step 5**, pass the nucleopore membrane filters through ethanol gradient (50 %, 80 %, 100 %, 3 min each). Let the filter discs dry and place them on microscope slides. Follow the FISH staining protocol as previously described.

43. Carnoy's or formalin-fixed tissues can be used for staining. Mucus is best preserved in Carnoy's fixed tissues.

44. SDS in FISH hybridization buffer can precipitate out of the solution. Ensure that FISH hybridization buffer is warmed up with continuous stirring to clear the solution before usage.

45. Avoid prolonged exposure to light as the probes are light-sensitive.

46. This staining method can also be used for visualizing commensal bacterial populations (*Bacteriodes*, *Firmicutes*, *Lactobacillus*). Design specific DNA probes complementary to the target sequence (16S rDNA sequence).

References

1. Vulcano AB, Tino-De-Franco M, Amaral JA, Ribeiro OG et al (2014) Oral infection with enteropathogenic Escherichia coli triggers immune response and intestinal histological alterations in mice selected for their minimal acute inflammatory responses. Microbiol Immunol 58:352–359

2. Mundy R, Girard F, FitzGerald AJ, Frankel G (2006) Comparison of colonization dynamics and pathology of mice infected with enteropathogenic Escherichia coli, enterohaemorrhagic E. coli and Citrobacter rodentium. FEMS Microbiol Lett 265:126–132

3. Borenshtein D, McBee ME, Schauer DB (2008) Utility of the Citrobacter rodentium infection model in laboratory mice. Curr Opin Gastroenterol 24:32–37

4. Petty NK, Bulgin R, Crepin VF, Cerdeno-Tarraga AM et al (2010) The Citrobacter rodentium genome sequence reveals convergent evolution with human pathogenic Escherichia coli. J Bacteriol 192:525–538

5. Deng W, Vallance BA, Li Y, Puente JL, Finlay BB (2003) Citrobacter rodentium translocated intimin receptor (Tir) is an essential virulence factor needed for actin condensation, intestinal colonization and colonic hyperplasia in mice. Mol Microbiol 48:95–115

6. Wiles S, Clare S, Harker J, Huett A et al (2004) Organ specificity, colonization and clearance dynamics in vivo following oral challenges with the murine pathogen Citrobacter rodentium. Cell Microbiol 6:963–972

7. Bhinder G, Sham HP, Chan JM, Morampudi V, Jacobson K, Vallance BA (2013) The Citrobacter rodentium mouse model: studying pathogen and host contributions to infectious colitis. J Vis Exp 72:e50222

8. Prescher JA, Contag CH (2010) Guided by the light: visualizing biomolecular processes in living animals with bioluminescence. Curr Opin Chem Biol 14:80–89

9. Meighen EA (1991) Molecular biology of bacterial bioluminescence. Microbiol Rev 55:123–142

10. Martinez-Garcia E, Aparicio T, de Lorenzo V, Nikel PI (2014) New transposon tools tailored for metabolic engineering of gram-negative microbial cell factories. Front Bioeng Biotechnol 2:46

11. Kasai GJ, Burrows W (1966) The titration of cholera toxin and antitoxin in the rabbit ileal loop. J Infect Dis 116:606–614

12. Luperchio SA, Schauer DB (2001) Molecular pathogenesis of Citrobacter rodentium and transmissible murine colonic hyperplasia. Microbes Infect 3:333–340

13. Koroleva EP, Halperin S, Gubernatorova EO, Macho-Fernandez E, Spencer CM, Tumanov AV (2015) Citrobacter rodentium-induced colitis: a robust model to study mucosal immune responses in the gut. J Immunol Methods 421:61–72

14. Wlodarska M, Willing B, Keeney KM, Menendez A et al (2011) Antibiotic treatment alters the colonic mucus layer and predisposes the host to exacerbated Citrobacter rodentium-induced colitis. Infect Immun 79:1536–1545

15. Chen J, Waddell A, Lin YD, Cantorna MT (2015) Dysbiosis caused by vitamin D receptor deficiency confers colonization resistance to Citrobacter rodentium through modulation of innate lymphoid cells. Mucosal Immunol 8:618–626

16. Lupp C, Robertson ML, Wickham ME, Sekirov I, Champion OL, Gaynor EC, Finlay BB (2007) Host-mediated inflammation disrupts the intestinal microbiota and promotes the overgrowth of Enterobacteriaceae. Cell Host Microbe 2:119–129

17. Bergstrom KS, Kissoon-Singh V, Gibson DL, Ma C et al (2010) Muc2 protects against lethal infectious colitis by disassociating pathogenic and commensal bacteria from the colonic mucosa. PLoS Pathog 6:e1000902

18. Diner EJ, Garza-Sanchez F, Hayes CS (2011) Genome engineering using targeted oligonucleotide libraries and functional selection. Methods Mol Biol 765:71–82

19. Datsenko KA, Wanner BL (2000) One-step inactivation of chromosomal genes in Escherichia coli K-12 using PCR products. Proc Natl Acad Sci U S A 97:6640–6645

20. Wiles S, Dougan G, Frankel G (2005) Emergence of a 'hyperinfectious' bacterial state after passage of Citrobacter rodentium through the host gastrointestinal tract. Cell Microbiol 7:1163–1172

21. Vallance BA, Deng W, Jacobson K, Finlay BB (2003) Host susceptibility to the attaching and effacing bacterial pathogen Citrobacter rodentium. Infect Immun 71:3443–3453

Chapter 22

Murine Trinitrobenzoic Acid-Induced Colitis as a Model of Crohn's Disease

John F. Kuemmerle

Abstract

Inflammatory Bowel Diseases, Crohn's disease and ulcerative colitis, result from the uncontrolled inflammation that occurs in genetically susceptible individuals and the dysregulation of the innate and adaptive immune systems. The response of these immune systems to luminal gut microbiota and their products results in altered intestinal permeability, loss of barrier function, and mucosal inflammation and ulceration. Animal models of experiment intestinal inflammation have been developed that leverage the development of spontaneous inflammation in certain mouse strains, e.g. Samp1/Yit mice, or induction of inflammation using gene-targeting e.g. IL-10 null mice, administration of exogenous agents e.g. DSS, or adoptive transfer of T-cells into immunodeficient mice, e.g. CD4$^+$ CD45Rb^{Hi} T-cell transfer. Colitis induced by rectal instillation of the haptenizing agent, 2,4,6 trinitrobenzene sulfonic acid, is one of the most commonly used and well-characterized models of Crohn's disease in humans.

Key words T Cell-mediated colitis, Inflammatory bowel disease

1 Introduction

Crohn's disease and ulcerative colitis are the two main forms of inflammatory bowel disease. Both diseases are characterized by uncontrolled inflammation in the GI tract. Recent advances in the pathogenesis of inflammatory bowel disease inflammatory bowel disease have identified over 160 genetic variants conferring risk of disease [1]. Analysis of the intestinal microbiome has also revealed diminished diversity in the composition of commensal and pathogenic bacteria comprising the gut microbiome [2–4]. The combined effects of genetic polymorphisms and dysbiosis combine to result in altered activation and regulation of the intestine's innate immune and adaptive immune systems that result in sustained inflammation. Ulcerative colitis is characterized by mucosal inflammation. Crohn's disease is a transmural process with patients expressing an inflammatory phenotype, Montreal B1, fibrostenotic phenotype, Montreal B2, or a penetrating phenotype, Montreal B3 [5].

Andrei I. Ivanov (ed.), *Gastrointestinal Physiology and Diseases: Methods and Protocols*, Methods in Molecular Biology, vol. 1422, DOI 10.1007/978-1-4939-3603-8_22, © Springer Science+Business Media New York 2016

In order to investigate the immunopathogenesis of inflammatory bowel disease, animal models of experimental enterocolitis have been developed that can be used to understand the pathogenesis and examine the efficacy of putative anti-inflammatory strategies. A number of animal models have been developed using spontaneous inflammation in mouse strains, or induction of inflammation using gene-targeting, administration of exogenous agents, or transfer of T-cells into immunodeficient mice [6].

One of the most commonly used models is rectal administration of the haptenizing agent, 2,4,6 trinitrobenzene sulfonic acid (TNBS). Colitis can be induced in strains of mice susceptible to TNBS-induced colitis once the mucosal barrier is disrupted by the co-administration of ethanol vehicle. TNBS-induced haptenization of colonic microbiota proteins or autologous proteins renders them immunogenic to the host immune system. This process is mediated by $CD4^+$ T-cells, and as such, the TNBS model is useful to understand the role of $CD4^+$ T-cell-mediated immune responses. The initial injury induced by TNBS/ethanol is followed by an acute inflammatory phase, after 7 days, which can be followed by chronic inflammation and development of fibrosis using appropriate TNBS administration protocols.

In mice, the specific inflammatory process that develops in response to TNBS is highly strain-dependent and requires individual optimization of TNBS concentrations, especially when used to generate chronic colitis and fibrosis. Murine TNBS-induced colitis was initially described in SIL/J mice with high susceptibility and developed a Th1 predominant inflammatory milieu [7–9]. BALB/c mice are also readily susceptible to TNBS-induced colitis, and with chronic administration, develop a sequential Th1, Th2, and Treg-patterned immunologic response [10, 11]. C57BL/6J mice are more resistant to TNBS, but with administration of higher escalating doses develop the concomitant Th1, Th17, and Treg-patterned immunologic response that is seen clinically in patients with Crohn's disease and recapitulates the development of transmural fibrosis in the 30–50 % of patients susceptible to fibrostenosis [12, 13]. The differences in susceptibility to TNBS have been shown to result from a genetic difference in IL-12 response [14].

TNBS-induced colitis can be used in three types of investigation. (1) Acute TNBS-induced colitis can be used to investigate the initial priming of Th1-patterned nonspecific inflammatory response. (2) Intermediate duration colitis can be used to investigate the delayed hypersensitivity response and the specific Th1 the responses that result. (3) Chronic TNBS-induced colitis can be used to investigate the mechanisms of fibrosis that involve specific Th1, Th17, and Treg responses and share a similar pathobiology as that in fibrostenotic Crohn's disease [15–18].

2   Materials

2.1   Reagents

1. 2,4,6-Trinitrobenzenesulfonic acid (TNBS) stock solution: 5 % w/v in H_2O (Sigma-Aldrich, St. Louis, MO). Caution: handle with appropriate safety equipment.

2. Absolute EtOH.

3. 50 % EtOH v/v in distilled water.

4. Isoflurane or other inhaled anesthetic. Caution: use in appropriate vented hood with scrubber.

5. Phosphate buffered saline, pH 7.2.

6. Calcium- and Magnesium-free Hank's balanced salt solution, pH 7.2.

7. Bouin's solution (Sigma-Aldrich).

8. Sirius Red Collagen Assay (Sircol) (Chondrex, Inc, Redmond, WA).

9. Tissue protein extraction reagent (T-PER) buffer (Thermal Science, Amarillo, TX).

10. Hematoxylin and eosin staining solution (Sigma-Aldrich).

11. Iron hematoxylin Stock Solution A: dissolve 1 g of Hematoxylin in 99 ml of 95 % EtOH.

12. Iron hematoxylin Stock Solution B: Mix 4 ml of 29 % w/v solution of ferric chloride in water with 95 ml of distilled water, and 1 ml of concentrated HCl.

13. Weigert's Iron Hemotoxylin Solution: prior each experiment, mix equal volumes of Stock solutions A and B.

14. Biebrich Scarlet-Acid Fuchsin Solution: mix 90 ml of 1 % aqueous Biebrich scarlet solution with 10 ml of 1 % aqueous acid fuchsin solution and 1 ml of glacial acetic acid.

15. Phosphomolybdic-Phosphotungstic Acid Solution: mix 25 ml of 5 % phosphomolybdic acid solution with 25 ml of 5 % phosphotungstic acid solution in distilled water.

16. Aniline Blue Solution: dissolve 2.5 g of Aniline Blue in 90 ml of distilled water, add 2 ml of glacial acetic acid, adjust final volume to 100 ml. Alternatively, the pre-maid Masson's Trichrome stain solution can be purchased from Sigma Aldrich.

17. Acetic acid solution, 1 % (v/v): mix 1 ml of glacial acetic acid with 99 ml of distilled water.

2.2   Animals

1. 6–8-week old sex- and age-matched mice of your strain of choice (SIL/J, BALB/c, or C57BL/6J). All animals must be housed and handled according to local, national, and international animal care and use guidelines.

2.3 Equipment

1. Vortex mixer.

2. 3.5 Fr polyethylene catheter tubing.

3. 1 ml syringe.

4. Anesthesia apparatus with scrubber with rodent mask (SurgiVet, Smith Medical, Dublin OH).

3 Methods

1. Prepare TNBS solution: Mix appropriate volume of 5 % TNBS (w/v) in H_2O TNBS solution with the appropriate volume of absolute EtOH to obtain the desired TNBS dosing. Caution: Handle with appropriate safety equipment.

2. Adequately anesthetize each mouse (*see* **Note 1**).

3. Fit the 3.5 Fr tubing to the 1 ml syringe and fill with TNBS working solution or 50 % EtOH solution.

4. Insert the tubing 4–6 cm into the colon via the rectum taking care to advance the catheter gently to avoid perforating the colon.

5. Slowly deliver 0.1 ml of solution of your intended concentration into the colon (*see* **Note 2**).

6. Remove the catheter slowly while keeping the animal in a vertical head down position for 1 min. For consistent result, it is important that all of the TNBS solution be retained in the colon.

7. Return mouse to its cage and assure it recovers from anesthesia.

8. Repeat the instillation of TNBS solution or EtOH solution on a weekly basis for 8 weeks using escalating doses of TNBS to establish chronic fibrosis (*see* **Note 3**).

9. Record daily or every other day weights (*see* **Note 4**).

10. Record the presence of loose stools and presence of blood (*see* **Note 5**).

11. At the desired end point: 1 week for acute colitis, or 3–8 weeks for chronic fibrosis, the animals can be euthanized adhering strictly to local, national, and international requirements for animal care and use.

12. At the time point of choice after mice have been euthanized, harvest the colon via a midline laparotomy.

13. Place the isolated colon into a petri dish filled with cold phosphate buffered saline or calcium and magnesium-free Hank's balanced solution.

14. The layers of the colon and specific cell types can be isolated by dissection and enzymatic digestion for preparation of protein lysates, mRNA, or initiation of primary cells if desired [12, 13, 19, 20] (*see* **Notes 6** and **7**).

15. Remove the mesenteric lymph nodes and remove to cold buffer [20].

16. Pass the collected mesenteric lymph nodes through a 100 nm nylon mesh into a 15 ml tube filled with calcium and magnesium-free Hank's balanced solution with 20 % fetal calf serum. Disrupt the lymph nodes mechanically on the mesh with a pair of forceps. Wash the collected cells and store them in the same buffer on ice until use.

17. Determine and record the gross macroscopic damage scores (Table 1).

18. Prepare histologic sections prepared for either frozen section or formalin-fixed paraffin-embedded sections to assess microscopic damage scores with hematoxylin and eosin staining (Table 2 and Fig. 1).

Table 1
Gross macroscopic damage score [13, 15, 22]

Symptom	Score
Presence and severity of adhesions (none—0, few—1, many—2)	0–2
Maximum bowel wall thickness	Millimeters
Presence or absence of diarrhea (none—0, diarrhea ± blood—1)	0–1

Table 2
Microscopic damage score [13, 15, 22]

Histological parameter	Score
Extent of destruction of normal architecture	0–3
Presence and degree of cellular infiltration	0–3
Extent of muscle thickening	0–3
Presence or absence of goblet cell mucus	0–1
Presence or absence of crypt abscesses	0–1
Maximum score	11

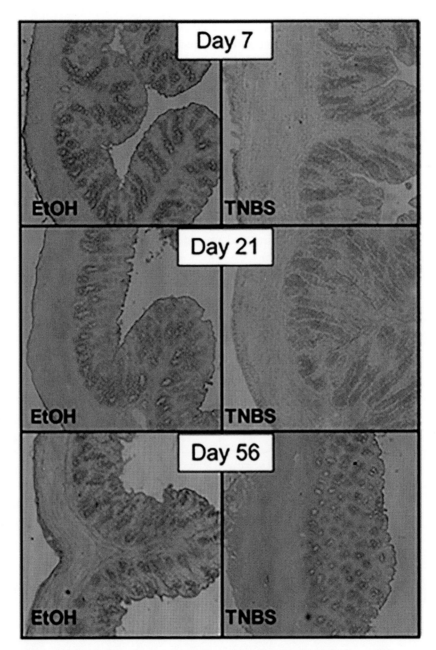

Fig. 1 Progressive development of colonic thickening and fibrosis over the course of TNBS-induced colitis in C57BL/6J mice

19. The average thickness of the subepithelial space and the muscularis propria can be measured directly using image scanning micrometry at five villous bases per section and reported in µm (*see* **Note 8**).

20. Collagen deposition can determined in Masson's Trichrome-stained sections (Fig. 2). This stain is used for the detection of

collagen fibers on formalin-fixed, paraffin-embedded sections or frozen sections as well. The collagen fibers will be stained blue and the nuclei will be stained black, while the background is stained red. Fixation in Bouin's solution improves the quality of the stain. Deparaffinize PFA-fixed tissue sections and rehydrate them passing through 100 % alcohol, 95 % alcohol, 70 % alcohol, or use frozen sections. Stain in Weigert's iron hematoxylin working solution for 10 min. Rinse in running tap water for 10 min. Wash in distilled water. Stain in Biebrich scarlet-acid fuchsin solution for 10–15 min. Wash in distilled water. Differentiate in phosphomolybdic-phosphotungstic acid solution for 10–15 min or until collagen is not red. Transfer sections without washing to Aniline Blue solution and stain for 5–10 min. Rinse briefly in distilled water and differentiate in 1 % acetic acid solution for 2–5 min. Wash in distilled water. Dehydrate very quickly through 95 % ethyl alcohol, absolute ethyl alcohol (these step will wipe off Biebrich scarlet-acid fuchsin staining), and clear in xylene. Mount tissue sections.

21. Quantify the area positive for collagen staining in microscopic images by red-green-blue (RGB) segmentation using fixed threshold values [13, 15, 21].

22. Collagen content could be quantitated using a Sirius red collagen (Sircol) assay [15]. Homogenize mesenchymal tissue of mouse colon in T-PER buffer, incubate on ice for 15 min, and centrifuge for 5 min at $10,600 \times g$ at 4 °C. Dilute each protein sample in 0.5 M acetic acid to a final concentration of 100 mg/ml. Follow instructions in the Sircol kit and read optical density at 530 nm. Calculate the results as amount of collagen per 100 mg of total protein [15].

EtOH   **TNBS**

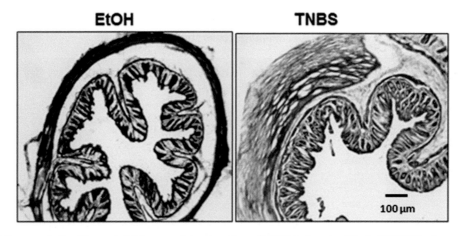

100 µm

Fig. 2 Representative Masson's trichrome-stained section of TNBS-induced colitis in C57Bl/6J mice

4   Notes

1. Plan for additional mice in the TNBS arm of your experiment. In general, the younger the animals, the greater is their susceptibility to TNBS. However, high mortality can be expected in mice younger than 4 weeks. While both genders develop colitis, male mice tend to develop a more chronic colitis in response to TNBS administration [9].

2. The dosing of TNBS to achieve the desired effect, acute or chronic colitis, is dependent on the mouse strain being used and on local vivarium factors, i.e. microbiome. Investigators with established TNBS dosing protocols when moving to a new vivarium may encounter difficulties with inadequate or excessive colitis necessitating the need to re-titrate dosing to the desired effect. Suggested dose ranges for various commonly used mouse strains are outlined in Table 3.

3. Troubleshooting for TNBS colitis is summarized in Table 4.

4. The progress and severity of TNBS-induced colitis can be easily monitored by the weight loss that reflects disease severity.

5. Diarrhea and rectal bleeding caused by TNBS administration represent other characteristic symptoms that reflect severity of the disease.

6. To separate the tissue layers of mouse colon, slip 5 cm colon sections over a 3–5 mm glass rod, and with gentle tangential stroking with a moistened wipe, peel off the muscularis propria. The mucosal and submucosal section still remain

Table 3
TNBS doses required to induce colitis in different mouse strains [23]

Mouse strain	TNBS dose (mg/kg, ~30 g mouse)	Day administered	Day of death
BALB/C	Variable 1–5 mg	0	7
BALB/C	Escalating 1.5–2.5 mg Subsequent dose: 3.375 mg	Weekly	49
C57	3–5 mg	0	7
C57	Escalating 4–6 mg	Weekly	56
SJL/J	0.5–5	0	7
SJL/J	0.5	0	56

Table 4
Troubleshooting TNBS colitis

Problem	Possible reason	Possible solution
No or weak colitis	Mouse strain has low susceptibility to TNBS colitis	Increase TNBS dose
Excessive colitis and death	Mouse strain has high susceptibility to TNBS colitis Colonic perforations are occurring	Decrease TNBS dose Take care to avoid colonic injury
Variable extent of colitis develops between groups	TNBS is not retained in the colon TNBS is outdated	Remove catheter more gently and maintain inversion for 60 s Replace TNBS with fresh

on the glass rod and can be used for isolation of epithelial cell, subepithelial myofibroblasts, or lamina propria mononuclear cells [13].

7. Preparation of protein lysates or mRNA extraction from whole thickness colon results in data that cannot be attributed to either specific regions or cell types, or may yield mixed results that tend to negate each other.

8. Generally, the muscularis propria is ~80 μm-thick in naïve or EtOH vehicle-treated mice, and its thickness can increase up to twofold over the course of 8 weeks of chronic TNBS-induced colitis.

Acknowledgments

This work was sponsored by NIH DK4961 and the Harrison Family Trust.

References

1. Jostins L, Ripke S, Weersma RK, Duerr RH et al (2012) Host-microbe interactions have shaped the genetic architecture of inflammatory bowel disease. Nature 491:119–124

2. Ott S, Musfeldt J, Wenderoth M, Hampe DFJ et al (2004) Reduction in diversity of the colonic mucosa associated bacterial microflora in patients with active inflammatory bowel disease. Gut 53:685–693

3. Joossens M, Huys G, Cnockaert M, De Preter V, Verbeke K, Rutgeerts P, Vandamme P, Vermeire S (2011) Dysbiosis of the faecal microbiota in patients with Crohn's disease and their unaffected relatives. Gut 60:631–637

4. Andoh A, Imaeda H, Aomatsu T, Inatomi O et al (2011) Comparison of the fecal microbiota profiles between ulcerative colitis and Crohn's disease

using terminal restriction fragment length polymorphism analysis. J Gastroenterol 46:479–486

5. Satsangi J, Silverberg MS, Vermeire S, Colombel J-F (2006) The Montreal classification of inflammatory bowel disease: controversies, consensus, and implications. Gut 55:749–753

6. Blumberg RS, Saubermann LJ, Strober W (1999) Animal models of mucosal inflammation and their relation to human inflammatory bowel disease. Curr Opin Immunol 11:648–656

7. Neurath MF, Fuss I, Kelsall BL, Stuber E, Strober W (1995) Antibodies to interleukin 12 abrogate established experimental colitis in mice. J Exp Med 182:1281–1290

8. Neurath MF, Fuss I, Pasparakis M, Alexopoulou L et al (1997) Predominant pathogenic role of tumor necrosis factor in experimental colitis in mice. Eur J Immunol 27:1743–1750

9. Scheiffele F, Fuss IJ (2001) Induction of TNBS colitis in mice. Curr Protoc Immunol Chapter 15:Unit 15.19

10. Fichtner-Feigl S, Fuss IJ, Young CA, Watanabe T, Geissler EK, Schlitt HJ, Kitani A, Strober W (2007) Induction of IL-13 triggers TGF-beta1-dependent tissue fibrosis in chronic 2,4,6-trinitrobenzene sulfonic acid colitis. J Immunol 178:5859–5870

11. Fichtner-Feigl S, Young CA, Kitani A, Geissler EK, Schlitt H-J, Strober W (2008) IL-13 signaling via IL-13Rα2 induces major downstream fibrogenic factors mediating fibrosis in chronic TNBS colitis. Gastroenterology 135:2003–2013

12. Hazelgrove KB, Flynn RS, Qiao LY, Grider JR, Kuemmerle JF (2009) Endogenous IGF-I and αvβ3 integrin ligands regulate increased smooth muscle growth in TNBS-induced colitis. Am J Physiol Gastrointest Liver Physiol 296:G1230–G1237

13. Mahavadi S, Flynn RS, Grider JR, Qiao L-Y, Murthy KS, Hazelgrove KB, Kuemmerle JF (2011) Amelioration of excess collagen IαI, fibrosis, and smooth muscle growth in TNBS-induced colitis in IGF-I(+/−) mice. Inflamm Bowel Dis 17:711–719

14. Bouma G, Kaushiva A, Strober W (2002) Experimental murine colitis is regulated by two genetic loci, including one on chromosome 11 that regulates IL-12 responses. Gastroenterology 123:554–565

15. Li C, Flynn S, Grider JR, Murthy KS, Kellum JM, Akbari HM, Kuemmerle JF (2013) Increased activation of latent TGF-β1 by αVβ3 in human Crohn's disease and fibrosis in TNBS colitis can be prevented by cilengitide. Inflamm Bowel Dis 19:2829–2839

16. Lawrance IC, Wu F, Leite AZA, Willis J, West GA, Fiocchi C, Chakravarti S (2003) A murine model of chronic inflammation-induced intestinal fibrosis down-regulated by antisense NF-κB. Gastroenterology 125:1750–1761

17. Rieder F, Kessler S, Sans M, Fiocchi C (2012) Animal models of intestinal fibrosis: new tools for the understanding of pathogenesis and therapy of human disease. Am J Physiol Gastrointest Liver Physiol 303:G786–G801

18. Li C, Kuemmerle JF (2014) Mechanisms that mediate the development of fibrosis in patients with Crohn's disease. Inflamm Bowel Dis 20:1250–1258

19. Kuemmerle JF (1998) Synergistic regulation of NOS II expression by IL-1β and TNF-α in cultured rat colonic smooth muscle cells. Am J Physiol 274:G178–G185

20. Koscielny A, Wehner S, Engel DR, Kurts C, Kalff J (2011) Isolation of T cells and dendritic cells from peripheral intestinal tissue, Peyer's patches and mesenteric lymph nodes in mice after intestinal manipulation. Protocol Exchange (Online).

21. Ortolan EVP, Spadella CT, Caramori C, Machado JLM, Gregorio EA, Rabello K (2008) Microscopic, morphometric and ultrastructural analysis of anastomotic healing in the intestine of normal and diabetic rats. Exp Clin Endocrinol Diabetes 116:198–202

22. Wirtz S, Neufert C, Weigmann B, Neurath MF (2007) Chemically induced mouse models of intestinal inflammation. Nat Protoc 2:541–546

23. te Velde AA, Verstege MI, Hommes DW (2006) Critical appraisal of the current practice in murine TNBS-induced colitis. Inflamm Bowel Dis 12:995–999

Chapter 23

Oxazolone-Induced Colitis as a Model of Th2 Immune Responses in the Intestinal Mucosa

Benno Weigmann and Markus F. Neurath

Abstract

Murine models of intestinal inflammation have been widely used in biomedical research. Similarities in anatomy and physiology between such murine models and patients with inflammatory bowel diseases may allow a better understanding of the pathogenesis of Crohn's disease and ulcerative colitis. Additionally, models of intestinal inflammation may be used for the analysis of potentially new therapeutic agents. One key class of models consists of chemically induced inflammation models. Within this group, colitis induced by the haptenizing agent oxazolone is an important model that results in induction of acute or chronic inflammation of the large bowel. Here, we describe the induction and the analysis of this experimental colitis model.

Key words Inflammatory bowel disease, Ulcerative colitis, Experimental colitis models, Intestinal mucosa, Oxazolone

1 Introduction

Crohn's disease and ulcerative colitis are chronic inflammatory disorders of the intestine [1]. The etiology of IBD is still not exactly understood, but there is strong evidence that immunocompetent cells, especially T-cells in the lamina propria, contribute to disease initiation and progression. Much of the recent progress in the understanding of mucosal immunity has been achieved by analyzing experimental animal models of intestinal inflammation [2, 3]. One group of these models is based on a chemical induction of the inflammatory response, which resembles human gut inflammation. It should be noted that these models do not fully represent the complexity of the human disease. However, they are valuable tools for understanding the pathophysiological mechanisms. Intestinal immune cell populations and cytokines have been characterized in humans over a long time period and studies of animal models emphasized the role of immune cell activation resulting in the induction of mucosal inflammation.

Andrei I. Ivanov (ed.), *Gastrointestinal Physiology and Diseases: Methods and Protocols*, Methods in Molecular Biology, vol. 1422, DOI 10.1007/978-1-4939-3603-8_23, © Springer Science+Business Media New York 2016

T-Lymphocytes, in particular CD4+ T-cells, play an important role in all immune regulatory processes in the gastrointestinal tract [4]. Production of proinflammatory cytokines such as IFN-γ, TNF-α, and IL-12p40 has been observed in the inflamed mucosa of Crohn's disease patients, which is consistent with a Th1-related cytokine response. Interestingly, the cytokine profile in ulcerative colitis is characterized by the increased production of several Th2 cytokines such as IL-5, IL-13, and IL-9 [4–6]. However, IL-4 production in ulcerative colitis is low, demonstrating that not all classical Th2 cytokines are induced in this disorder. Therefore, the cytokine response in human ulcerative colitis has been described as a "modified Th2 response". Consequently, an animal model with exaggerated production of Th2 cytokines would be ideal to study the pathogenesis of ulcerative colitis.

Interestingly, a Th2-type cytokine response has been described in the oxazolone-induced colitis model. This model was first characterized by the development of an acute T cell-dependent inflammation limited to colonic mucosa and submucosa, especially the distal colon [7]. Rectal administration of oxazolone dissolved in 30–50 % ethanol was found to cause marked colitis associated with weight loss and diarrhea. Histopathological analysis of the colon obtained from mice with active disease revealed superficial colonic inflammation, epithelial cell erosions, ulcerations, infiltration of lymphocytes, and an elevated production of IL-13 by NK T-cells [8]. Clinical manifestations of inflammation included the body weight loss and diarrhoea. Investigation of molecular mechanisms underlying this model of colitis revealed that treatment with neutralizing anti-IL-4 or anti-TNF-α antibodies or with a decoy IL-13R2α-Fc protein attenuate development of mucosal inflammation in oxazolone-treated animals [7, 8]. Therefore, oxazolone-induced colitis recapitulated many important features of human ulcerative colitis.

Similarly to other rodent models of inflammation, development of the oxazolone-induced colitis depends on mice strain and the commensal microflora at different animal facilities. Initially, the oxazolone-induced colitis model was performed in SJL/J or C57/BL10 mice, but later it was found that other mice stains like BALB/c are susceptible as well [9]. Furthermore, depending on the experimental setup, oxazolone colitis may also result in a mixed Th1/Th2-dependent colitis [10] or inflammation in immunodeficient NOD-SCID-IL2R mice [11].

A number of studies have demonstrated the major advantage of the oxazolone-mediated colitis as a simple and reproducible model of colonic inflammation. The presented protocol describes an acute model of the diseases that can be easily transformed into chronic inflammation via repetitive administration of low doses of oxazolone [12].

2   Materials

All solutions should be freshly prepared under sterile conditions at the beginning of the experiment (unless indicated otherwise).

2.1   Reagents

1. Olive oil.
2. Acetone.
3. Oxazolone (4-Ethoxymethylene-2-phenyl-2-oxazolin-5-one).
4. The sensitization solution: Mix acetone and olive oil in a 4:1 v/v ratio by vortexing. Dissolve a 60 mg oxazolone powder in 2 ml of this solution to obtain a 3 % (w/v) oxazolone sensitization solution. Mix the solution by carefully vortexing (*see* **Note 1**).
5. The oxazolone challenge solution: Dissolve 20 mg oxazolone in 2 ml of 50 % ethanol to obtain a 1 % (w/v) solution. Mix the solution by careful vortexing. The oxazolone powder should be completely dissolved before use (*see* **Note 2**).

2.2   Equipment

1. Vortex.
2. Electric clipper for animals.
3. Balances.
4. Centrifuge.
5. Flexible catheter (3.5 F, length 20 cm) with soft tip.
6. Syringe (1 ml).
7. Murine colonoscopy system.

2.3   Animals

1. Mice; usually 8–16 weeks old.

3   Methods

The procedure is divided in three phases: the sensitization, the challenge, and finally the analysis of the intestinal inflammation. The described protocol requires up to 9 days for the induction of acute colitis, starting with animal sensitization on day 0. Chronic colitis takes usually 4–5 weeks (Fig. 1, *see* **Note 3**).

3.1   Sensitization

1. On day 0, shave an approximately 2 cm × 2 cm field on the skin of the back of the mouse by using an electric clipper. Be careful to avoid open wounds. Animals will then be given oxazolone as detailed below. Some mice should be treated with sensitization solution without oxazolone as a control (*see* **Note 4**).
2. To apply the solution, hold the mouse with one hand and use the other hand to a 150 µl of the oxazolone sensitization

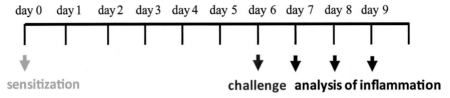

day 0   day 1   day 2   day 3   day 4   day 5   day 6   day 7   day 8   day 9

sensitization   challenge analysis of inflammation

Fig. 1 Time schedule for the acute oxazolone colitis model. The experiment starts at day 0 with animal sensitization, following by the challenge step on day 6 and subsequent analysis of the inflammatory response

solution on the shaved skin. The solution is absorbed by the skin quickly.

3. Return the animals to their cages.

3.2 Challenge

1. On day 6, weigh and label the sensitized animals.

2. Anesthetize the mice before oxazolone challenge. To achieve this, use an intraperitoneal injection of 80 μl of a ketamine/xylazine solution per 10 g body weight. Alternatively, use an isoflurane inhalation anesthetic system (*see* **Note 5**).

3. Connect a 3.5 F catheter to a 1 ml syringe and fill the syringe with the oxazolone challenge solution.

4. Before oxazolone administration, carefully position the catheter into the murine colon (insertion depth about 3–4 cm proximal to the anus). If necessary, remove stool pellets by flushing the colon carefully with a pipette filled with water (*see* **Note 6**).

5. Administer approximately 100 μl of the oxazolone challenge solution into the colonic lumen via the catheter. Administration should be performed slowly and should take 10–30 s (*see* **Note 7**).

6. Remove the catheter and keep the mouse in a vertical position (head down) for 60 s to avoid spilling out the injected solution.

7. Return the animals to their cages.

3.3 Analysis

1. Measure animals body weight using balances and record the weight daily in order to monitor for body weight changes (Fig. 2a).

2. Examine the appearance of the colonic mucosa by using the murine colonoscopy system [13]. This technique allows to

Fig. 2 (continued) pictures illustrating erosion of the colonic mucosa in two different mouse strains (C57BL/6 and BALB/C) on day 2 post-oxazolone challenge; (**c**) the decrease in the colon length; (**d**) histological evidence of tissue damage and inflammation in H&E-stained colonic sections, on day 3 post-oxazolone challenge

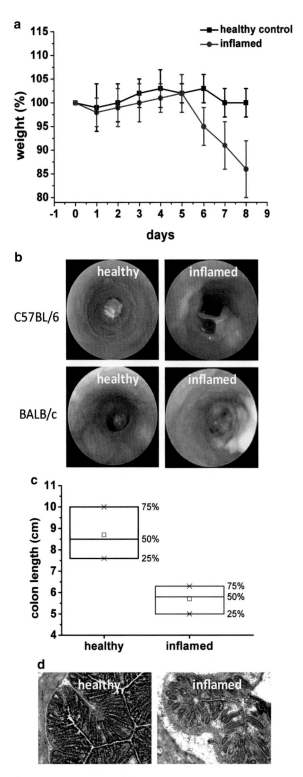

Fig. 2 Analysis of mucosal inflammation during oxazolone-induced colitis. The figures show different clinical and histological manifestation of colonic inflammation in oxazolone-treated mice that include: (**a**) body weight loss; (**b**) endoscopic

visualize the architecture of colonic mucosa and to detect early symptoms of mucosal inflammation (Fig. 2b, *see* **Note 8**).

3. Euthanize the animals at the end of the experiment. The development of colonic inflammation starts immediately after the oxazolone challenge and reaches its nadir approximately on day 8 post-sensitization. Open the abdominal cavity by using surgical scissors and remove the entire colon. This can be done by cutting the colon directly below the cecum and around the anus.

4. Lay the entire colon carefully on a paper and measure the length of the colon without stretching. Signs of inflammation are the shortening of the colon and thickening of the bowel wall (Fig. 2c).

5. Carefully cut the colon into several segments and clean them from the fecal material by squeezing the pellets out. Snap freeze some segments in liquid nitrogen. These segments can be used to obtain colonic cryo-sections for the histopathological analysis (e.g., calculation of the tissue injury score using hematoxylin and eosin-stained sections) (Fig. 2d, *see* **Note 9**).

6. Store the remaining colonic segments either in the RNA stabilization solutions for further analysis of mRNA expression, or in cold phosphate-buffered saline for subsequent isolation of different cell types from the intestinal mucosa.

7. Collect the blood to measure inflammatory cytokine level in the blood serum. This procedure should be performed shortly after euthanizing the animal in order to avoid rapid blood coagulation. The best approach for the blood collection is a cardiac puncture using a 10 ml syringe with the 18 G needle. This cardiac puncture allows collecting 1–2 ml of blood. Place a tube with collected blood in an upright position at room temperature followed by incubation for 30–45 min (no longer than 60 min) to allow clotting. If using a clot-activator tube, invert carefully 5–6 times to mix clot activator and blood before incubation.

8. Centrifuge the blood samples for 15 min at $400 \times g$ at 4 °C. Do not use brake to stop centrifugation.

9. Carefully aspirate the supernatant (serum) at room temperature and pool into a centrifuge tube, taking care not to disturb the cell layer or transfer any cells. Use a clean pipette for each tube.

10. Inspect serum for turbidity. Turbid samples should be centrifuged and aspirated again to remove remaining insoluble matter.

11. Aliquot the serum into cryo-vials and store at −80 °C for the subsequent analysis of cytokine levels by ELISA.

4   Notes

1. Both acetone and oxazolone are chemical hazards. Therefore, caution is needed to avoid potential toxic effects of these reagents.

2. To completely dissolve oxazolone crystals in 50 % ethanol, warm up the ethanol to 35 °C/95 °F and shake the solution on a tumbler till the entire pellet is dissolved. This may take some time.

3. Typically, signs of the acute inflammation will develop on day 8 of the protocol (2 days post-oxazolone administration), whereas the chronic inflammation develops within 3–4 weeks. To induce chronic colitis, the sensitization step with 3 % oxazolone solution is required, followed by three challenge steps with 0.5 % oxazolone challenge solutions [6]. Considering the sensitization step as day 0, the challenge steps should be performed on days 6, 20, and 34, allowing for approximately 2 weeks recovery phases between them. After each challenge, analyzing signs of inflammation (body weight or colonoscopy) will be helpful in order to evaluate the efficiency of colitis induction. Each investigator should experimentally determine the working concentration of oxazolone, given the differential sensitivity of different mouse strains and distinct composition of the commensal microflora of the local animal facilities.

4. All animal experiments must be performed in accordance with approved institutional animal care and use protocols that are based on national and international guidelines about humane care and use of the laboratory animals.

5. The isoflurane inhalation anesthetic system should not be used together with cyclosporine A, because cyclosporine influences the blood level of isoflurane.

6. Proceed very carefully to avoid damage or perforation of the colon wall by the instillation of the catheter. Colon perforation may result in lethal peritonitis.

7. In order to obtain reproducible results, it is important to ensure that the oxazolone solution remains trapped in the colon lumen.

8. The severity of inflammation is commonly judged based on a MEICS (murine endoscopic index of colitis severity) score. This score includes five different features of the colonic mucosa evaluated by endoscopy: translucency, granularity, fibrin, vascularity, and stool [13].

9. A blinded analysis should be performed for histological evaluation of H&E-stained colonic sections. The Tissue Injury Index should be calculated by scoring the following parameters:

mucosal erosion and ulceration, loss of crypt architecture, submucosal spread and transmural involvement, loss of goblet cells, and infiltration of neutrophils. The lowest score (grade 0) should be given for the normal tissue architecture. The grade 1 is associated with minimal scattered mucosal inflammatory cell infiltrates, with or without minimal epithelial hyperplasia. The grade 2 is given when a mild scattered to diffuse inflammatory cell infiltrates, sometimes extending into the submucosa and associated with erosions, with minimal to mild epithelial hyperplasia and minimal to mild mucin depletion from goblet cells is observed. The grade 3 involves a mild to moderate inflammatory cell infiltrate, sometimes transmural and often associated with ulceration, with moderate epithelial hyperplasia and mucin depletion. The grade 4 is linked with a marked inflammatory cell infiltrate that is often transmural and associated with ulceration, with a marked epithelial hyperplasia and mucin depletion. Finally, the grade 5 is given when a marked transmural inflammation with severe ulceration, loss of intestinal glands, and massive infiltration of neutrophils is observed.

Acknowledgment

This work was supported by DFG grant WE4656/2-2 and SFB 1181-B02.

References

1. Strober W, Fuss I, Mannon P (2007) The fundamental basis of inflammatory bowel disease. J Clin Invest 117:514–521

2. Strober W, Fuss IJ, Blumberg RS (2002) The immunology of mucosal models of inflammation. Annu Rev Immunol 20:495–549

3. Wirtz S, Neurath MF (2000) Animal models of intestinal inflammation: new insights into the molecular pathogenesis and immunotherapy of inflammatory bowel disease. Int J Colorectal Dis 15:144–160

4. Ina D, Pallone F (2008) What is the role of cytokines and chemokines in IBD? Inflamm Bowel Dis 14:S117–S118

5. Fuss IJ, Neurath M, Boirivant M et al (1996) Disparate CD4+ lamina propria (LP) lymphokine secretion profiles in inflammatory bowel disease. Crohn's disease LP cells manifest increased secretion of IFN-γ, whereas ulcerative colitis LP cells manifest increased secretion of IL-5. J Immunol 157:1261–1270

6. Gerlach K, Hwang Y, Nikolaev A, Atreya R et al (2014) TH9 cells that express the transcription factor PU.1 drive T cell-mediated colitis via IL-9 receptor signaling in intestinal epithelial cells. Nat Immunol 15:676–686

7. Boirivant M, Fuss IJ, Chu A, Strober W (1998) Oxazolone colitis: a murine model of T helper cell type 2 colitis treatable with antibodies to interleukin 4. J Exp Med 188:1929–1939

8. Heller F, Fuss IJ, Nieuwenhuis EE, Blumberg RS, Strober W (2002) Oxazolone colitis, a Th2 colitis model resembling ulcerative colitis, is mediated by IL-13-producing NK-T cells. Immunity 17:629–638

9. Daniel C, Sartory NA, Zahn N, Schmidt R et al (2007) FTY720 ameliorates oxazolone colitis in mice by directly affecting T helper type 2 functions. Mol Immunol 44:3305–3316

10. Iijima H, Neurath MF, Nagaishi T, Glickman JN et al (2004) Specific regulation of T helper cell 1-mediated murine colitis by CEACAM1. J Exp Med 199:471–482

11. Nolte T, Zadeh-Khorasani M, Safarov O, Rueff F et al (2013) Oxazolone and ethanol induce

colitis in non-obese diabetic-severe combined immunodeficiency interleukin-2Rγ(null) mice engrafted with human peripheral blood mononuclear cells. Clin Exp Immunol 172:349–362

12. Schiechl G, Bauer B, Fuss I, Lang SA et al (2011) Tumor development in murine ulcerative colitis depends on MyD88 signaling of

colonic F4/80+ CD11b(high)Gr1(low) macrophages. J Clin Invest 121:1692–1708

13. Becker C, Fantini MC, Wirtz S, Nikolaev A et al (2005) In vivo imaging of colitis and colon cancer development in mice using high resolution chromoendoscopy. Gut 54: 950–954

Chapter 24

The Mongolian Gerbil: A Robust Model of Helicobacter pylori-Induced Gastric Inflammation and Cancer

Jennifer M. Noto, Judith Romero-Gallo,
M. Blanca Piazuelo, and Richard M. Peek

Abstract

The Mongolian gerbil is an efficient, robust, and cost-effective rodent model that recapitulates many features of *H. pylori*-induced gastric inflammation and carcinogenesis in humans, allowing for targeted investigation of the bacterial determinants and environmental factors and, to a lesser degree, host constituents that govern *H. pylori*-mediated disease. This chapter discusses means through which the Mongolian gerbil model has been used to define mechanisms of *H. pylori*-inflammation and cancer as well as the current materials and methods for utilizing this model of microbially induced disease.

Key words *Helicobacter pylori*, Mongolian gerbil, Gastric inflammation, Gastric cancer

1 Introduction

1.1 Helicobacter pylori and Gastric Cancer

Gastric adenocarcinoma is the third leading cause of cancer-related death worldwide, accounting for over 700,000 deaths each year [1]. There are two main histologically distinct forms of gastric adenocarcinoma, diffuse- and intestinal-type. Diffuse-type cancer is characterized by non-cohesive neoplastic cells that infiltrate the stroma and is not associated with histological precancerous lesions, while intestinal-type cancer progresses through a series of well-defined pathological steps from normal gastric mucosa to chronic superficial gastritis, atrophic gastritis, intestinal metaplasia, and finally dysplasia and adenocarcinoma [2] (Fig. 1).

Chronic gastric inflammation induced by *Helicobacter pylori* is the strongest known risk factor for the development of gastric premalignant lesions and cancer, and *H. pylori* eradication significantly reduces the intensity of premalignant lesions and the subsequent incidence of gastric adenocarcinoma [3–5]. Despite more than half of the world's population being colonized with *H. pylori*, only a fraction of individuals ever develop gastric dysplasia or adenocarcinoma.

Andrei I. Ivanov (ed.), *Gastrointestinal Physiology and Diseases: Methods and Protocols*, Methods in Molecular Biology, vol. 1422, DOI 10.1007/978-1-4939-3603-8_24, © Springer Science+Business Media New York 2016

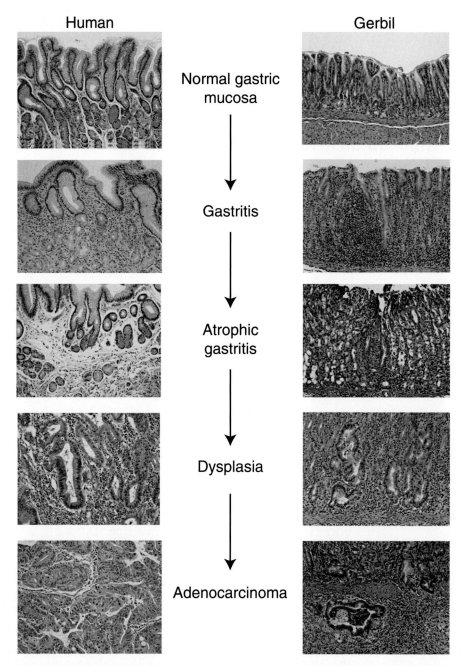

Fig. 1 Chronological steps in the progression to gastric cancer in humans and gerbils. Representative hematoxylin and eosin (H&E)-stained human (*left*) or gerbil (*right*) gastric tissue sections showing the various stages in the progression and transformation from normal gastric mucosa to gastric adenocarcinoma, including gastritis, atrophic gastritis, dysplasia, and adenocarcinoma at 20× magnification

The discrepancy in overall *H. pylori* infection rates versus disease outcomes is governed by specific relationships among host inflammatory responses, strain-specific bacterial virulence determinants, and environmental factors, which ultimately influence interactions between *H. pylori* and its human host.

Table 1
Characteristics of the Mongolian gerbil model

Characteristic	Description
Family	*Cricetidae*
Genus species	*Meriones unguiculatus*
Coat	Agouti
Weight	70–100 g as adult
Activity	Nocturnal, burrowing
Water	~4 mL/day
Food	~8 g/day
Breeding	Monogamous, pair 10–12 weeks of age
Disposition	Friendly, not prone to fighting or biting

1.2 Mongolian Gerbils

The Mongolian gerbil (*Meriones unguiculatus*) is a small rodent member of the *Cricetidae* family (Table 1). The gerbil has been increasingly used in research focused on *H. pylori* pathogenesis as it represents an efficient and cost-effective rodent model that recapitulates many features of *H. pylori*-induced gastric inflammation and carcinogenesis in humans (Fig. 1). The first published description of this model in 1991 reported that *H. pylori* colonized gerbils and induced mild gastritis following a 2-month infection [6]. Numerous other studies demonstrated that *H. pylori*-infected gerbils developed gastric ulcers, duodenal ulcers, and intestinal metaplasia following experimental challenge [7–11]. Subsequent studies showed that gerbils develop gastric carcinoma following co-administration of *H. pylori* and chemical carcinogens such as N-methyl-N-nitrosourea (MNU) or N-methyl-N-nitro-N--nitrosoguanidine (MNNG) [12–15]. In 1998, two seminal studies demonstrated that Mongolian gerbils developed gastric adenocarcinoma following long-term *H. pylori* infection in the absence of additional chemical carcinogens [16, 17], which was subsequently confirmed by other groups [18, 19]. Carcinomas that developed in *H. pylori*-infected gerbils typically occurred in the distal stomach and the pyloric region and contained well-differentiated intestinal-type epithelium, reflecting many features of intestinal-type gastric adenocarcinoma in humans. Consistent with reports in humans, *H. pylori* eradication in the gerbil model significantly reduces the severity of gastritis, premalignant lesions, and incidence of gastric adenocarcinoma [8, 20–25].

Similar to the human disease, the colonization of the gerbil gastric mucosa by *H. pylori* elicits a mixed inflammatory infiltrate in the lamina propria consisting of neutrophils and mononuclear leukocytes. Severe inflammation may be accompanied by pit abscesses, formation of lymphoid follicles, and epithelial hyperplasia. In the Mongolian gerbil, inflammation is usually more pronounced in the transitional mucosa between the corpus and the antrum and spreads proximal and distally as the disease progresses. Over time, there is loss of parietal and chief cells that usually starts in distal corpus and progresses towards the proximal mucosa. The loss of these specialized cells in the oxyntic mucosa is usually accompanied by hyperplasia of mucous neck cells (sometimes called mucous metaplasia) and the base of fundic glands may show features of spasmolytic polypeptide expressing metaplasia (also known as pseudopyloric metaplasia). In more advanced stages of *H. pylori* infection, epithelial dysplastic changes may be observed, which usually start in the deep portion of the glands in the transitional mucosa. These changes are characterized by epithelial nuclear crowding and by irregularity, budding, and branching of the glands. Dysplastic glandular epithelium tends to spread laterally, parallel to the muscularis mucosa. The penetration of dysplastic glands through the muscularis mucosa into the submucosa is characteristic of invasive adenocarcinoma. However, one caveat of using the Mongolian gerbil model of gastric cancer is that there has been some difficulty in distinguishing herniation of non-neoplastic mucosa from invasive carcinoma, particularly within the setting of inflammation. To address this issue, guidelines have been established for pathological interpretation of these lesions [26]. Although these guidelines were originally formulated for evaluation of intestinal tumors, they have been applied to gastric lesions by numerous investigators [27–29], and in many studies of *H. pylori*-induced cancer in gerbils, several features have been used to distinguish invasive carcinoma from mucosal herniation (Table 2).

1.3 Host Constituents

Although limited due to their outbred nature, Mongolian gerbils have been used to study the role of a subset of host constituents on the development of gastric cancer. IL-1β is a Th1-type

Table 2
Features that distinguish invasive carcinoma from herniated mucosa

Feature	Description
1	Invasive cells differ from overlying mucosa with atypia exceeding low-grade dysplasia
2	Desmoplasia, unassociated with a predominant inflammatory infiltrate
3	Irregular, sharp, or angulated glands in the invasive component
4	Invading glands spread laterally deep to the surface mucosa
5	Loss of epithelial cells from invading glands
6	Greater than 2 invading glands in submucosa
7	Absence of basement membrane around invading glands

cytokine that inhibits acid secretion and is increased within gastric mucosa of *H. pylori*-infected individuals [30]. In humans, polymorphisms in the IL-1β gene cluster are associated with increased IL-1β production and a significantly increased risk for hypochlorhydria, gastric atrophy, and distal gastric adenocarcinoma, but only among *H. pylori*-infected individuals [31–33]. These relationships were investigated in gerbils infected with *H. pylori* by quantifying changes in gastric acid secretion that were mediated by IL-1β. Compared to uninfected animals, gerbils infected with *H. pylori* exhibited elevated levels of IL-1β expression within the gastric mucosa, which was accompanied by decreased gastric acidity. Treatment of *H. pylori*-infected gerbils with an IL-1β antagonist abolished loss of acid secretion, implicating IL-1β in the development of achlorhydria within an *H. pylori*-infected gerbil stomach [34].

Other groups have also developed novel reagents specific for analyses in gerbils. Numerous primers targeting chemokines and cytokines have been developed based on species-specific gerbil cDNA to assess *H. pylori*-induced inflammatory responses in gerbils [35–41]. In addition to chemokines and cytokines, *H. pylori* has been shown to alter expression of other inflammatory mediators in gerbils including iNOS and COX2 [37, 42]. The role of NF-κB activation has also been assessed in gerbils within the context of *H. pylori*-induced inflammation. One group developed a gastric cell line derived from a gerbil gastric cancer specimen and demonstrated that *H. pylori* can activate NF-κB signaling in an in vitro, species-specific model [43], while another group demonstrated that activation of NF-κB was essential for *H. pylori*-induced inflammation in gerbils [44]. *H. pylori* infection of gerbils has also been shown to increase serum levels of gastrin, which can promote cell growth, and increased gastrin levels were directly related to heightened gastric epithelial cell proliferation [45, 46]. Furthermore, a recent study demonstrated that a gastrin antagonist prevents *H. pylori*-induced gastritis in gerbils [47]. Overall, despite the fact that the gerbil model is not genetically tractable, many host responses have been successfully investigated and targeted in this system.

1.4 Bacterial Determinants

In contrast to constraints regarding the study of host factors, the gerbil model is particularly robust for studying cancer-associated microbial determinants. The *cag* pathogenicity island (*cag* PAI), one of the most well-studied *H. pylori* virulence determinants, encodes a bacterial type IV secretion system (T4SS) that allows for the delivery of the bacterial effector protein, CagA, into host gastric epithelial cells [48, 49]. Transgenic mice that overexpress CagA develop gastric epithelial cell hyperproliferation and gastric adenocarcinoma [50], implicating this molecule as a bacterial oncoprotein. One limitation of using murine models of *H. pylori* infection is that *cag*+ strains frequently lose function of the *cag* island following chronic infection [51, 52]. In contrast, *cag*[+] strains of *H. pylori* efficiently colonize gerbils and maintain a functional *cag* T4SS secretion system [45], which allows for examination of the role of this key virulence determinant in the context of *H. pylori*-induced inflammation and cancer. Compared to gerbils infected with *cag*– *H. pylori*, gerbils challenged with *cag*+ strains develop significantly more severe gastritis [18, 53–56], indicating the importance of the *cag* island in *H. pylori*-mediated inflammation. Rieder et al. investigated alterations not only in the intensity but also in the topography of inflammation in gerbils infected with wild-type *H. pylori* compared to isogenic *cagA*– mutants. Loss of *cagA* resulted in an inflammatory response primarily restricted to the gastric antrum, and which did not significantly involve the acid-secreting corpus [57]. Cumulatively, these results indicate that a functional *cag* T4SS is required for *H. pylori*-induced corpus-predominant gastritis, a precursor in the progression to intestinal-type gastric adenocarcinoma, which can be easily evaluated in the Mongolian gerbil model.

As previously demonstrated by Watanabe et al., gerbils develop gastric cancer in response to *H. pylori* [16]; however, the prolonged time-course required for transformation in these early studies precluded large-scale analyses that comprehensively evaluate mediators that are critical to gastric carcinogenesis. Since serial passage of *H. pylori* in rodents can increase colonization efficiency, Franco et al. investigated whether in vivo adaptation of a human *H. pylori* strain (B128) would enhance its carcinogenic potential [58]. A single gerbil was infected with a human *H. pylori* isolate B128 and then sacrificed 3 weeks post-challenge. A single colony output derivative (*H. pylori* strain 7.13) was isolated and used to infect an independent population of gerbils. Although the levels of inflammation induced by both the parental strain B128 and output strain 7.13 were similar, gastric dysplasia and adenocarcinoma only developed in gerbils infected with the output strain 7.13 [58]. These findings indicated that in vivo adaptation of *H. pylori* strains can increase the virulence potential of *H. pylori* strains in the gerbil model.

1.5 Environmental Factors

Host constituents and/or *H. pylori* virulence determinants are not solely responsible for the development of gastric cancer in humans or animal models, as environmental factors, such as diet, are also important risk factors for disease [59]. However, epidemiologic studies of diet in humans are subject to many limitations, including a reliance on patient reporting and difficulty in ascertaining diets that were consumed decades prior to the development of gastric cancer. Moreover, it is difficult to determine whether dietary parameters are causally linked to the development of gastric cancer or merely represent markers for other factors that are important in gastric cancer pathogenesis. To further investigate potential relationships between diet and gastric cancer risk, several studies have examined the role of dietary factors in the gerbil model of *H. pylori*-induced gastric carcinogenesis.

Increased salt consumption has been shown to increase the risk for gastric cancer in humans [60], and the effects of high-salt diets on *H. pylori* infection and gastric cancer have been investigated using the gerbil model. One study reported that *H. pylori* infection and a high-salt diet could independently induce gastric atrophy and intestinal metaplasia in Mongolian gerbils [61]. Other studies have demonstrated a synergistic effect of *H. pylori* infection and high-salt diets on gastric carcinogenesis in the presence of a chemical carcinogen, MNU [62–64]. To further investigate the direct effect of a high-salt diet on *H. pylori*-induced carcinogenesis in gerbils, Gaddy et al. maintained Mongolian gerbils on high-salt or normal-salt diets and then challenged gerbils with *H. pylori*. Compared to infected gerbils maintained on normal-salt diets, the incidence of gastric adenocarcinoma was significantly increased among *H. pylori*-infected animals maintained on a high-salt diet [65].

Another dietary factor that increases the risk for gastric cancer is iron deficiency [66]. To address the effect of dietary iron depletion on *H. pylori*-induced gastric carcinogenesis, Noto et al. maintained gerbils on iron-replete or iron-depleted diets and then challenged gerbils with *H. pylori* [67]. *H. pylori* induced more severe gastritis and increased the incidence and frequency of gastric dysplasia and gastric adenocarcinoma among gerbils maintained on iron-depleted diets compared to gerbils maintained on iron-replete diets [67], phenotypes that were abrogated in animals infected with an *cagA*– isogenic mutant strain. *H. pylori cagA*– strains also exhibited a significant decrease in colonization density [68] and altered colony morphology [69], but only among gerbils maintained on iron-depleted diets. Cumulatively, these data demonstrate that dietary factors, including high salt and low iron, significantly increase the severity and frequency of *H. pylori*-induced premalignant and malignant lesions in the gerbil model of gastric cancer.

2 Materials

2.1 *Helicobacter pylori* Culture

1. Personal protective equipment (PPE).
2. *Helicobacter pylori* strains (*see* **Note 1**).
3. Sterile cotton-tipped applicators.
4. Trypticase soy agar plates with 5 % sheep blood.
5. Baffled-bottom flasks.
6. Sterile Brucella broth with 10 % fetal bovine serum and vancomycin (10 μg/mL final concentration).
7. Spectrophotometer.

8. 37 °C incubator with 5 % CO_2 with platform shaker.

9. Glass slides.

10. Gram stain reagents.

11. Oxidase test.

12. Catalase test.

13. Microscope.

2.2 Gerbil Challenge with H. pylori

1. 50 mL conical tubes.

2. Tabletop centrifuge.

3. 1 mL syringes with 18 G feeding needles.

4. Sterile Brucella broth (control).

5. *H. pylori* resuspended in Brucella broth (2×10^9 CFU/mL).

6. Biosafety cabinet.

7. Male Mongolian gerbils (*see* **Note 2**).

2.3 Gerbil Dissection

1. Dissecting tray with pins.

2. 70 % ethanol.

3. 1 mL syringes with 27 G needles.

4. Sterile blunt forceps.

5. Sterile blunt scissors.

6. Sterile sharp forceps.

7. Sterile sharp scissors.

8. Sterile scalpel.

9. Sterile surgical blades.

10. 100 % ethanol for tool sterilization.

11. Sterile petri dishes.

12. Sterile 1× phosphate-buffered saline (PBS).

13. Pencil or solvent-resistant marker for labeling histology cassettes.

14. Histology cassettes.

15. Sponges for histology cassettes.

16. 10 % formalin for histology.

17. Microcentrifuge tubes.

18. Freezer tubes.

19. Dry ice.

2.4 Sample Preparation After Dissection

1. Microcentrifuge.

2. Digital scale.

3. Sterile tissue homogenizer.

4. Tryptic soy agar with 5 % sheep blood.

Table 3
Antifungal/antibiotic concentrations for isolation of *H. pylori* by quantitative culture

Antifungal/antibiotic	Stock concentration (mg/mL)	Final concentration (µg/mL)
Amphotericin	20	2
Bacitracin	30	30
Nalidixic acid	10	10
Vancomycin	20	20

5. Antifungal/antibiotics (amphotericin, bacitracin, nalidixic acid, vancomycin; Table 3).

6. GasPak™ Campy Container sachets.

7. 100 % ethanol for histology.

8. Equipment for paraffin embedding and sectioning.

9. Hematoxylin & eosin (H&E).

10. Microscope.

3   Methods

3.1 Helicobacter pylori Culture

1. Recover minimally passaged *H. pylori* strains from −80 °C freezer stocks onto trypticase soy agar plates with 5 % sheep blood. Incubate at 37 °C with 5 % CO_2 for 2–4 days (*see* **Note 3**).

2. Expand *H. pylori* strains onto 1–2 trypticase soy agar plates with 5 % sheep blood and incubate at 37 °C with 5 % CO_2 for 24–48 h.

3. Continue expanding *H. pylori* strains onto trypticase soy agar plates with 5 % sheep blood and incubate at 37 °C with 5 % CO_2 for 24 h until desired amount is obtained (*see* **Note 4**).

4. Collect *H. pylori* from the 24-h plates using sterile cotton-tipped applicators. Using sterile technique, inoculate Brucella broth culture (*see* **Note 5**) with a starting $OD_{600} > 0.15$ (*see* **Note 6**). Incubate on a platform shaker at 37 °C with 5 % CO_2, ~150 rpm for 16–18 h.

5. The next day, harvest *H. pylori* broth cultures. Remove 1 mL of *H. pylori* broth culture to measure OD_{600} and a small sample to perform Gram stain, oxidase, and catalase tests. *H. pylori* should be in the log phase of growth and typically $OD_{600} > 1.0$ (range = 1.0–2.0) (*see* **Note 6**).

6. Transfer broth culture into 50 mL conicals and subject to centrifugation at 4000 rpm for 10 min.

7. Discard supernatants and resuspend bacterial pellets to a concentration of 2×10^9 CFU/mL. To do so, multiply the final

OD_{600} measurement to determine CFU/mL (1 $OD_{600} = 5.5 \times 10^8$ CFU/mL). Multiply total CFU/mL by the total culture volume to determine total CFU. Next, divide total CFU by the desired concentration of bacteria needed per gerbil (2×10^9 CFU/mL). This calculation will provide the volume required to resuspend the bacterial pellet so that there are 2×10^9 CFU/mL. The following example calculation would result in enough inoculum to infect 41 gerbils.

Example calculation:

50 mL culture, final $OD_{600} = 1.50$

$OD_{600} \times 5.5 \times 10^8$ CFU/mL $\times$ total volume (mL)

$1.50 \times 5.5 \times 10^8$ CFU/mL $\times 50$ mL $= 4.125 \times 10^{10}$ CFU

4.125×10^{10} CFU/2×10^9 CFU/mL $= 20.6$ mL

3.2 Gerbil Challenge with H. pylori

1. Order male Mongolian gerbils (36–49 days old or 41–50 g) (*see* **Note 2**). Adapt gerbils to new environment in the animal housing facility for at least 1 week prior to challenge.

2. Fast the gerbils for 8–12 h prior to first challenge (*see* **Note 7**).

3. Orally gavage gerbils with 0.5 mL of Brucella broth as a control or 0.5 mL of *H. pylori* (1×10^9 CFU) using sterile feeding needles (one feeding needle per strain of *H. pylori*). For oral gavage, hold the gerbil by the ears and support and straighten the body. Guide the feeding needle along the roof of the mouth to the back of the throat and pass it through the esophagus into the stomach. Dispense 0.5 mL of bacteria (1×10^9 CFU). After challenging, fast the gerbils for an additional 2–4 h (*see* **Note 8**). Reintroduce food and wait 24–48 h before the second challenge (*see* **Note 9**).

4. Fast the gerbils for 8–12 h prior to second challenge.

5. Orally gavage gerbils with 0.5 mL of Brucella broth as a control or 0.5 mL of *H. pylori* (1×10^9 CFU) using sterile feeding needles. After challenging, fast the gerbils for an additional 2–4 h and then reintroduce food.

6. Monitor gerbils throughout the course of infection to address any health concerns, such as changes in body weight (*see* **Note 10**).

3.3 Preparation for Gerbil Dissection

1. Prepare and autoclave tryptic soy agar. Prepare enough plates to have approximately 3–4 per gerbil (*see* **Note 11**).

2. Once media has adequately cooled to ~56 °C, add 5 % sheep blood and antifungal/antibiotics (Table 3 and *see* **Note 12**).

3. Pour plates and allow them to solidify overnight. Store plates at 4 °C for short-term storage.

4. Prepare collection tubes and histology cassettes for gerbil dissection and tissue collection (Table 4).

Table 4
Collection tubes and histology cassettes required for gerbil dissection and tissue collection

Collection tube/cassette	Purpose
Tube 1	Microcentrifuge tube for blood collection
Tube 2[a]	Microcentrifuge tube with 1 mL of sterile 1× PBS with antifungal/antibiotics for gastric tissue collection for quantitative culture[a]
Tube 3	Freezer tube for gastric tissue collection for long-term storage at −80 °C
Cassette	Histology cassette labeled with pencil or solvent-resistant marker for gastric tissue collection for histology
Tube 4	Microcentrifuge tube with 900 µL sterile 1× PBS with antifungal/antibiotics for 1:10 serial dilution
Tube 5	Microcentrifuge tube with 900 µL sterile 1× PBS with antifungal/antibiotics for 1:100 serial dilution
Tube 6	Microcentrifuge tube with 900 µL sterile 1× PBS with antifungal/antibiotics for 1:1000 serial dilution

[a]Weigh tube 2 prior to collection of tissue to ultimately determine CFU per gram of gastric tissue.

3.4 Gerbil Dissection

1. Euthanize gerbils according to the practices and procedures designated by the Institutional Animal Care and Use Committee (IACUC) and individual animal housing facility. Following euthanasia, perform cervical dislocation.

2. Secure gerbil in the supine position on the dissecting tray. Drench the anterior chest and abdomen in 70 % ethanol.

3. Using a 27-G syringe, collect blood by cardiac puncture and place in first collection tube, Tube 1 (Table 4 and *see* **Note 13**).

4. Using sterile blunt forceps and scissors, open the top layer of skin to expose the peritoneum.

5. Using sharp forceps and scissors, open the peritoneum to expose the stomach.

6. Excise the stomach, including a portion of the duodenum. Discard the forestomach lined with squamous epithelium and place the stomach in a sterile petri dish (Fig. 2).

7. Cut the stomach along the greater curvature so that it lies flat (Fig. 2) and remove any remaining food.

8. Cut the stomach, extending from the squamocolumnar junction through the proximal duodenum, in half longitudinally and place one-half into a histology cassette (Fig. 2) and submerge in 10 % formalin (*see* **Note 14**).

9. Cut the remaining half of the stomach in half, creating two one-quarter sections (Fig. 2). Place one-quarter of the stomach

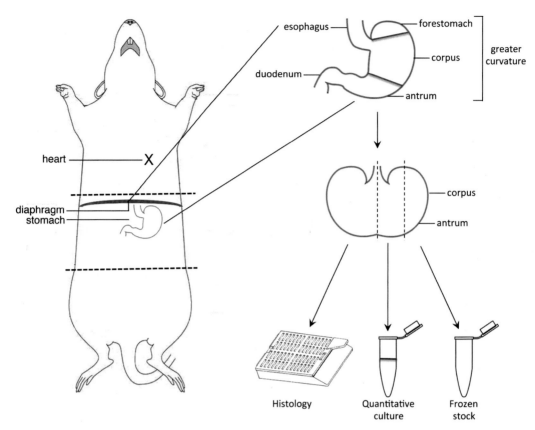

Fig. 2 Gerbil dissection and isolation of gastric tissue. Perform cardiac puncture for blood collection. Open the peritoneum to expose the stomach. Excise the stomach, including a portion of the duodenum. Cut the stomach along the greater curvature so that it lies flat. Cut the stomach, extending from the squamocolumnar junction through the proximal duodenum, in half longitudinally and place one-half into a histology cassette. Cut the remaining half of the stomach in half, creating two one-quarter sections. Place one-quarter of the stomach into a collection tube for quantitative culture. Place the other one-quarter of the stomach into a collection tube for long-term storage

into the second collection tube with sterile 1× PBS and antifungal/antibiotics (Tube 2) for quantitative culture. Place the other one-quarter of the stomach into the third collection tube (Tube 3) and place on dry ice.

3.5 Blood Samples and Frozen Gastric Tissue Samples

1. For blood samples, subject samples to centrifugation for 5 min at 5000 rpm. Transfer serum/supernatant to a new microcentrifuge tube and repeat centrifugation for 5 min at 5000 rpm or 2400 ×g. Transfer serum/supernatant to a fresh microcentrifuge tube and store at –20 °C until needed (*see* **Note 15**).

2. For gastric tissue samples temporarily stored on dry ice, transfer and store gastric tissue samples at –80 °C for long-term storage until needed (*see* **Note 16**).

3.6 Gastric Tissue Samples for Quantitative Culture

1. Weigh the microcentrifuge tube containing gastric tissue to determine the weight of each gastric tissue section. Subtract the initial tube weight to determine weight of each piece of tissue to ultimately determine CFU per gram of gastric tissue by quantitative culture.

2. Transfer each gastric tissue sample into a sterile tissue homogenizer containing 1 mL sterile 1× PBS + antifungal/antibiotics.

3. Homogenize the gastric tissue and plate 100 μl of homogenate onto tryptic soy agar plates containing 5 % sheep blood and antifungal/antibiotics.

4. Perform serial dilutions and plate 100 μL of each serial dilution out to a 1:1000 dilution (Tubes 4, 5, and 6).

8. Incubate plates at 37 °C with 5 % CO_2 for 5–7 days in GasPak™ Campy Container sachets and then count colonies to determine CFU per gram of gastric tissue.

9. Isolate and expand *H. pylori* single colony isolates. Perform Gram stain, oxidase, and catalase tests on single colonies and then freeze for future analyses of in vivo-derived strains (*see* **Note 17**).

3.7 Gastric Tissue Samples for Histology

1. Incubate histology cassettes in 10 % formalin overnight and then transfer to 100 % ethanol.

2. Paraffin embed and section gastric tissue.

3. Stain gastric tissue sections with H&E and have a pathologist assess acute and chronic inflammation, and determine disease diagnosis based on the most advanced lesions present within the gastric tissue sections.

4 Notes

1. Some strains of *Helicobacter pylori* are commercially available through ATCC.

2. Due to the higher incidence of disease among males, male gerbils are typically used for assessing *H. pylori*-induced inflammation and gastric disease. Mongolian gerbils are classified as a USDA animal and are subject to different restrictions than murine models. Mongolian gerbils are commercially available from Charles River Laboratories.

3. Wait until single colonies are clearly visible on plate. Passing plates too early will lead to insufficient growth and not enough inoculum for broth cultures.

4. At least 2–3 plates per *H. pylori* strain are required for a 50 mL broth culture. Some laboratory strains of *H. pylori* grow very well, while clinical strains frequently do not, so adjust the

number of plates depending on the growth rates of individual *H. pylori* strains.

5. Choose the appropriate volume of broth culture depending on the number of gerbils in the experiment. A 50 mL volume should be enough for approximately 20 animals, depending on the final OD_{600}. *See* example calculation (Subheading 3.1, **step 7**).

6. Optical density readings can differ between spectrophotometer instruments. For reference: 1 $OD_{600} = 5.5 \times 10^8$ CFU. If the starting OD_{600} is too low, *H. pylori* will not grow. Determine the best starting OD_{600} based on individual spectrophotometers and CFU. Some laboratory strains of *H. pylori* grow very well, while other clinical strains do not, so adjust the starting OD_{600} based on spectrophotometer readings as well as growth rates of individual *H. pylori* strains.

7. Fasting gerbil prior to infection allows for more efficient *H. pylori* colonization.

8. Oral gavage is usually performed when the gerbils are awake and not sedated; however, the gerbils can be sedated for this procedure, but it is a much more time-intensive process.

9. The amount of time between challenges is due to the requirement for fasting. Fasting the gerbils for extended periods of time can have adverse effects on the health; therefore, adequate time is needed for re-feeding between infections.

10. Typically a 12-week infection with a gerbil-adapted *H. pylori* strain, such as 7.13, should be adequate to observe gastric dysplasia and adenocarcinoma. Longer infection times may be necessary for less well-adapted *H. pylori* strains.

11. One liter should yield approximately 30–40 plates.

12. If media is not adequately cooled prior to the addition of blood or antifungal/antibiotics, blood will lyse and antifungal/antibiotics will be inactivated.

13. Regarding cardiac puncture, collect blood with a single attempt to avoid lysis of red blood cells. However, if unsuccessful, more attempts can be made or blood can be collected immediately after the animal has been dissected, exposing the heart. Remove the needle following blood collection for dispensing into the microcentrifuge tube. This will prevent further red blood cell lysis.

14. For histology, harvest both the gastric antrum and corpus in each section because indices of inflammation are scored in both the antrum and corpus. The strips of tissue for paraffin embedding should be no greater than 3 mm wide.

15. The serum should be clear to yellow in color. Serum that is pink or red indicates lysis of red blood cells in the sample.

16. Frozen samples can be used for a variety of purposes, including but not limited to flow cytometry, quantitative PCR, and Western blotting.

17. If no colonies are isolated from serial dilutions, collectively sweep each plate onto a new tryptic soy agar plate with 5 % sheep blood and antifungal/antibiotics to determine if *H. pylori* can be isolated. This is not quantitative, but will provide information on whether the animal was successfully colonized. PCR can also be performed to determine *H. pylori* colonization.

References

1. Ferlay J, Soerjomataram I, Dikshit R, Eser S et al (2014) Cancer incidence and mortality worldwide: sources, methods, and major patterns in GLOBOCAN 2012. Int J Cancer 136:E359–386

2. Correa P, Chen VW (1994) Gastric cancer. Cancer Surv 20:55–76

3. Wong BC, Lam SK, Wong WM, Chen JS et al (2004) Helicobacter pylori eradication to prevent gastric cancer in a high-risk region of China: a randomized controlled trial. JAMA 291:187–194

4. Mera R, Fontham ET, Bravo LE, Bravo JC, Piazuelo MB, Camargo MC, Correa P (2005) Long term follow up of patients treated for Helicobacter pylori infection. Gut 54:1536–1540

5. Mera R, Fontham ET, Bravo LE, Bravo JC, Piazuelo MB, Camargo MC, Correa P (2007) Re: long term follow up of patients treated for Helicobacter pylori infection. Gut 56:436

6. Yokota K, Kurebayashi Y, Takayama Y, Hayashi S et al (1991) Colonization of Helicobacter pylori in the gastric mucosa of Mongolian gerbils. Microbiol Immunol 35:475–480

7. Hirayama F, Takagi S, Kusuhara H, Iwao E, Yokoyama Y, Ikeda Y (1996) Induction of gastric ulcer and intestinal metaplasia in Mongolian gerbils infected with Helicobacter pylori. J Gastroenterol 31:755–757

8. Matsumoto S, Washizuka Y, Matsumoto Y, Tawara S, Ikeda F, Yokota Y, Karita M (1997) Induction of ulceration and severe gastritis in Mongolian gerbil by Helicobacter pylori infection. J Med Microbiol 46:391–397

9. Honda S, Fujioka T, Tokieda M, Gotoh T, Nishizono A, Nasu M (1998) Gastric ulcer, atrophic gastritis, and intestinal metaplasia caused by Helicobacter pylori infection in Mongolian gerbils. Scand J Gastroenterol 33:454–460

10. Ikeno T, Ota H, Sugiyama A, Ishida K, Katsuyama T, Genta RM, Kawasaki S (1999) Helicobacter pylori-induced chronic active gastritis, intestinal metaplasia, and gastric ulcer in Mongolian gerbils. Am J Pathol 154:951–960

11. Ohkusa T, Okayasu I, Miwa H, Ohtaka K, Endo S, Sato N (2003) Helicobacter pylori infection induces duodenitis and superficial duodenal ulcer in Mongolian gerbils. Gut 52:797–803

12. Tatematsu M, Yamamoto M, Shimizu N, Yoshikawa A et al (1998) Induction of glandular stomach cancers in Helicobacter pylori-sensitive Mongolian gerbils treated with N-methyl-N-nitrosourea and N-methyl-N′-nitro-N-nitrosoguanidine in drinking water. Jpn J Cancer Res 89:97–104

13. Sugiyama A, Maruta F, Ikeno T, Ishida K et al (1998) Helicobacter pylori infection enhances N-methyl-N-nitrosourea-induced stomach carcinogenesis in the Mongolian gerbil. Cancer Res 58:2067–2069

14. Shimizu N, Inada K, Nakanishi H, Tsukamoto T et al (1999) Helicobacter pylori infection enhances glandular stomach carcinogenesis in Mongolian gerbils treated with chemical carcinogens. Carcinogenesis 20:669–676

15. Tokieda M, Honda S, Fujioka T, Nasu M (1999) Effect of Helicobacter pylori infection on the N-methyl-N′-nitro-N-nitrosoguanidine-induced gastric carcinogenesis in Mongolian gerbils. Carcinogenesis 20:1261–1266

16. Watanabe T, Tada M, Nagai H, Sasaki S, Nakao M (1998) Helicobacter pylori infection induces gastric cancer in Mongolian gerbils. Gastroenterology 115:642–648

17. Honda S, Fujioka T, Tokieda M, Satoh R, Nishizono A, Nasu M (1998) Development of Helicobacter pylori-induced gastric carcinoma in Mongolian gerbils. Cancer Res 58:4255–4259

18. Ogura K, Maeda S, Nakao M, Watanabe T et al (2000) Virulence factors of Helicobacter pylori responsible for gastric diseases in Mongolian gerbil. J Exp Med 192:1601–1610

19. Zheng Q, Chen XY, Shi Y, Xiao SD (2004) Development of gastric adenocarcinoma in Mongolian gerbils after long-term infection with *Helicobacter pylori*. J Gastroenterol Hepatol 19:1192–1198

20. Shimizu N, Ikehara Y, Inada K, Nakanishi H et al (2000) Eradication diminishes enhancing effects of *Helicobacter pylori* infection on glandular stomach carcinogenesis in Mongolian gerbils. Cancer Res 60:1512–1514

21. Keto Y, Ebata M, Okabe S (2001) Gastric mucosal changes induced by long term infection with *Helicobacter pylori* in Mongolian gerbils: effects of bacteria eradication. J Physiol 95:429–436

22. Nozaki K, Shimizu N, Tsukamoto T, Inada K et al (2002) Reversibility of heterotopic proliferative glands in glandular stomach of *Helicobacter pylori*-infected Mongolian gerbils on eradication. Jpn J Cancer Res 93:374–381

23. Hirayama F, Takagi S, Yokoyama Y, Yamamoto K, Iwao E, Haga K (2002) Long-term effects of *Helicobacter pylori* eradication in Mongolian gerbils. J Gastroenterol 37:779–784

24. Brzozowski T, Konturek PC, Kwiecien S, Konturek SJ et al (2003) Triple eradication therapy counteracts functional impairment associated with *Helicobacter pylori* infection in Mongolian gerbils. J Physiol Pharmacol 54:33–51

25. Nozaki K, Shimizu N, Ikehara Y, Inoue M et al (2003) Effect of early eradication on *Helicobacter pylori*-related gastric carcinogenesis in Mongolian gerbils. Cancer Sci 94:235–239

26. Boivin GP, Washington K, Yang K, Ward JM et al (2003) Pathology of mouse models of intestinal cancer: consensus report and recommendations. Gastroenterology 124:762–777

27. Houghton J, Stoicov C, Nomura S, Rogers AB et al (2004) Gastric cancer originating from bone marrow-derived cells. Science 306:1568–1571

28. Rogers AB, Taylor NS, Whary MT, Stefanich ED, Wang TC, Fox JG (2005) *Helicobacter pylori* but not high salt induces gastric intraepithelial neoplasia in B6129 mice. Cancer Res 65:10709–10715

29. Hagiwara T, Mukaisho K, Nakayama T, Sugihara H, Hattori T (2011) Long-term proton pump inhibitor administration worsens atrophic corpus gastritis and promotes adenocarcinoma development in Mongolian gerbils infected with *Helicobacter pylori*. Gut 60: 624–630

30. Noach LA, Bosma NB, Jansen J, Hoek FJ, van Deventer SJ, Tytgat GN (1994) Mucosal tumor necrosis factor-alpha, interleukin-1 beta, and interleukin-8 production in patients with *Helicobacter pylori* infection. Scand J Gastroenterol 29:425–429

31. El-Omar EM, Carrington M, Chow WH, McColl KE et al (2000) Interleukin-1 polymorphisms associated with increased risk of gastric cancer. Nature 404:398–402

32. Figueiredo C, Machado JC, Pharoah P, Seruca R et al (2002) *Helicobacter pylori* and interleukin 1 genotyping: an opportunity to identify high-risk individuals for gastric carcinoma. J Natl Cancer Inst 94:1680–1687

33. Santos JC, Ladeira MS, Pedrazzoli J Jr, Ribeiro ML (2012) Relationship of IL-1 and TNF-alpha polymorphisms with *Helicobacter pylori* in gastric diseases in a Brazilian population. Braz J Med Biol Res 45:811–817

34. Takashima M, Furuta T, Hanai H, Sugimura H, Kaneko E (2001) Effects of *Helicobacter pylori* infection on gastric acid secretion and serum gastrin levels in Mongolian gerbils. Gut 48:765–773

35. Yamamoto N, Sakagami T, Fukuda Y, Koizuka H et al (2000) Influence of *Helicobacter pylori* infection on development of stress-induced gastric mucosal injury. J Gastroenterol 35:332–340

36. Crabtree JE, Court M, Aboshkiwa MA, Jeremy AH, Dixon MF, Robinson PA (2004) Gastric mucosal cytokine and epithelial cell responses to *Helicobacter pylori* infection in Mongolian gerbils. J Pathol 202:197–207

37. Matsubara S, Shibata H, Takahashi M, Ishikawa F, Yokokura T, Sugimura T, Wakabayashi K (2004) Cloning of Mongolian gerbil cDNAs encoding inflammatory proteins, and their expression in glandular stomach during *H. pylori* infection. Cancer Sci 95:798–802

38. Yamaoka Y, Yamauchi K, Ota H, Sugiyama A et al (2005) Natural history of gastric mucosal cytokine expression in *Helicobacter pylori* gastritis in Mongolian gerbils. Infect Immun 73:2205–2212

39. Toyoda T, Tsukamoto T, Takasu S, Shi L, Hirano N, Ban H, Kumagai T, Tatematsu M (2009) Anti-inflammatory effects of caffeic acid phenethyl ester (CAPE), a nuclear factor-kappaB inhibitor, on *Helicobacter pylori*-induced gastritis in Mongolian gerbils. Int J Cancer 125:1786–1795

40. Sugimoto M, Ohno T, Graham DY, Yamaoka Y (2009) Gastric mucosal interleukin-17 and -18 mRNA expression in *Helicobacter pylori*-induced Mongolian gerbils. Cancer Sci 100:2152–2159

41. Sugimoto M, Ohno T, Graham DY, Yamaoka Y (2011) *Helicobacter pylori* outer membrane proteins on gastric mucosal interleukin 6 and 11 expression in Mongolian gerbils. J Gastroenterol Hepatol 26:1677–1684

42. Sakai T, Fukui H, Franceschi F, Penland R et al (2003) Cyclooxygenase expression during *Helicobacter pylori* infection in Mongolian gerbils. Dig Dis Sci 48:2139–2146

43. Nozaki K, Tanaka H, Ikehara Y, Cao X et al (2005) *Helicobacter pylori*-dependent NF-kappa B activation in newly established Mongolian gerbil gastric cancer cell lines. Cancer Sci 96:170–175

44. Yanai A, Maeda S, Shibata W, Hikiba Y et al (2008) Activation of IkappaB kinase and NF-kappaB is essential for *Helicobacter pylori*-induced chronic gastritis in Mongolian gerbils. Infect Immun 76:781–787

45. Peek RM, Wirth HP, Moss SF, Yang M et al (2000) *Helicobacter pylori* alters gastric epithelial cell cycle events and gastrin secretion in Mongolian gerbils. Gastroenterology 118:48–59

46. Konturek PC, Brzozowski T, Konturek SJ, Kwiecień S et al (2003) Functional and morphological aspects of *Helicobacter pylori*-induced gastric cancer in Mongolian gerbils. Eur J Gastroenterol Hepatol 15:745–754

47. Sordal O, Waldum H, Nordrum IS, Boyce M, Bergh K, Munkvold B, Qvigstad G (2013) The gastrin receptor antagonist netazepide (YF476) prevents oxyntic mucosal inflammation induced by *Helicobacter pylori* infection in Mongolian gerbils. Helicobacter 18:397–405

48. Akopyants NS, Clifton SW, Kersulyte D, Crabtree JE et al (1998) Analyses of the *cag* pathogenicity island of *Helicobacter pylori*. Mol Microbiol 28:37–53

49. Censini S, Lange C, Xiang Z, Crabtree JE et al (1996) *cag*, a pathogenicity island of *Helicobacter pylori*, encodes type I-specific and disease-associated virulence factors. Proc Natl Acad Sci U S A 93:14648–14653

50. Ohnishi N, Yuasa H, Tanaka S, Sawa H et al (2008) Transgenic expression of *Helicobacter pylori* CagA induces gastrointestinal and hematopoietic neoplasms in mouse. Proc Natl Acad Sci U S A 105:1003–1008

51. Sozzi M, Crosatti M, Kim SK, Romero J, Blaser MJ (2001) Heterogeneity of *Helicobacter pylori cag* genotypes in experimentally infected mice. FEMS Microbiol Lett 203:109–114

52. Philpott DJ, Belaid D, Troubadour P, Thiberge JM, Tankovic J, Labigne A, Ferrero RL (2002) Reduced activation of inflammatory responses in host cells by mouse-adapted *Helicobacter pylori* isolates. Cell Microbiol 4:285–296

53. Akanuma M, Maeda S, Ogura K, Mitsuno Y et al (2002) The evaluation of putative virulence factors of *Helicobacter pylori* for gastroduodenal disease by use of a short-term Mongolian gerbil infection model. J Infect Dis 185:341–347

54. Saito H, Yamaoka Y, Ishizone S, Maruta F et al (2005) Roles of *virD4* and *cagG* genes in the *cag* pathogenicity island of *Helicobacter pylori* using a Mongolian gerbil model. Gut 54:584–590

55. Ohnita K, Isomoto H, Honda S, Wada A et al (2005) *Helicobacter pylori* strain-specific modulation of gastric inflammation in Mongolian gerbils. World J Gastroenterol 11:1549–1553

56. Shibata W, Hirata Y, Maeda S, Ogura K et al (2006) CagA protein secreted by the intact type IV secretion system leads to gastric epithelial inflammation in the Mongolian gerbil model. J Pathol 210:306–314

57. Rieder G, Merchant JL, Haas R (2005) *Helicobacter pylori cag*-type IV secretion system facilitates corpus colonization to induce precancerous conditions in Mongolian gerbils. Gastroenterology 128:1229–1242

58. Franco AT, Israel DA, Washington MK, Krishna U et al (2005) Activation of beta-catenin by carcinogenic *Helicobacter pylori*. Proc Natl Acad Sci U S A 102:10646–10651

59. Cover TL, Peek RM Jr (2013) Diet, microbial virulence, and *Helicobacter pylori*-induced gastric cancer. Gut Microbes 4(6):482–493

60. Tsugane S, Sasazuki S (2007) Diet and the risk of gastric cancer: review of epidemiological evidence. Gastric Cancer 10:75–83

61. Bergin IL, Sheppard BJ, Fox JG (2003) *Helicobacter pylori* infection and high dietary salt independently induce atrophic gastritis and intestinal metaplasia in commercially available outbred Mongolian gerbils. Dig Dis Sci 48(3):475–485

62. Nozaki K, Shimizu N, Inada K, Tsukamoto T et al (2002) Synergistic promoting effects of *Helicobacter pylori* infection and high-salt diet on gastric carcinogenesis in Mongolian gerbils. Jpn J Cancer Res 93:1083–1089

63. Kato S, Tsukamoto T, Mizoshita T, Tanaka H et al (2006) High salt diets dose-dependently promote gastric chemical carcinogenesis in *Helicobacter pylori*-infected Mongolian gerbils associated with a shift in mucin production from glandular to surface mucous cells. Int J Cancer 119:1558–1566

64. Gamboa-Dominguez A, Ubbelohde T, Saqui-Salces M, Romano-Mazzoti L et al (2007) Salt and stress synergize *H. pylori*-induced gastric lesions, cell proliferation, and p21 expression in Mongolian gerbils. Dig Dis Sci 52:1517–1526

65. Gaddy JA, Radin JN, Loh JT, Zhang F et al (2013) High dietary salt intake exacerbates *Helicobacter pylori*-induced gastric carcinogenesis. Infect Immun 81:2258–2267

66. Pra D, Rech Franke SI, Pegas Henriques JA, Fenech M (2009) A possible link between iron deficiency and gastrointestinal carcinogenesis. Nutr Cancer 61:415–426

67. Noto JM, Gaddy JA, Lee JY, Piazuelo MB et al (2013) Iron deficiency accelerates *Helicobacter pylori*-induced carcinogenesis in rodents and humans. J Clin Invest 123:479–492

68. Tan S, Noto JM, Romero-Gallo J, Peek RM Jr, Amieva MR (2011) *Helicobacter pylori* perturbs iron trafficking in the epithelium to grow on the cell surface. PLoS Pathog 7(5), e1002050

69. Noto JM, Lee JY, Gaddy JA, Cover TL, Amieva MR, Peek RM Jr (2015) Regulation of *Helicobacter pylori* virulence within the context of iron deficiency. J Infect Dis 211:1790–1794

Chapter 25

A Rapid Screenable Assay for Compounds That Protect Against Intestinal Injury in Zebrafish Larva

Jason R. Goldsmith, Sarah Tomkovich, and Christian Jobin

Abstract

This chapter describes a method to assay compounds modulating NSAID-induced intestinal injury in zebrafish larvae. The assay employs the NSAID glafenine, which causes intestinal epithelial cell damage and death by inducing organelle stress responses (endoplasmic reticulum and mitochondrial) and blocking the unfolded protein response pathway. This epithelial damage includes sloughing of intestinal cells into the lumen and out the cloaca of the zebrafish larvae. Exposing larvae to acridine orange highlights this injury when visualized under fluorescence microscope; injured fish develop intensely red-staining intestines, as well as a "tube" or cord of red color extending through the intestine and out the cloaca. Using this rapid visually screenable method, various candidate compounds were successfully tested for their ability to prevent glafenine-induced intestinal injury. Because this assay involves examination of larval zebrafish intestinal pathology, we have also included our protocol for preparation and analysis of zebrafish histology. The protocol includes numerous steps to generate high-quality zebrafish histology slides, as well as protocols to establish accurate anatomic localization of any given tissue cross-section-processes that are made technically difficult by the small size of zebrafish larvae.

Key words Zebrafish, Intestinal injury, NSAIDs, ER stress, UPR, Drug screen

1 Introduction

An important component of intestinal homeostasis is the maintenance of a functional barrier composed of a single layer of intestinal epithelial cells (IEC) that separates the host from the highly antigenic luminal milieu [1]. Conditions leading to an impaired mucosal barrier function are diverse and include erosive gastritis and enteritis from medications (non-steroidal anti-inflammatory drugs or NSAID use being the most common culprit [2]), radiation exposure [3], ischemic episodes [4], and inflammatory bowel diseases (IBD) [1, 5]. Notably, genes that disrupt the unfolded protein response (UPR) and autophagy following cellular stress have recently been implicated in the pathogenesis of IBD [6, 7]. The disease etiology remains unclear; however, genetic evidence and experimental models suggest that intestinal cell death due to

Andrei I. Ivanov (ed.), *Gastrointestinal Physiology and Diseases: Methods and Protocols*, Methods in Molecular Biology, vol. 1422, DOI 10.1007/978-1-4939-3603-8_25, © Springer Science+Business Media New York 2016

improper responses to endoplasmic reticulum (ER) stress leads to impaired barrier function [8]. Because of the importance of the epithelium in maintaining intestinal homeostasis, the identification of compounds that promote epithelial restitution could lead to new therapeutic strategies for IBD, as well as NSAID-induced gastroenteritis, chemotherapy-induced diarrhea, and ischemic colitis.

Zebrafish (*Danio rerio*) possess several features that make it an attractive model for in vivo investigation of intestinal injury. They are transparent through early adulthood, allowing for the use of in vivo imaging techniques, and the anatomy and physiology of their digestive tract is similar to mammals, with a pancreas, liver, gall bladder, and intestine [9–11]. The zebrafish intestinal epithelium displays proximal-distal functional specification and contains most of same cell lineages found in mammals including absorptive enterocytes, goblet cells, and enteroendocrine cells [9, 11]. Their digestive tract develops rapidly to permit feeding and digestive function by 5 days post-fertilization (dpf) and zebrafish also possess innate and adaptive immune systems homologous to those of mammals [12, 13], enabling interrogation of host–diet–microbiome interactions [14], especially given the development of gnotobiotic zebrafish [12].

This model employs the NSAID glafenine to induce intestinal injury in zebrafish larvae, and then uses the vital dye acridine orange (AO) to visualize this injury. NSAIDs are known to disrupt the intestinal epithelium, leading to ulceration and inflammation in both humans and mice [2, 15]. Additionally, zebrafish have homologs for both cyclooxygenase (COX) isoforms (the targets of NSAIDs) that function similarly and display the same responses to prototypical pharmacological inhibitors as seen in mammals [16]. In this model, injury is thought to occur through enhancement of ER stress, with blockade of downstream compensatory pathways, specifically the unfolded protein response (UPR) [17]. This injury leads to apoptosis of the IECs, a fatal effect known to be caused by other NSAIDs [18]. These dead cells can be easily visualized using AO. The model of injury is responsive to several intestinal-protective compounds, including the long-acting prostaglandin dmPGE2, R-spondin (a β-catenin activator via Lgr5), and the caspase inhibitor Q-VD-OPh [17]. Co-administration of the NSAID and an intestinal-protective drug lead to a lack of positive AO signal [17], making this model appealing for rapid drug screens. Using our drug screen, we found that the mu opioid agonist DALDA prevented intestinal injury, with further investigation demonstrating a mechanism of action involving re-establishment of the UPR [17].

The protocol involves rearing zebrafish larvae in the standard fashion to 5.5–6 days post-fertilization. Administration of glafenine and any therapeutic interventions occur concurrently for 12 h. Acridine orange is then statically added to the zebrafish culture dish and then the larvae are visualized using a fluorescence microscope with a dsRed filter. In average trained hands, six compounds per hour can be screened with sufficient power to achieve statistical significance (*see* **Note 1**).

2 Materials

2.1 Reagents

1. Instant Ocean Stock solution: a sea salt aquarium mix, available from www.instantocean.com, product SS15-10.

2. Bullseye 7.0: a pH regulator for aquariums, manufactured by Wardley Wellness, and available from numerous online vendors, including Amazon® and Petco®.

3. Gnotobiotic Zebrafish Medium (GZM): Mix 7.5 mL Instant Ocean stock solution and 1.25 mL Bullseye 7.0, adjust to 1 L with dH₂O and sterilize by autoclaving [19]. Store at room temperature.

4. 0.017 % Tricaine: 0.017 % tricaine (w/v) in sterile GZM media, filtered with a 0.22 μm filter. Store at room temperature.

5. 0.083 % Tricaine: 0.083 % tricaine (w/v) in sterile GZM media, filtered with a 0.22 μm filter. Used for euthanization. Store at room temperature.

6. Acridine Orange Solution, 20 mg/mL: dissolve 20 mg of Acridine orange in 1 mL of sterile Phosphate-Buffered Saline (PBS, pH 7.2) (*see* **Note 2**). Store at –20 °C covered in an Eppendorf tube-covered aluminum foil.

7. Glafenine Solution, 50 mM: dissolve 20.5 mg glafenine hydrochloride in 1 mL of DMSO (*see* **Note 2**). Store at –20 °C.

8. 3 % methylcellulose: Dissolve 3 g of methylcellulose in 100 mL of house fish water (*see* **Note 3**), place on a rocker overnight to dissolve (*see* **Note 4**). Aliquot into 50 mL tubes and freeze the solution until needed. Store the solution in a light-opaque tube, or cover the tube in aluminum foil at room temperature.

9. 4 % paraformaldehyde in PBS: Add 4 g of paraformaldehyde to 100 mL of pre-made sterile PBS (pH 7.2), heat gently until the paraformaldehyde is dissolved and then store at 4 °C to cool and until ready for use. The solution should be made the same day it is used.

10. Agarose, low-melting point.

2.2 Equipment and Supplies

1. Non-sterile incubator capable of maintaining fish at 28.0 ± 0.5 °C. Does not need an extra water bath for humidity or precise CO_2-tension.

2. Fluorescence stereomicroscope with a dsRed filter and camera for image capture.

3. Light microscope of choice for H&E histology.

4. Metal dissection probe.

5. Water bath, set to 60 °C.

6. Aluminum alloy block (typically used to hold eppendorf tubes in a sand block heater), approximately $4 \times 4 \times 10$ cm in size. The exact size and specifications are not important; rather the block must be easily heated and moved around in a water bath and it must have a flat side large enough for a $15 \times 15 \times 5$ mm plastic cryomold to rest on it.

7. Disposable $15 \times 15 \times 5$ mm cryomold.

8. Cloth towel.

9. Small forceps.

10. Glass pipette

11. Razor blade or scalpel

3   Methods

3.1   Zebrafish Husbandry

These protocols uses wild-type TL strain zebrafish (*see* **Note 5**) maintained under a 14 h light/10 h dark cycle at a constant temperature of 28.0 ± 0.5 °C using standard, published protocols for zebrafish husbandry [20].

1. For each experiment, setup breeding pairs using standard protocols [20].

2. Place fertilized eggs into 10 cm diameter petri dishes at a density no greater than 50 eggs/plate using standard house media to grow until ready for the experiment.

3.2   Zebrafish Plating

This assay is performed on zebrafish that are 5.5–6 days post-fertilization (dpf).

1. The morning before the experiment (*see* **Note 6**), anesthetize zebrafish with 1–2 drops of tricaine solution and transfer them to a new 10 cm diameter petri dish containing 20 mL of GZM, to a maximum density of 20 fish/plate (*see* **Note 7**).

2. Store the fish in a 28.0 ± 0.5 °C incubator used for experimental plates of fish for the remainder of the experiment (*see* **Note 8**).

3.3   Glafenine and Drug Administrations

For this assay, both glafenine (the intestinal-injury-inducing compound) and all therapeutics are administered simultaneously, with 12 h of incubation time.

1. Add 10 μL of glafenine stock to each 20 mL plate, for a final concentration of 25 μM. After administration, swirl the plate in both the clockwise and counter-clockwise directions to ensure adequate mixing of the glafenine into the fish media.

2. Add other drugs as desired [17]; dosing volumes should be kept to under 20 µL per compound to minimize effects/toxicity from the drug vehicles. Swirl each plate after the administration of any compound, as above.

3. Create appropriate plates of zebrafish with isovolumetric vehicle controls (*see* **Note 9**).

3.4 Assessment of Intestinal Injury via Acridine Orange "Tube Assay"

1. After the 12 h incubation, remove the plates with the zebrafish from the incubator and add 1 µL of acridine orange solution to each plate (*see* **Note 10**), for a final concentration of 1 µg/mL, using the same swirling protocol as described above to mix in the reagent. The fish should be exposed to acridine orange for at least 10 min before proceeding to the next step.

2. Anesthetize the fish with 0.017 % tricaine and mount them in 3 % methylcellulose (*see* **Note 11**) for visualization using the Leica fluorescence stereomicroscope with a dsRed filter. All fish from one plate should be mounted simultaneously and oriented in the same manner (*see* **Note 12**).

3. Analyze fish by eye one-by-one to determine if a "fluorescent tube" is present (*see* **Note 13**). A zebrafish larvae is considered positive if there is an AO-positive tube extending from the intestine out of the cloaca, or if a strong, red signal is present in intestinal segment 2 (the distal half of the intestine). Figure 1 shows an image of a zebrafish with both of these traits, as well as an uninjured fish for comparison. Any images can also be captured if desired, provided the microscope has a camera.

3.5 Zebrafish Larvae Fixation for Histology

1. To prepare the larvae for histology, euthanize 5.5–6 dpf zebrafish with 0.083 % tricaine and quickly fix in 4 % paraformaldehyde in PBS overnight at 4 °C, typically in a 15 mL conical tube.

2. After fixing overnight in paraformaldehyde (minimum of 12 h, but no more than 24), transfer larvae to 70 % ethanol for another 24 h at 4 °C.

3.6 Paraffin Embedding of Zebrafish Larvae Tissue

Because of their small size, zebrafish cannot simply be embedded with paraffin as they will be lost in most embedding machines. Thus, they must first be embedded in 1 % agarose, and then this agarose block can go through the typical paraffin-embedding process like normal tissue. Embedding zebrafish in 1 % agarose is similar in many ways to mounting them in methylcellulose, but must be done with some expedience before the agarose has a chance to cool.

1. Make fresh 1 % agarose (low-melting point) in PBS (w/v, pH 7.2); once the solution is heated, keep it stored in its flask in the 60 °C water bath (*see* **Note 14**). Do not proceed until the solution has cooled to reach a temperature of 60 °C, as higher temperatures could damage the tissue.

Uninjured zebrafish

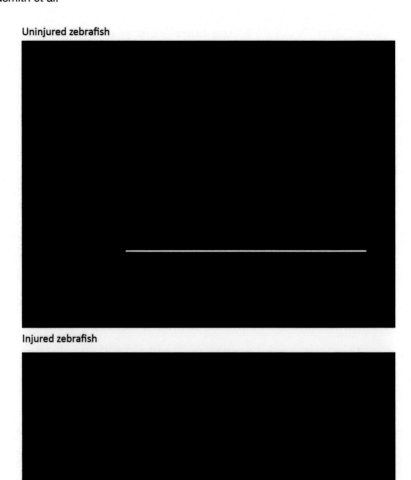

Injured zebrafish

Fig. 1 Uninjured and injured zebrafish stained with acridine orange. The uninjured fish represents what is typically seen on control or in a zebrafish successfully protected from injury by a drug. In the injured zebrafish, *white arrows* indicate contiguous areas of positive red staining, extending from segment 2 of the intestine to a tube extruding out of the cloaca of the fish. Scale bar is 1 mm

2. Remove the heated metal block from the hot water bath and quickly dry it off with a towel.

3. Place the cryomold on the metal block (mold side up) and pour agarose into the cryomold until it is 80 % full.

4. Transfer zebrafish using a glass pipette to the mold in columns in the standard orientation (*see* **Note 12**), using the dissection probe to position the fish (*see* **Note 15**). The depth of the zebrafish in a given column does not need to be uniform, but you want the fish flat on the z-axis (not angling across multiple depths). Thus, it is often easiest to push the fish to the bottom of the mold. In a given column, the zebrafish need only 1 mm of space between each fish. However, each column of fish should have a minimum of 5 mm of space between them and often 1 cm is more ideal working space to prevent movement of one fish from perturbing the position of other columns of fish. *See* Fig. 2a for an example.

5. Allow the metal block and agarose to cool in the air. Once the agarose starts to solidify, you can gently remove the cryomold from the metal block and place the block back in the water bath to reheat.

6. Carefully cut each column of fish from the cryomold using the razor blade/scalpel and forceps (Fig. 2b). Make sure there is a border of 1–2 mm of solid agarose on each edge that is cut to ensure integrity of each block.

7. Using your columns of fish, a new, final agarose block is constructed. For axial images, the column is re-oriented so that the heads are facing upwards and the tails are placed down into the cryomold. Multiple columns of fish are stacked side-by-side next to each other, and then liquid agarose is added to these stacked columns to bind them together and make a new solid block. A little space between each solid block is necessary so that the liquid agarose can flow between them and bind the strips together (Fig. 2c). Extra agarose can be added to the top of the block to ensure cohesion. For longitudinal/sagittal views (which can be included in a block of other fish oriented axially or in their own block), the column of fish are simply kept in their original orientation and placed in a new mold with the addition of liquid agarose as needed to make a new, final block.

8. Once the final blocks are formed and the agarose has solidified, carefully remove the agarose block and place it in a large histological cassette. Label the cassette and place it in a container with 70 % ethanol for 24 h.

9. Embed tissues with paraffin using your standard lab or core protocol.

3.7 Zebrafish Larvae Histology Tissue Sectioning and Orientation

Beyond the issues of scale, a second difficulty with zebrafish larvae gut histology is knowing your anatomic location when viewing the intestine; specifically if you are in segment one, two, or three. This can be sorted out by knowing how far from the cloaca any given slice of intestine is, which is the objective of this sub-protocol. This protocol assumes axially oriented zebrafish.

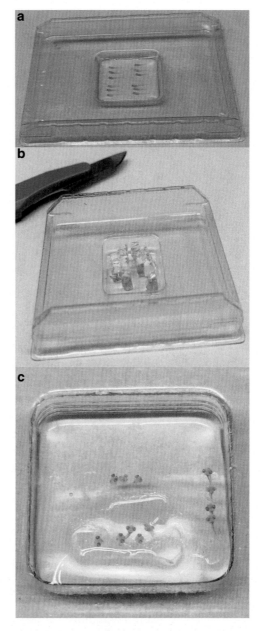

Fig. 2 Preparation of larval zebrafish for histological staining. (**a**) Fixed larvae are laid into liquid, but not hot, 1 % agarose with all heads oriented in the same direction, in parallel columns, using a cryomold. (**b**) The agarose is allowed to cool and these columns of fish are cut out. (**c**) The rows of fish are reoriented head-up in a new cryomold and more agarose is added to form the final block

1. Start sectioning each block of zebrafish from the head. Take serial 10 μm slices, with each slide holding three slices of tissue. Stain every third slide, including the first, with H&E. Thus, each set of three slides (1 with H&E staining and 2 unstained) is

90 μm of tissue advancement. Section each block until all tissue is captured on slides. Make sure all slides are serially numbered.

2. Using the light microscope, screen your H&E slides to find a slide that captures all (or most) of one column of zebrafish with all their heads in the field of view. Make a note that captures the slide number, the relative orientation of this row of fish relative to the tissue block, and assign a number (or letter) to each fish.

3. Now skip forward about 750 μm of slides and find the same row of fish. You should be seeing images that capture the intestine at this point and are likely in segment 2. More slowly, scan caudally with your H&E slides until you find the last slide for any given fish that has the intestine captured in the H&E stain. On your note/diagram from step 2, write this slide number down. Now, for any given fish, a slide 200–300 μm proximal to this slide number will be in segment 2 (*see* **Note 16**). Remember that each section of each H&E slide is 90 μm apart from the corresponding section of another H&E slide. Thus, going to the third H&E slide proximal to this "last intestine" slide will get you to the caudal portion of segment 2.

4. Analyze the slides for intestinal damage (*see* Fig. 3 for examples of uninjured—control or treated—fish and injured fish); you can use the scoring system that we validated during the development of this method [17], or any other intestinal scoring system. Proper scoring involves a blinded scorer. For IHC or immunofluorescence staining, standard protocols apply, although there may be some limitations given the paucity of antibodies available for zebrafish studies. Of note, 5-ethynyl-2′-deoxyuridine (EdU) staining can be performed using a final EdU concentration of 15 μg/mL and an exposure time of ~12 h.

4 Notes

1. Because this model is designed to be a rapid screen, promising compounds typically need verification of protective effects. One of the most powerful methods to confirm tissue injury is histological assessment, and a protocol for zebrafish larvae histology preparation is included. Because of their small size, orientation of the zebrafish for histological sectioning can be difficult. Our method involves the use of 1 % agarose as a temporary mounting media to help orient the zebrafish larvae prior to paraffin embedding. A second difficulty is accurate localization of a given slide—specifically if a cross-sectional slide is in segment 1, 2, or 3 of the intestine. While there are anatomical landmarks (specifically the liver) that can help differentiate between proximal segment 1 and more distal

Uninjured zebrafish

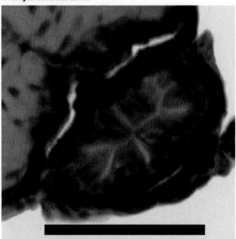

Injured zebrafish

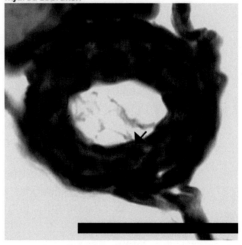

Fig. 3 Histology of uninjured and injured zebrafish. *Arrow heads* point to apical, disorganized enterocyte nuclei that are typical of intestinal damage in this model [17]. More rarely, sloughing intestinal epithelial cells can be seen (not shown) [17]. Scale bar is 12.5 μm

segments, distinguishing between various segments of the intestine by eye alone can be challenging. Thus, it is important to ensure that any observed changes between treatment groups are due to a therapeutic effect and not simply an incorrect comparison between different intestinal segments.

2. Keep frozen at –20 °C between uses, withstands multiple freeze–thaw cycles.

3. This is the circulating fish water used in the zebrafish aquarium.

4. You can also heat the solution to 60 °C and then add the methylcellulose to help it dissolve.

5. Glafenine-induced intestinal injury has also been observed in a variety of transgenic strains on the TL background [17], as well as the AB background, specifically AB-background *Tg(mpo:GFP)* zebrafish.

6. The experimental setup is designed such that chemicals are added 12 h before analysis. Typically, this is done at night. Thus, fish are plated in the morning, treated with chemicals in the evening, and then analyzed the next morning.

7. Zebrafish are best transferred using a glass pipette; they tend to stick to plastic pipettes.

8. This is a separate incubator from where the larvae are grown up prior to being treated with various agents. This prevents any accidental contamination of experimental larvae stock with experimental reagents.

9. Thus, a no-glafenine control would receive 10 μL of DMSO. A glafenine-only group would similarly receive whatever volume of vehicle that the treatment compound was dissolved in. For experiments with multiple different therapeutic interventions, simplicity can be maintained by attempting to use the same vehicle for all agents (typically sterile water, GZM media, or DMSO).

10. The acridine orange solution typically takes 10 min to thaw, so it is advisable to remove the frozen stock 10 min ahead of removing the fish from the incubator.

11. To mount fish, first use a disposable plastic pipette to place a postage stamp-sized patch of methylcellulose onto an inverted lid of a 10 cm petri dish (the raised rim of the petri dish lid is pointing upward, and can thus contain liquid). For each fish, gently transfer the anesthetized fish using a glass pipette from its petri dish into the top of the methylcellulose. The fish should be barely submerged into the methylcellulose. It should not be on top of the methylcellulose and able to slide around, nor should the fish be deeply submerged in the methylcellulose.

12. Convention in zebrafish research is to have the fish oriented head towards the left and dorsum "upwards." In other words, if the plate was a compass, the head would point towards the west and the spine would be northwards, with the belly of the fish southwards.

13. For proper data analysis, the scorer should be blinded as to the nature of each group being scored. In our studies, 20 fish per group was more the sufficient to achieve adequate power and distinguish between healthy-control, glafenine-only, and treatment groups. In our hands, healthy-controls had a false-positive tube rate of about 15 %, glafenine injured fish had a tube rate of 80–90 %, and treatment groups had tube rates of

50 % or lower. Mounting and scoring a plate of 20 fish typically took 10 min, with most of the time spent mounting and orienting (or re-orienting) the fish to get an adequate view. Tubes could be easily visualized at 3.2× magnification, and with training could be regularly identified at 2× magnification.

14. Use a ring-weight made for water baths to keep the flask of agarose steady in the water bath. Typically, we use a 125 mL Erlenmeyer flask to make 50 mL of 1 % agarose.

15. To keep things clean and prevent too much liquid runoff from occurring, we recommend that you decant all but 3 mL of the ethanol from each conical flask before transferring fish to the agarose.

16. Segment 2 extends more proximally than this by several hundred micrometers, the 200–300 μm guide just places you in the caudal portion of segment 2.

References

1. Sartor RB (2008) Microbial influences in inflammatory bowel diseases. Gastroenterology 134:577–594

2. Morteau O, Morham SG, Sellon R et al (2000) Impaired mucosal defense to acute colonic injury in mice lacking cyclooxygenase-1 or cyclooxygenase-2. J Clin Invest 105:469–478

3. Packey CD, Ciorba MA (2009) Microbial influences on the small intestinal response to radiation injury. Curr Opin Gastroenterol 26:88–94

4. Kinross J, Warren O, Basson S et al (2009) Intestinal ischemia/reperfusion injury: defining the role of the gut microbiome. Biomark Med 3:175–192

5. Williams KL, Fuller RC, Dieleman LA et al (2001) Enhanced survival and mucosal repair after dextran sodium sulfate-induced colitis in transgenic mice that overexpress growth hormone. Gastroenterology 120:925–937

6. Kaser A, Martinez-Naves E, Blumberg RS (2010) Endoplasmic reticulum stress: implications for inflammatory bowel disease pathogenesis. Curr Opin Gastroenterol 26:318–326

7. Stappenbeck TS, Rioux JD, Mizoguchi A et al (2010) Crohn disease: a current perspective on genetics, autophagy and immunity. Autophagy 7:355–374

8. Kaser A, Lee AH, Franke A et al (2008) XBP1 links ER stress to intestinal inflammation and confers genetic risk for human inflammatory bowel disease. Cell 134:743–756

9. Ng AN, de Jong-Curtain TA, Mawdsley DJ et al (2005) Formation of the digestive system in zebrafish: III. Intestinal epithelium morphogenesis. Dev Biol 286:114–135

10. Pack M, Solnica-Krezel L, Malicki J et al (1996) Mutations affecting development of zebrafish digestive organs. Development 123: 321–328

11. Wallace KN, Akhter S, Smith EM, Lorent K, Pack M (2005) Intestinal growth and differentiation in zebrafish. Mech Dev 122:157–173

12. Kanther M, Rawls JF (2010) Host-microbe interactions in the developing zebrafish. Curr Opin Immunol 22:10–19

13. Meeker ND, Trede NS (2008) Immunology and zebrafish: spawning new models of human disease. Dev Comp Immunol 32:745–757

14. Goldsmith JR, Sartor RB (2014) The role of diet on intestinal microbiota metabolism: downstream impacts on host immune function and health, and therapeutic implications. J Gastroenterol 49:785–798

15. Maiden L, Thjodleifsson B, Theodors A, Gonzalez J, Bjarnason I (2005) A quantitative analysis of NSAID-induced small bowel pathology by capsule enteroscopy. Gastroenterology 128:1172–1178

16. Grosser T, Yusuff S, Cheskis E, Pack MA, FitzGerald GA (2002) Developmental expression of functional cyclooxygenases in zebrafish. Proc Natl Acad Sci USA 99: 8418–8423

17. Goldsmith JR, Cocchiaro JL, Rawls JF, Jobin C (2012) Glafenine-induced intestinal injury in zebrafish is ameliorated by mu-opioid signaling via enhancement of Atf6-dependent cellular stress responses. Dis Model Mech 6: 146–159

18. Franceschelli S, Moltedo O, Amodio G, Tajana G, Remondelli P (2011) In the Huh7 hepatoma cells diclofenac and indomethacin activate differently the unfolded protein response and induce ER stress apoptosis. Open Biochem J 5:45–51

19. Pham LN, Kanther M, Semova I, Rawls JF (2008) Methods for generating and colonizing gnotobiotic zebrafish. Nat Protoc 3: 1862–1875

20. Westerfield M (2000) The zebrafish book. A guide for laboratory use of zebrafish (*Danio rerio*). University of Oregon Press, Eugene, OR

Part V

Animal Models of Gastrointestinal Cancer

Chapter 26

AOM/DSS Model of Colitis-Associated Cancer

Bobak Parang, Caitlyn W. Barrett, and Christopher S. Williams

Abstract

Our understanding of colitis-associated carcinoma (CAC) has benefited substantially from mouse models that faithfully recapitulate human CAC. Chemical models, in particular, have enabled fast and efficient analysis of genetic and environmental modulators of CAC without the added requirement of time-intensive genetic crossings. Here we describe the Azoxymethane (AOM)/Dextran Sodium Sulfate (DSS) mouse model of inflammatory colorectal cancer.

Key words Colitis-associated cancer, Colon cancer, AOM, DSS, Inflammatory carcinogenesis

1 Introduction

Colorectal cancer (CRC) is the fourth most common cancer in the world [1]. It is well established that colitis predisposes individuals to colorectal tumorigenesis [2–4]. Patients with inflammatory bowel disease, for example, are at an elevated risk for developing colon cancer, although the magnitude of this risk has recently come under debate [5–11]. While the molecular pathogenesis of colitis-associated cancer (*CAC*) remains incompletely understood, significant advances have been made from studying murine models of CAC. Here we outline the application of the Azoxymethane (*AOM*)/Dextran sodium sulfate (*DSS*) model of CAC. The AOM/DSS model is a powerful, reproducible, and relatively inexpensive initiation-promotion model that utilizes chemical induction of DNA damage followed by repeated cycles of colitis [12–15].

AOM (Methyl-methylimino-oxidoazanium, $CH_3N = N(\rightarrow O)$ CH_3) is a procarcinogen that is metabolized by cytochrome p450, isoform CYP2E1, converting it into methylazocymethanol (MAM), a highly reactive alkylating species that induces O^6 methylguanine adducts in DNA resulting in $G \rightarrow A$ transitions [16]. After excretion into the bile, it is taken up by colonic epithelium and induces mutagenesis. DSS is a heparin-like polysaccharide that is dissolved in the drinking water and inflicts colonic epithelial

Andrei I. Ivanov (ed.), *Gastrointestinal Physiology and Diseases: Methods and Protocols*, Methods in Molecular Biology, vol. 1422, DOI 10.1007/978-1-4939-3603-8_26, © Springer Science+Business Media New York 2016

damage, inducing colitis mimicking some of the features of IBD [17]. Combining AOM and DSS provides a two-step tumor model of CAC.

Key features of the AOM/DSS model include its relatively short timeline and accurate modeling of CAC. Tumor development can occur in as short as 10 weeks [12]. Moreover, the histopathology of AOM/DSS-induced tumors recapitulates key facets of human CAC such as distally located tumors and invasive adenocarcinomas [13]. Application of the AOM/DSS model has been critical in unraveling the pathogenesis of CAC: from the role of signaling pathways (e.g. Toll-like receptor 4, IKKβ, and IL-6 [18–20]) and antioxidant machinery (e.g. glutathione peroxidase [21]) to the influence of the microbiota [22] and transcriptional corepressors (e.g. Myeloid translocation genes [23]). Thus, the AOM/DSS model is a powerful platform to employ when studying the pathogenesis of inflammatory colorectal cancer.

2 Materials

1. Azoxyemethane solution: 1 mg/ml. Dissolve 10 mg of AOM (Sigma-Aldrich, Cat# A4586) in 10 ml of sterile Phosphate-buffered saline (PBS). Filter the solution using a 0.45 μm cellulose acetate filter and aliquot into 1 ml sterile Eppendorf tubes. Aliquots can be stored at −20 °C for up to a year.

2. 0.5 ml Tuberculin Syringe with $28^{1/2}$ G needle.

3. Dextran Sodium Sulfate solution: 3 % (w/v). Weigh 30 g of DSS (Affymetrix Cat# 14489, MW 40–50 kDa) and dissolve into 1 l of water. Once the DSS is dissolved, filter-sterilize the solution using 0.45 μm cellulose acetate filter.

4. Scale for weighing mice (Model SP402, Ohaus Scout Balance).

5. 10 % Buffered Formalin.

6. 70 % Ethanol. Dilute 190 proof ethanol to 70 % ethanol with sterile, deionized water.

7. Isoflurane, USP (Phoenix Pharmaceuticals, Inc.).

8. Tissue Pathology Macrosette Cassettes.

9. 20 G Straight feeding needle.

10. Dissection Scissors Sharp/Blunt Tip (VWR International, Cat# 82027-588).

11. Waugh Forceps (VWR International, Cat# 82027-428).

12. $27^{1/2}$ G Precision Glide Needle.

13. Whatman Blotting Paper.

14. Carbon Fiber Composites Digital Caliper.

15. Nalgene Surfactant-Free Cellulose Acetate (SFCA) Filter (Cole-Palmer, Cat# EW-06731-2).

16. RNA*later* solution (Life Technologies).

17. RIPA buffer (Thermo Scientific).

18. Sterile Phosphate-Buffered Saline (PBS).

3 Methods

3.1 Treating mice with AOM/DSS (See Fig. 1 for an Example of a Typical Experimental Timeline)

Day 1: Injecting Azoxymethane

1. Ensure experimental groups are age- and gender-matched with control mice (*see* **Note 1**). Weigh 8- to 12-week old C57BL/6 (*see* **Note 2**) mice and record weights. Accurate weights are required in order to ensure uniform dosing of AOM (*see* **Note 3**). We recommend weighing each mouse three times to increase precision. Calculate the volume of AOM (1 mg/ml) to inject to achieve a dose of 12 mg/kg. For example: a 25 g mouse would receive a 300 µl injection of 1 mg/ml AOM solution. It may be necessary to reduce the dosage if substantial toxicity is observed (*see* **Note 4**).

2. Once you have recorded weights and injection volumes, anesthetize mice using isoflurane in accordance with your institution's IACUC protocols. Using a $28^{1/2}$ G tuberculin syringe, inject each mouse intraperitoneally with the appropriate volume of AOM.

3. Place the mice back in their cages. Weigh and monitor them over the next 48 h.

4. If your mouse facility provides lixit drinking valves or other automatic watering systems, be sure to cap or disengage this water supply to ensure each cage only has one water supply. It is important for mice to become accustomed to drinking only from a water bottle, as this will be the source of DSS (*see* **Note 5**). We recommend disengaging automatic watering systems for the duration of the experiment (*see* **Note 6**).

Day 3: Start DSS cycle 1

5. Replace drinking water in cages with 3 % DSS formula.

Day 3–8: Monitoring Animals: DSS cycle 1

6. Weigh mice daily to evaluate response to DSS-induced colitis.

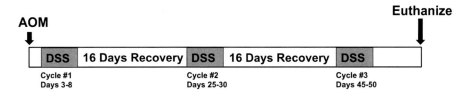

Fig. 1 Schematic timeline for AOM/DSS-induced inflammatory carcinogenesis

7. During treatment with DSS, mice can lose significant body weight depending on strain and genotype (Fig. 2). If mice lose substantial body weight (between 10 and 20 % weight loss relative to the day prior to DSS administration), it is advisable to administer up to 1 ml of sterile saline via IP injection or provide wet food (*see* **Note 7**).

8. If mice lose greater than 20 % body weight, demonstrate hunched posture, or move in a limited fashion, then it may be necessary to euthanize the animal. Be sure to follow all appropriate IACUC protocols.

Day 8: End DSS Cycle 1

9. Replace 3 % DSS with sterile drinking water.

Day 9–12: Initial Recovery

10. It is important to continue to monitor the mice, especially in the 3–5 days after replacing the 3 % DSS with water. It is not unusual for mice to continue to lose weight several days after 3 % DSS administration.

Day 13–24: Recovery

11. Weigh mice every 2–3 days.

Day 25: Start DSS cycle 2

12. Replace water with 3 % DSS and weigh mice daily.

Day 25–30: Monitoring Animals: DSS cycle 2

13. Monitor and weigh mice as exactly as detailed in Day 3–8.

Day 30: End DSS cycle 2

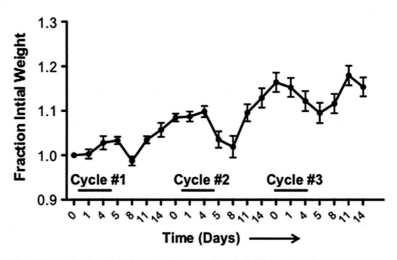

Fig. 2 Example of weight loss during repeat cycle DSS treatment

14. Replace 3 % DSS with water.

15. Tumor burden can be safely monitored via endoscopy (Fig. 3a) throughout the duration of the experiment. We recommend using endoscopy 1 week after completion of the second cycle of DSS (*see* **Note 8**).

Day 30–44: Recovery

16. Weigh mice every 2–3 days.

Day 45: Start DSS cycle 3

17. Replace drinking water in cages with 3 % DSS formula.

Day 45–50: Monitoring Animals: DSS cycle 3

18. Monitor and weigh mice as exactly as detailed in Day 3–8.

Day 50: End DSS Cycle 3

19. Replace 3 % DSS with sterile drinking water.

Day 51–65: Recovery

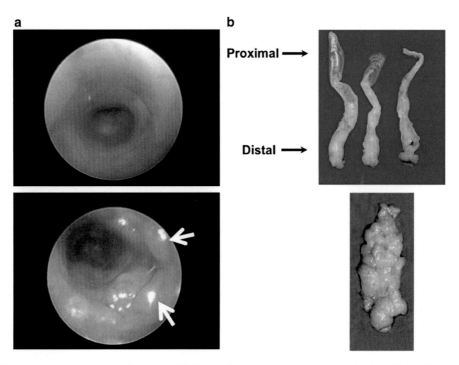

Fig. 3 Endoscopic analysis of murine colon. (**a**) *Above*: Endoscopy image of normal colon. *Below*: Endoscopy of colon after AOM injection followed by two cycles of DSS. *White arrows* indicate tumors. (**b**) *Above*: Colons harvested and oriented with the distal end toward the dissector. *Below*: Example of gross tumor burden. Images are reproduced from prior report [23]

3.2 Sacrificing Mice

1. Weigh mice before euthanizing. Euthanize mice by a combination of inhalational isoflurane overdose and cervical dislocation or other institutionally, IACUC-approved protocols. Expose the ventral side of the mouse by placing the mouse on a surgical dissection table with its abdomen facing up. Secure legs for unobstructed access to the abdomen. Cover the abdomen with 70 % ethanol to prevent fur from interfering with dissection.

2. Using forceps pinch and pull the abdomen up at the midline (thus forming a "tent"). Using scissors incise the pinched abdominal tissue to access the peritoneum. Then extend the incision to the xyphoid process at the midline (away from the dissector) and to the costal margins bilaterally (toward the dissector). Gently push peritoneal fat and small intestine to the side and locate the cecum (*see* **Note 9**).

3. Once the cecum is identified, cut immediately distal to isolate proximal colon. Follow the colon using forceps and gently dissect away the mesentery. Cut through the pelvis to allow removal of the distal colon including the anus (*see* **Note 10**). Because DSS-induced colitis damages the distal colon, it is critical to remove the entire colon to accurately assess tumor burden.

4. Flush the colon with PBS using a 10 ml syringe. Place the colon lengthwise on Whatman paper (Fig. 3b) with the distal end (anus end) nearest to the dissector and the proximal (cecum end) furthest away. Cut the colon longitudinally along the proximal–distal axis so that the colon is splayed open length-wise and the distal most portion of the colon is located nearest to the dissector.

5. Assess and record tumor burden grossly. Tumor size can be measured using digital calipers. If desired, isolate tumor tissue or adjacent tissue for RNA or protein analysis using a scalpel. Place tissue for RNA analysis directly into RNA*later*; for protein analysis, place tissue directly in lysis buffer with protease and phosphatase inhibitors. We recommend doubling the normal concentration of protease and phosphatase inhibitors to preserve protein integrity (*see* **Note 11**).

6. Using two fine-tipped forceps held in two hands, grasp both lateral sides of the distal edge of the colon and roll the colon. The end product should be a rolled colon resembling a Swiss roll and the distal colon will be in the center and the proximal colon will be the outermost layer (Fig. 4 and *see* **Note 12**). Once the colon is rolled, place a 27½ G needle through the roll to secure it. Place the Swiss-rolled colon into a labeled tissue cassette.

7. Immerse the cassette into 10 % buffered formalin for 24 h.

8. Process samples for histological analysis according to your lab's preferred method (*see* **Note 13**).

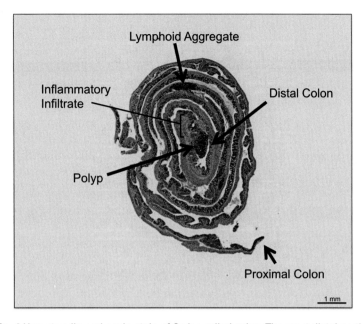

Fig. 4 Hematoxylin and eosin stain of Swiss-rolled colon. The most distal portion of the colon is located in the innermost segment of the roll. A large polyp can be seen in the distal colon surrounded by inflammatory infiltrate

4   Notes

1. If possible, it is ideal to use control and experimental mice that are littermates. This will control for any environmental differences, including microbiota variability. If this is not possible, then it is strongly recommended that the control and experimental mice be *housed* in the same room.

2. Different strains of mice will have different sensitivities to AOM/DSS treatment. Tumor penetrance and multiplicity as well as colitis damage can all vary based on strain [15]. Thus, it is critical to adjust your AOM and DSS doses according to your genetic strain.

3. As mentioned, AOM is metabolized by cytochrome p450, isoform CYP2E1 [16]. Consider the possibility that if using genetically modified mice and the gene is expressed in the liver, activation of AOM may be impaired thus confounding the results.

4. AOM concentration is an important variable to adjust. Depending on the lab, mouse facilities, and mouse background, AOM can have varying effects. In preparation for your experiment, we suggest performing trial experiments using three different doses of AOM (7.5, 10, and 12.5 mg/kg) on three different cohorts (3–5 mice) of wild-type mice. Mortality from AOM is often observed between 24 and 72 h after the injection.

5. Consistent DSS dosing is critical and the volume of DSS-containing water should be monitored to ensure uniform exposure across all cages. This can be done by measuring the initial volume of 3 % DSS placed into the cages on Day 1 and measuring the final volume on Day 5 before replacing with water.

6. We recommend that the investigator disengage automatic watering systems or cap lixit valves for all of the mice to be used as soon as they are weaned. This allows them to become accustomed to only one water source.

7. If mice lose significant body weight, they will often become too weak to access their water or food supply. If a mouse demonstrates signs of discomfort or weakness such as hunched posture, lethargy, or decreased grooming as indicated by soiled or rough hair coat, we recommend administering up to 1 ml of sterile saline by IP injection after weighing the animal. In our experience, this is an insufficient volume to affect mouse weights 24 h later. Alternatively, wet chow is a good way to provide food and hydration. A medium-sized weigh boat can be filled with standard rodent chow, soaked in water for 30 s, and then drained and placed in the cage.

8. Endoscopy can be performed to visualize tumor incidence during the experiment. We recommended conducting this 1 week after completion of the second cycle of DSS (Fig. 3a). This allows sufficient time for tumor development. Moreover, allowing the mice to recover for 1 week after DSS reduces inflammation, making tumors more visible. In our experience, mice do not need to be given an oral purgative or laxative to evacuate the colonic contents. When performing endoscopy, encountered stool can be gently pushed toward the proximal regions of the colon, so as not to obscure the luminal view. If obstruction persists, 1 ml of sterile PBS can be administered as an enema to expel contents.

9. The cecum is the junction at which the small intestine ends and the colon begins. The cecum can be easily identified as a large intestinal pouch containing stool and located in the right lower quadrant of the mouse.

10. AOM/DSS produces tumors primarily located in the distal colon (see Fig. 3b for an example of a distal colon with a high tumor burden). Thus, it is critical to cut through the pelvis in order to remove the entire colon. This will allow you to isolate the colon with the anus intact and provide the most accurate accounting of tumor number.

11. When isolating tissue from AOM/DSS-treated colons, it is important to work as quickly as possible to preserve tissue integrity. Preparation of all reagents and recording documents should be performed prior to sacrificing the mice. When har-

vesting tumor or colonic tissue, place the tissue directly into 350 µl of RNA*later* in a pre-labeled eppendorf tube for RNA analysis. For protein analysis, we recommend preparing 500 µl of RIPA buffer with twice the amount of protease and phosphatase inhibitors as recommended. When isolating tissue for RNA or protein, it is imperative to place the tubes immediately on ice. As soon as you have completed sacrificing the mouse, recording tumor number and size, and rolled the colon, place the preserved tissue into –80 °C (for protein) or –20 °C (for RNA*later*).

12. While Swiss rolling can be technically challenging, especially in the presence of tumors, it is important to roll colons properly in order to obtain well-oriented samples for histological analysis. A video demonstration of proper rolling technique is available if needed [24].

13. We recommend Hematoxylin and eosin (H&E) staining to assess crypt and tumor pathology (Figs. 4 and 5). H&E stain-

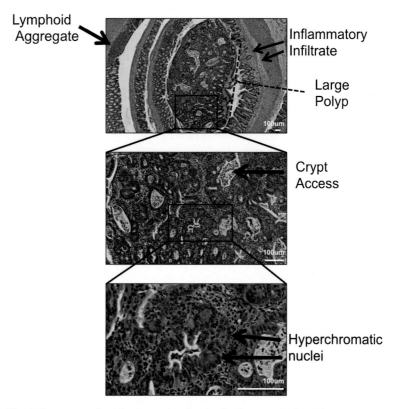

Fig. 5 Representative histology of a distal colonic tumor isolated from a mouse treated with the AOM/DSS protocol. Pathology features indicating injury include inflammatory infiltrates that are observed at low power (*top panel*) and intratumoral crypt abscesses (*middle panel*). Features of neoplasia such as hyperchromatic nuclei and increased nuclei/cytoplasmic ratios are identified at higher magnification (*middle* and *bottom panels*)

ing of a well-aligned colon rolls allows microscopic examination of tumors and their location within the colon. The severity of inflammatory injury can be observed with identification of inflammatory infiltrates and presence of crypt abscess. Tumor pathology such as hyperchromatic nuclei, increased nuclei/cytoplasmic ratio, and eccentric nuclei placement are all features of dysplasia and can be seen in the representative images.

Acknowledgments

This work was supported by the National Institutes of Health grants DK080221 (C.S.W.), 1F30DK096718-01A1 (B.P.), T32 GM07347 (NIH/NIGMS) (B.P.), Merit Review Grants from the Office of Medical Research, Department of Veterans Affairs 1I01BX001426 (C.S.W.), and ACS-RSG 116552 (C.S.W.).

References

1. Siegel R, Ma J, Jemal A (2014) Cancer statistics, 2014. CA Cancer J Clin 64:9–29

2. Danese S, Malesci A, Vetrano S (2011) Colitis-associated cancer: the dark side of inflammatory bowel disease. Gut 60:1609–1610

3. Danese S, Mantovani A (2010) Inflammatory bowel disease and intestinal cancer : a paradigm of the Yin – Yang interplay between inflammation and cancer. Oncogene 29:3313–3323

4. Terzić J, Grivennikov S, Karin E, Karin M (2010) Inflammation and colon cancer. Gastroenterology 138:2101–2114

5. Eaden JA, Abrams KR, Mayberry JF (2001) The risk of colorectal cancer in ulcerative colitis : a meta analysis. Gut 48:526–535

6. Ekbom A, Helmick C, Zack M, Adami H (1990) Ulcerative colitis and colorectal cancer. N Engl J Med 323:1228–1233

7. Hovde O, Kempski-Monstad I, Småstuen MC, Solberg IC, Henriksen M, Jahnsen J et al (2013) Mortality and causes of death in Crohn's disease: results from 20 years of follow-up in the IBSEN study. Gut 63:771–775

8. Beaugerie L, Svrcek M, Seksik P, Bouvier AM, Simon T, Allez M et al (2013) Risk of colorectal high-grade dysplasia and cancer in a prospective observational cohort of patients with inflammatory bowel disease. Gastroenterology 145:166–175

9. Gyde SN, Prior P, Allan RN, Stevens A, Jewell DP, Truelove SC et al (1988) Colorectal cancer in ulcerative colitis: a cohort study of primary referrals from three centres. Gut 29:206–217

10. Lakatos PL, Lakatos L (2008) Risk for colorectal cancer in ulcerative colitis: changes, causes and management strategies. World J Gastroenterol 14:3937–3947

11. Jess T, Rungoe C, Peyrin-Biroulet L (2012) Risk of colorectal cancer in patients with ulcerative colitis: a meta-analysis of population-based cohort studies. Clin Gastroenterol Hepatol 10:639–645

12. Tanaka T, Kohno H, Suzuki R, Yamada Y, Sugie S, Mori H (2003) A novel inflammation-related mouse colon carcinogenesis model induced by azoxymethane and dextran sodium sulfate. Cancer Sci 94:965–973

13. De Robertis M, Massi E, Poeta ML, Carotti S, Morini S, Cecchetelli L et al (2011) The AOM/DSS murine model for the study of colon carcinogenesis: from pathways to diagnosis and therapy studies. J Carcinog 10:9

14. Chen J, Huang XF (2009) The signal pathways in azoxymethane-induced colon cancer and preventive implications. Cancer Biol Ther 14:1313–1317

15. Suzuki R, Kohno H, Sugie S, Nakagama H, Tanaka T (2006) Strain differences in the susceptibility to azoxymethane and dextran sodium sulfate-induced colon carcinogenesis in mice. Carcinogenesis 1:162–169

16. Sohn OS, Fiala ES, Requeijo SP (2001) Differential effects of CYP2E1 status on the metabolic activation of the colon carcinogens azoxymethane. Cancer Res 61:8435–8440

17. Okayasu I, Hatakeyama S, Yamada M, Ohkusa T, Inagaki Y, Nakaya R (1999) A novel method in the induction of reliable experimental acute and chronic colitis in mice. Gastroenterology 98:694–702

18. Fukata M, Chen A, Vamadevan AS, Cohen J, Breglio K, Krishnareddy S et al (2007) Toll-like receptor-4 promotes the development of colitis-associated colorectal tumors. Gastroenterology 133:1869–1881

19. Greten FR, Eckmann L, Greten TF, Park JM, Li ZW, Egan LJ et al (2004) IKKβ links inflammation and tumorigenesis in a mouse model of colitis-associated cancer. Cell 118:285–296

20. Grivennikov S, Karin E, Terzic J, Mucida D, Yu GY, Vallabhapurapu S et al (2009) IL-6 and Stat3 are required for survival of intestinal epithelial cells and development of colitis-associated cancer. Cancer Cell 15:103–113

21. Barrett CW, Ning W, Chen X, Smith JJ, Washington MK, Hill KE et al (2012) Tumor suppressor function of the plasma glutathione peroxidase Gpx3 in colitis-associated carcinoma. Cancer Res 73:1245–1255

22. Uronis JM, Mühlbauer M, Herfarth HH, Rubinas TC, Jones GS, Jobin C (2009) Modulation of the intestinal microbiota alters colitis-associated colorectal cancer susceptibility. PLoS One 4, e6026

23. Barrett CW, Fingleton B, Williams A, Ning W, Fischer M, Washington MK et al (2011) MTGR1 is required for tumorigenesis in the murine AOM/DSS colitis-associated carcinoma model. Cancer Res 71:1302–1312

24. Whitten C, Williams A, Williams CS (2010) Murine colitis modeling using dextran sulfate sodium. J Vis Exp 35:5–8

Chapter 27

Characterization of Colorectal Cancer Development in *Apc*^{min/+} Mice

ILKe Nalbantoglu, Valerie Blanc, and Nicholas O. Davidson

Abstract

The $Apc^{min/+}$ mouse provides an excellent experimental model for studying genetic, environmental, and therapeutic aspects of intestinal neoplasia in humans. In this chapter, we will describe techniques for studying colon cancer development in $Apc^{min/+}$ mice on C57BL/6J (B6) background, focusing on the roles of environmental modifiers, including Dextran Sulfate Sodium (DSS), high fat diet, and bile acid supplementation in the context of experimental colorectal cancer. This chapter also includes protocols describing extraction and purification of DSS-contaminated RNA, as well as sampling, harvesting, and tissue processing. The common pathologic lesions encountered in these animals are described in detail.

Key words $Apc^{min/+}$, Colorectal cancer, Adenoma, Dextran sulfate sodium, High fat, Bile acids

1 Introduction

Colorectal cancer (CRC) is the third most commonly diagnosed cancer as well as the third most common cause of death from cancer in the United States [1]. An estimated 132,700 men and women will be newly diagnosed with the disease and 49,700 will die from colorectal cancer in 2015 [2]. Approximately 30 % of colorectal cancers have a familial or genetic component, yet less than 10 % are related to well-defined syndromes [3]. In addition to genetic alterations, several environmental factors, including consumption of red meats, high fat diet, obesity, insulin resistance [4], chronic inflammation [5], and smoking [6] have been associated with a higher risk for developing sporadic colorectal cancer.

Among the pathways to colorectal cancer is the adenoma-carcinoma sequence. Adenomas develop both in humans and mice as a result of chromosomal instability initiated by inactivation of APC (Adenomatous Polyposis Coli), a tumor suppressor gene located on chromosome 5q [7] and which is mutated in the overwhelming majority of sporadic CRC [8]. The APC gene is an essential component of β-catenin and Wnt signaling pathways,

Andrei I. Ivanov (ed.), *Gastrointestinal Physiology and Diseases: Methods and Protocols*, Methods in Molecular Biology, vol. 1422, DOI 10.1007/978-1-4939-3603-8_27, © Springer Science+Business Media New York 2016

regulating cell adhesion, differentiation, polarity, migration, and apoptosis [7]. Germline APC mutations cause FAP (Familial Adenomatosis Polyposis), making this an important target to study the genetic and environmental modifiers of CRC.

The laboratory mouse has proven useful in the study of CRC [9]. The Min (Multiple intestinal neoplasia) mutant arose following treatment of mice with ethylnitrosourea (ENU) [10]. Crossing ENU-treated mice with wild-type controls yielded offspring with the min phenotype. $Apc^{min/+}$ is an autosomal dominant allele characterized by development of multiple intestinal tumors and anemia [11]. $Apc^{min/+}$ mice in the C57BL/6J (B6) genetic background develop more than 50 tumors by 90 days of age, mostly throughout the small intestine, and although these tumors rarely become invasive, they cause death around 150 days of age as a result of intestinal obstruction and anemia due to bleeding from the larger polyps [11]. Other models, including conditional APC mutant alleles, have also been described [12].

$Apc^{min/+}$ mice share many genetic and phenotypic similarities to humans with FAP. The mouse and human APC orthologs are approximately 90 % identical [13], but colonic tumors are more common and have much greater malignant potential in humans, whereas small intestinal tumors are more prevalent in mice [13]. Unlike the human phenotype, desmoid tumors and epidermoid cysts are rarely observed in the mouse model [13].

The $Apc^{min/+}$ mouse provides an excellent experimental model for studying genetic, environmental, and therapeutic aspects of intestinal neoplasia in humans. In this chapter, we will describe techniques for studying colon cancer development in $Apc^{min/+}$ mice on a C57BL/6J (B6) genetic background, focusing on Dextran Sulfate Sodium-induced (DSS) [14–17], high fat diet-induced [18–20], and bile acid supplementation [21, 22] as environmentally modifiable models of colorectal cancer. This chapter also includes protocols describing extraction and purification of DSS-contaminated RNA [23], as well as sampling, harvesting, and tissue processing. The common pathologic lesions encountered in these animals are described in detail.

2 Materials

2.1 Dextran Sulfate Sodium Model of Colorectal Cancer in Apc^min/+ Mice

1. $Apc^{min/+}$ mice, 4 weeks old, commercially purchased from Jackson Laboratories (Bar Harbor, ME) (*see* **Notes 1–4**).

2. Dextran Sulfate Sodium (DSS), molecular weight of 40,000 (Affymetrix): 2–2.5 % (w/w) solution in water.

3. Teklad Mouse Breeder diet, 10.5 % fat.

4. Ketamine/Xylazine cocktail: 100 µl/10 kg solution containing 8 mg/ml ketamine and 3 mg/ml Xylazine.

5. Insulin syringe, 1 ml.

6. Scale, 0.1 (g) digits.

7. BrdU solution: 18 mg/ml BrdU and 1.8 mg/ml 5 Fluoro-2deoxyuridine in deionized water.

2.2 High Fat Diet and Colorectal Cancer in *Apc^min/+* Mice

1. High fat diet, 20.5 % protein, 36 % fat, and 35.7 % carbohydrate.

2. Contour TS Glucometer.

3. Heat lamp.

4. Rodent restrainer.

5. Scalpel blade.

6. Alcohol swap.

7. Insulin Syringe.

8. Heparin.

2.3 Bile Acids and Colorectal Cancer in *Apc^min/+* Mice

1. Sodium deoxycholate ($\geq$ 97 % titration): 0.2 % (w/v) solution in drinking water.

2.4 Tissue Collection and Processing

1. Agar.

2. PBS (Phosphate Buffered Saline, 1×): 137 mM NaCl, 2.7 mM KCl, 10 mMNa$_2$HPO$_4$, 1.8 mM KH$_2$PO$_4$, pH 7.4.

3. Dissecting PenWax (1 lb is enough for ten wax boards).

4. Square dish with grid (ten dishes per bag).

5. Ethanol, 70 %.

6. Scalpel blade Scissors (5 in. surgical scissors).

7. Razor blade.

8. Pins.

9. Formalin solution, 10 % buffered.

10. Fisherbrand, Tru-Flow tissue cassette.

11. Solvent resistant cassette marking pen.

12. Eppendorf tubes (1.7 ml).

13. Liquid nitrogen.

14. Nikon Stereomicroscope.

15. MetaVue software (Molecular Devices).

16. Photometrics CooLSNAP camera (ROPER Scientific).

2.5 Magnetic Bead-Based Poly-A Purification of DSS-Exposed mRNA

1 Tissue homogenizer.

2. Trizol Reagent.

3. UV spectrophotometry or Agilent 2100 Bioanalyzer.

4. Dynabeads mRNA purification kit (Invitrogen, Life Technology).

5. Magnet Dynal:DynalMPC-S.

6. High Capacity Reverse Transcription Kit (Applied Biosystems).

7. Binding Buffer: 20 mM Tris–HCI pH 7.5, 1 M LiCI, and 2 mM EDTA.

8. Washing buffer: 10 mM Tris–HCI pH 7.5, 0.15 M LiCI, and 1 mM EDTA.

2.6 Protein Extract Preparation

1. Tissue lysis buffer (TLB): 20 mM Tris pH 7.5, 1 mM sodium vanadate, 150 mM NaCl, 2 mM EDTA, 100 mM sodium fluoride, 50 mM beta-glycerophosphate, 5 % glycerol, protease inhibitors cocktail.

2. Detergent buffer (10×): 10 % Triton X-100, 1 % Sodium Dodecyl Sulfate in tissue lysis buffer.

3   Methods

3.1 Dextran Sulfate Sodium Model of Colorectal Cancer in Apc$^{min/+}$ Mice

1. House 4-week-old $Apc^{min/+}$ mice (male and female) in groups and maintain on a 12 h light/dark cycle (*see* **Notes 1–4**).

2. Feed mice with Teklad Mouse Breeder diet ad libitum.

3. Weigh all animals before the start of DSS exposure.

4. Expose mice (experimental group-4 week old) to 2–2.5 % DSS in drinking water for 1 week (*see* **Notes 5–7**).

5. Follow this cycle by normal drinking water for 2 weeks.

6. Give the control group tap water ad libitum throughout the experiment.

7. Weigh animals twice a week. If multiple cycles of DSS are being administered, taking weight measurements once a week during consecutive water and DSS cycles is sufficient.

8. Record the mortality rate and the weight of the animals throughout the experiment.

9. Sacrifice animals with weight loss of 20 % or more by injecting 100 μl/10 kg of a Ketamine/Xylazine solution, followed by cervical dislocation (*see* **Note 8**).

10. Euthanize all other surviving animals at weeks 2, 3, 4, and ultimately 5 to follow the progression of the pathology.

11. Two hours before sacrification, inject 200 μl of BrdU solution (a marker of proliferation) per mouse.

12. Perform gross examination, imaging of the intestines including assessment of tumor number and size and tissue collection following sacrifice (*see* Subheading 4).

3.2 High Fat Diet and Colorectal Cancer in Apc$^{min/+}$

1. House 4-week-old $Apc^{min/+}$ mice (male and female) in groups and maintain on a 12 h light/dark cycle (*see* **Notes 1–4**).

2. Give regular tap water ad libitum.

3. Give the experimental group a high fat diet for 8 weeks ad libitum (*see* **Note 9**).

4. Give the control group Teklad breeder diet ad libitum.

5. Weigh both the experimental and control group animals weekly (*see* **Note 10**).

6. Collect blood samples from tail following 5 h fasting at 8 and 12 weeks of age for glucose concentrations (*see* **Note 11**).

7. At week 12, sacrifice all animals by injecting 100 μl/10 kg of a Ketamine/Xylazine cocktail solution followed by cervical dislocation (*see* **Note 8**).

8. At the time of sacrifice, obtain approximately 200–300 μl blood from inferior vena cava using a 1 ml heparinized syringe (*see* **Note 12**). Collect plasma by centrifugation ($1500 \times g$ for 10 min at 4 °C) and store at –80 °C (Optional).

9. Collect and weigh mesenteric, retroperitoneal, and epididymal fat pads.

10. Perform lipidomic profiling on the intestinal mucosa (optional) [24].

11. Perform gross examination, imaging of the intestines including assessment of tumor number and size and tissue collection following sacrifice (*see* Subheading 4).

3.3 Bile Acids and Colorectal Cancer in APC*^*min/+* *Mice

1. House 4-week-old *Apc*^*min/+*mice (male and female) and maintain on a 12 h light/dark cycle (*see* **Notes 1–4**).

2. Feed mice with Teklad Mouse Breeder diet ad libitum.

3. Weigh all animals before the start of the experiment.

4. Give 0.2 % sodium deoxycholate in drinking water to experimental group for 12 weeks ad libitum.

5. Give regular tap water to the control group ad libitum.

6. Monitor both the control and experimental groups daily and weigh them weekly.

7. Measure fecal bile acid output from stool (collected for 72 h from individually housed mouse) [15] if desired. This step is optional.

8. At week 16, sacrifice all animals by injecting 100 μl/10 kg of a Ketamine/Xylazine solution, followed by cervical dislocation (*see* **Note 8**).

9. Determine bile acid pool size from total content of small intestine, gallbladder, and liver following sacrifice (optional) if desired [15].

10. Perform gross examination, imaging of the intestines including assessment of tumor number and size and tissue collection following sacrifice (*see* Subheading 4).

3.4 Tissue Collection and Processing

3.4.1 Preparation of 2 % Agar

1. Melt 1 g of agar in 50 ml of water in microwave.

2. Place the flask containing agar in a Becker containing H_2O maintained at 55 °C.

   At this temperature, agar remains liquid but is not too hot and won't damage tissue.

3.4.2 Preparation of Wax Boards

1. Completely melt the block of PenWax in a container immersed in boiling water (*see* **Note 13**).

2. Once melted, pour the wax into square dishes up to 1 cm thick and let it solidify.

3. Wax plates can be kept at room temperature for unlimited time (*see* **Note 14**).

3.4.3 Tissue Collection and Image Analysis

1. Euthanize animals by Ketamine/Xylazine injection followed by cervical dislocation.

2. Swab the abdominal skin with alcohol and open the abdominal cavity with scissors using caution not to lacerate abdominal organs.

3. Dissect the spleen, located right under the stomach, from the surrounding adipose tissue and viscera with scissors. Record the dimensions and weight.

4. Place the tissue in an Eppendorf tube and snap freeze in liquid nitrogen.

5. Expose and dissect the small and large intestines.

6. Remove and discard the cecum, which is a small saccular structure between small and large intestine (*see* **Note 15**).

7. Separate small intestine from colon.

8. Measure and record the length of large and small intestine with a ruler.

9. Dissect off the mesenteric adipose tissue and attached vessels from the small and large intestines.

10. Cut small intestine in three equal sections (proximal, mid, distal). Depending on the size of the wax plate, further cut proximal, mid, and distal sections of small intestine into two equal pieces. Colon usually remains as a single piece, but can be cut into two equal pieces to fit on the wax plate.

11. Flush each segment with cold PBS, open longitudinally and pin down on a wax plate (Fig. 1 and *see* **Note 16**).

12. Perform the gross imaging using a Nikon SMZ800 microscope and photograph the tissues using a Photometrics CooLSNAP camera.

13. Place the intestines in 10 % formalin, enough to cover and submerge the tissue completely (1:10 ratio of tissue and formalin solution). Fix overnight at room temperature.

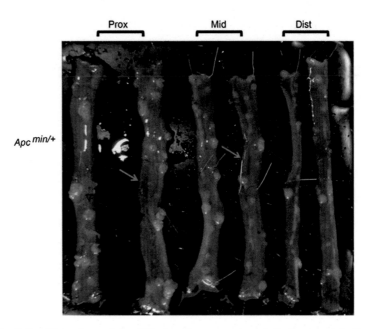

Fig. 1 Small intestines of *Apc* ^{min/+} pinned on a wax board. *Red arrows*: intestinal polyps. *Blue arrow*: Grossly normal mucosa. Note grossly normal mucosa may have adenomas on microscopic examination

14. Rinse the intestines with PBS and keep in PBS at 4 ° C until embedding (*see* **Note 17**).

15. Use MetaVue software for width, length, and total area measurements.

16. Before starting further dissection, make sure that all the materials are ready and are located within reach for further tissue sectioning and preparation (Fig. 2a). This will streamline the process and prevent tissue from drying.

17. Cut each pinned tissue segment into pieces approximately 2 cm in length (*see* **Note 18**).

18. Cut each segment longitudinally with a razor blade and separate each half (Fig. 2b, c).

19. Cover the mucosal sides of the cut sections with 55 ° C agar and set aside until the agar solidifies (Fig. 2d).

20. Cut excess agar around tissue (Fig. 2e).

21. Each 2 cm section results in two sister strips with mucosa facing each other (Figs. 2f and 3).

22. Maintain proximal–distal orientation (Fig. 3 and *see* **Note 19**).

23. While maintaining the proximal–distal orientation, align strips and cover with 55 °C agar to form a single block (Fig. 2g).

24. Label each tissue cassette with animal type/number, experiment, and tissue type (i.e. *Apc*^{min/+} DSS, small intestine) with special marker pen (*see* **Note 20**).

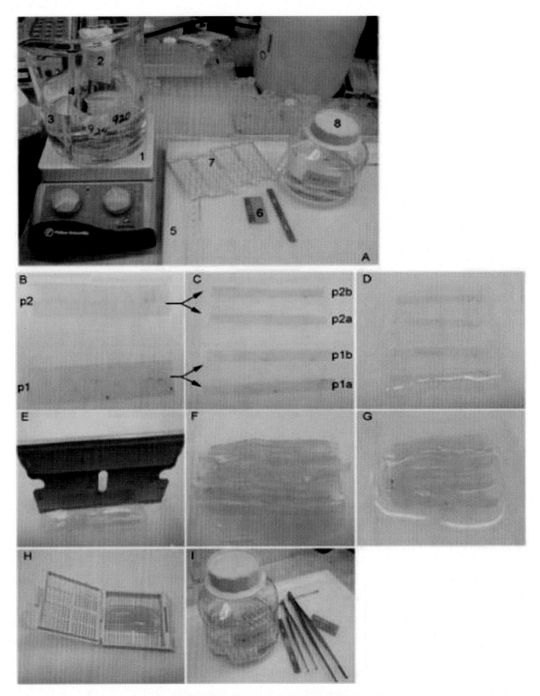

Fig. 2 Tissue embedding process. (**a**) Required material. (*1*) heat block maintaining the agar solution at 55 °C. The Becker containing the agar solution (*2*) is placed in a larger Becker (*3*) filled with water kept at 55 °C. The temperature is monitored with a thermometer (*4*). Cutting and embedding of tissue is done on a cutting board (*5*) with razor blades (*6*). Once processed, the embedded tissue is placed in a tissue cassette (*7*) and kept in 70 % ethanol (*8*) at room temperature. Embedding process (*B-I*). Each tissue section is cut transversally to generate a ~2 cm long piece (**b**). For each section, two pieces are thus generated (*p1* and *p2*). Each fragment is then cut longitudinally into two pieces (**c**) (*p1a, p1b, p2a, p2b*). The sections are then covered with agar solution (**d**). Once the agar has solidified, the excess agar is removed (**e**). Each section is positioned as shown in Fig. 3 and covered with agar (**f**). Once the block has solidified and the excess agar removed (**g**), the tissue block is placed into a tissue cassette (**h**) and kept in 70 % ethanol (**i**)

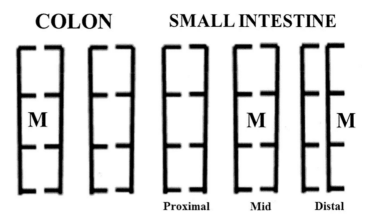

Fig. 3 Embedding scheme for small and large intestine. *M* mucosal side

25. Once solidification is completed, place the tissue blocks into a tissue cassette (Fig. 2h).

26. Keep the cassettes in 70 % Ethanol, enough to cover all the cassettes and keep at room temperature (Fig. 2i and *see* **Note 17**).

27. Submit the cassettes for tissue processing and hematoxylin and eosin (H&E) staining right after the tissue is placed in the cassette, or keep longer in 70 % Ethanol or formalin if submitting more than one batch.

28. Collect tissue from "lesional" and "normal" areas of colon and small intestine from separate animals within the same experimental group after the intestines are pinned on a formalin-free wax plate. Flash freeze this tissue in liquid nitrogen for RNA and protein analysis (*see* Subheading 3.5).

3.5 Magnetic Bead-Based Poly-A Purification of DSS-Exposed mRNA

This is a crucial step since any exposure of cells or tissues to DSS will severely impair the ability to extract RNA suitable for reverse transcription [23].

3.5.1 RNA Preparation

1. Homogenize 50–100 mg of harvested organs at maximum speed for 10 s using tissue homogenizer in 1 ml of Trizol agent.

2. Extract RNA in Trizol following Manufacturer's protocol. Five to ten microgram of total RNA will be used for binding to Dynabeads.

3. Adjust the volume of RNA to 100 μl with H_2O, heat the solution to 65 °C for 2 min to disrupt secondary structures.

4. Place the RNA solution on ice.

3.5.2 Preparation of Dynabeads

All the following steps are performed at room temperature.

1. Transfer 50 μl of beads to a 1.7 ml microcentrifuge tube.

2. Place the tube on the Magnet Dynal for 30 s.

3. Pipette off the supernatant, remove the tube from the magnet, and add 100 μl of Binding Buffer to equilibrate the beads.

4. Place the tube back to the magnet and remove the supernatant.

5. Remove the tube from the magnet and add 100 μl of Binding Buffer (*see* **Note 21**).

3.5.3   mRNA Purification

All the following steps are performed at room temperature.

1. Add the total RNA (100 μl) (Subheading 3.5, **step 5**) to the Dynabeads/Binding Buffer suspension.

2. Mix thoroughly manually until all the beads are re-suspended and rotate the tube on a wheel 3–5 min at room temperature to allow binding of the RNA to the oligo(dT) beads.

3. Place the tube on the magnet until the solution is cleared within seconds and remove supernatant.

4. Remove the tube from the magnet and wash the mRNA-beads complex twice with 200 μl Washing Buffer.

5. Remove the supernatant between each wash by placing the tube back on the magnet.

6. Elute the RNA with 20 μl of 10 mM Tris–HCl pH 7.5.

7. Heat the tube to 65–80 °C for 2 min and place it immediately on the magnet.

8. Transfer the eluted RNA to a new Eppendorf tube in ice.

3.5.4   cDNA Synthesis

1. Use the eluted RNA directly in the cDNA reaction carry out using the Applied Biosystems High Capacity Reverse transcription kit.

2. Thaw 10× Reverse Transcriptase (RT) buffer, 10× random primers and 25× dNTPs on ice to preserve reagents stability.

3. Prepare Master Mix on ice as follows: per reaction (10 μl): 2 μl 10× RT buffer, 0.8 μl 25× dNTPs, 2 μl 10× Random primers, 1 μl MultiScribe Reverse transcriptase, 4.2 μl H_2O.

4. Combine, RNA (~1 μg) with 10 μl Master Mix in thin-walled PCR tube and complete to 20 μl with autoclaved H_2O (*see* **Note 22**).

5. Perform the reaction as follows: 10 min at 25 °C, 120 min at 37 °C, 5 s at 85 °C.

6. Keep cDNAs at –80 °C until further used for analysis of RNA expression by PCR [24].

3.5.5   Protein Extract Preparation

1. Homogenize 100 mg of tissue in 600 μl of tissue lysis buffer for 10 s.

2. Place the homogenate in an Eppendorf tube and keep in ice.

3. Add 1/10 of the volume of 10× detergent to the homogenate.

4. Vortex for 20 s and keep on ice for 15 min.

5. Centrifuge the homogenate at maximum speed $(18,000 \times g)$ at 4 °C for 15 min in tabletop Eppendorf centrifuge.

6. Transfer the supernatant into a new Eppendorf tube and keep at −80 °C until further analysis by Western Blot.

3.6 Assessment of Histology and Immunohistochemistry

3.6.1 Hematoxylin & Eosin Assessment of Normal Mucosa and Common Pathologic Lesions

1. Hematoxylin and Eosin (H&E) staining is the standard method for evaluating the morphology of tissues. Make sure that the reviewer is blinded to the experimental and control groups. The normal wall structures from mucosa to serosa (serosal side towards exterior) as well as pathologic changes can be easily identified in well-oriented sections. If the sections are not oriented correctly, it may cause errors in histopathologic examination and lesions may be missed. All figures provided in this chapter represent well-oriented sections.

2. Identify wall structures and cellular components of small (Fig. 4a and *see* **Note 23**) and large intestine (Fig. 4b and

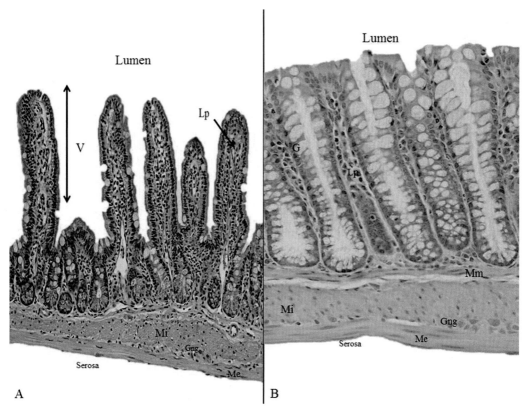

Fig. 4 Hematoxilin and Eosin staining of normal mouse small and large intestine. Note that small intestine mucosa has villi (**a**, 200×), whereas colonic mucosa is flat (**b**, 400×). The Goblet cells lining the surface epithelium are present in both small and large intestine. Lamina propria is the space between crypts. Muscularis interna and externa are thicker bundles of muscle and have ganglion cells and nerve bundles in between the muscle layers. *V* villus, *G* goblet cells, *Lp* lamina propria, *Mm* muscularis mucosae, *Mi* muscularis interna, *Gng* Ganglion cells, *Me* muscularis externa

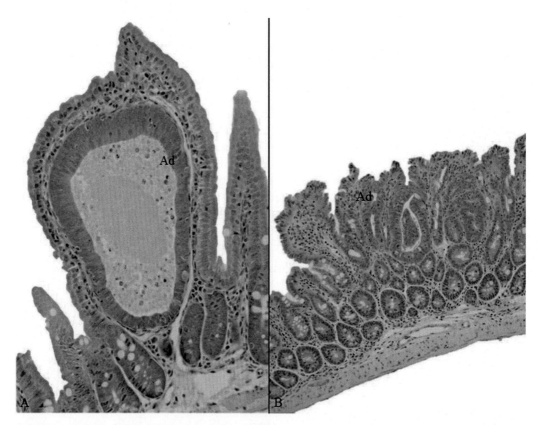

Fig. 5 Hematoxilin and Eosin staining of single crypt adenoma and tubular adenoma in mouse small intestine. Single crypt adenomas (Ad) are commonly seen on microscopic examination (**a**, 400×). A classic tubular adenoma in mouse small intestine (**b**, 200×). Note the single dilated adenomatous gland lined with pencillate nuclei. This one gland adenoma displays classic features of an adenoma with nuclear crowding, and pseudostratification. Apoptosis can also be seen (not shown in figure). Adenomas can reach bigger sizes. Note the larger adenoma in small intestine shows similar histologic features with involvement of the surface epithelium (**b**)

see **Note 24**). This will help recognize any pathologic lesions that may occur.

3. Identify and count the adenomas in both small and large intestine including single crypt adenomas (Figs. 5a, b, 6a and *see* **Note 25**). Remember that single crypt adenomas are small and are not visible by gross examination.

4. Identify and count the adenomas with high-grade dysplasia in both small and large intestine (Fig. 6b and *see* **Note 26**).

5. Identify and count intramucosal (not shown) and invasive carcinomas (Fig. 6c). Note the depth of invasion for invasive tumors (*see* **Note 27**).

*3.6.2 Immunohisto-
chemistry and Assessment*

1. Obtain 4-µm sections from each tissue on a charged slide (*see* **Note 28**).

2. Follow the antibody staining per manufacturer's protocol. Have positive and negative controls for each immunohistochemistry run (*see* **Note 29**).

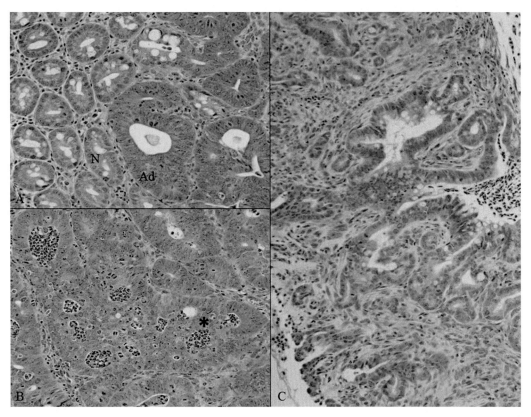

Fig. 6 Hematoxilin and Eosin staining of adenoma in mouse small intestine, adenoma with high-grade dysplasia, and adenocarcinoma in mouse small intestine. Adenomas are low-grade dysplastic lesions (**a**, 400×). Note the sharp transition between normal (N) and adenomatous crypts (Ad). Adenoma with high-grade dysplasia is characterized by cribriform glands (*, **b**) and loss of nuclear polarity (**b**, 400×). Prominent nuclear pleomorphism and luminal necrosis (*, **b**) are commonly seen. Invasive carcinomas are characterized by irregular glands and desmoplasic stroma (**c**, 200×)

3. The evaluation of immunohistochemistry depends on the type of antibody that is used. The manufacturer's manual is the best source for staining evaluation.

4. Ki67, BrdU, and TUNEL are nuclear markers; therefore count the number of positive nuclei.

5. Evaluate intestinal proliferation by scoring full longitudinal sections of crypts and report it as number of BrdU-positive cells normalized to the total number of cells per crypt (*see* Fig. 7a, b).

6. Beta-catenin normally stains cytoplasmic membrane, but when mutated nuclear staining is observed (*see* Fig. 7c, d).

7. Evaluate the apoptotic index similarly by counting the number of TUNEL-positive cells normalized to the total number of cells in crypts and villi (*see* Fig. 7e).

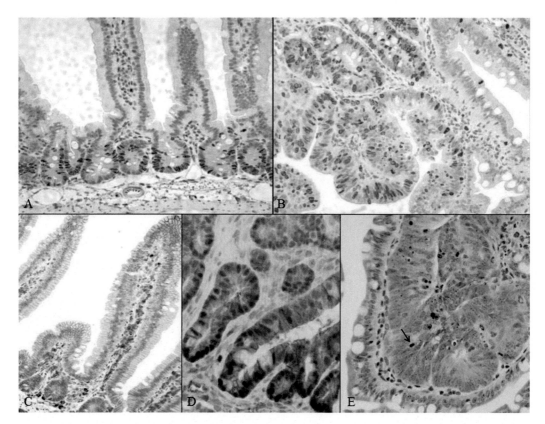

Fig. 7 Immunohistochemical staining of BrdU, β-catenin, and TUNEL in mouse intestine. Normal BrdU staining in *Apc^min/+* small intestine (**a**, 100×) and BrdU staining in *Apc^min/+* adenoma (**b**, 100×). Note the positive brown colored nuclear staining. Normal β-catenin staining in mouse small intestine shows cytoplasmic membranous staining (**c**, 400×), whereas β-catenin mutation in an adenoma displays dense nuclear positivity (**d**, 600×). TUNEL staining shows positive nuclei in a small intestine adenoma (*arrow*, brown nuclear staining) (**e**, 600×)

4   Notes

1. If the animals are going to be bred in house, maintain the *Apc ^min/+* mice as heterozygous breeding pairs. Homozygosity for this gene is lethal.

2. In the event that they are purchased commercially, acclimate the animals to the animal facility for at least a week so that the flora and intestinal microbiome of animals are similar. Therefore, purchase mice that are 2 weeks old and keep them in the animal facility for a week or two before the experiments begin.

3. Keep experimental and control groups to a minimum of 8–10 animals to ensure statistical power. It is understood that some animals will expire during the experiment, and having a larger group will yield enough data points and samples for additional studies.

4. Conduct experiments on both genders of mice to eliminate possible gender selection/susceptibility. In our experience, there is no gender effect in adenoma or carcinoma incidence.

5. Use the same batch of DSS (record lot number) throughout all experiments when possible as the potency of DSS may vary lot to lot.

6. The percentage of DSS depends on the experimental model and the additional genetic traits of the animals. Expose a small batch of animals to different percentages of DSS as an initial test run. If multiple cycles of DSS are involved in the experimental model [11], it is likely that most animals will die within the first two cycles. Depending on the mortality rate, either decreasing the DSS concentration or increasing the length of water cycle is suggested.

7. Induction of neoplasia in animals can be accelerated by injection of azoxymethane (10 mg/kg/kg body weight administered by intraperitoneal injection at 8 weeks of age) [15].

8. Another method for sacrificing mice is isofluorane overdose. In this method, soak a cotton ball with isofluorane and keep the animals in a closed chamber until they expire. Cervical dislocation can also be carried out after isofluorane anesthesia is established.

9. There are several commercially available high fat diet regimens. Adjusted calories diet (60 % saturated fatty acid, milk fat and butter fat, sucrose 34 % per weight and cholesterol 0.2 %, Harlan Teklad) is most commonly used in our laboratory.

10. If necessary, body composition can be measured at 4-, 8-, and 12-week endpoints by DEXA scan under isofluorane anesthesia.

11. Expose the animals to a heating lamp for 10 min. This will cause vascular congestion and the tail blood sampling will be much easier. Place the animal in a mouse restrainer, make a small nick at the 1/3 distal end of the tail at 12 o'clock position where the vessel is visible on the dorsal site. Clean with alcohol and make a small cut with the blade. Other methods for collecting blood are retro-orbital approach and cardiac puncture.

12. The heparinized syringes are commercially available and are more standardized. Another way to make a heparinized syringe is to wash a regular syringe with heparin. Pull a small amount of heparin into the injector and "wash" the barrel of the injector by moving the plunger back and forth a few times.

13. Melting the Penwax may take up to several hours. Dishes come in bag of ten and one case contains ten bags. One pound of wax is sufficient to prepare ten wax plates. One plate can be used several times. These can be kept at room temperature and can be melted and poured again if needed.

14. Keep at least one or two wax boards free of formalin contamination. This is necessary for gross examination/imaging and dissection of normal and tumor tissues that will be processed for protein and RNA extraction.

15. This step is optional. Cecum is a small sac-like structure at the intersection of small and large intestine. It is difficult to orient and pin on a wax plate. If the investigators would like to include this for histologic examination, marking it with tissue dyes is recommended, since it is histologically identical to the rest of the colon.

16. Use ten pins per intestinal section. Approximately 70 pins are used per animal. These pins are re-usable. Keep a set of pins free of formalin contamination.

17. The fixed tissue can also be stored in formalin solution.

18. A tissue cassette is approximately 2 cm in greatest dimension. Further section the intestines after fixation into 2 cm segments. This will enable the tissues to fit in the cassette and prevent tissue waste.

19. Tissue dyes can be utilized to mark proximal, mid, and distal intestine. Designate and apply different color dyes to serosal surface of the intestines (i.e. red-proximal, blue-distal, mid-unstained). This way, the investigators can tell which piece belongs to which segment. Apply the tissue dyes with a cotton-tip, make sure it does not overrun between the segments. Use small amounts of dye and blot the excess with a paper towel.

20. The labeling done with regular marker pens will fade during tissue processing. Therefore, use a solvent-resistant pen. Another cost-effective alternative is to use a regular pencil (not a sharp writer).

21. Obtain optimal hybridization conditions in Binding Buffer by adding in a 1:1 ratio relative to sample volume.

22. As little as 0.2 μg of RNA can be used in the reverse transcription reaction.

23. The wall of the small intestine in mouse is identical to humans and is composed of mucosa, muscularis mucosae, submucosa, muscularis interna, muscularis externa, and serosa [25] (Fig. 4a). It is important to recognize and document all the components of normal small intestinal mucosa, which is mainly composed of crypts lined by Goblet, absorptive, Paneth, and endocrine cells forming villiform structures protruding into the lumen. Lamina propria is the space between crypts that has minimal inflammation but contains vessels and lymphatics supported by a connective tissue (Fig. 4a). Muscularis mucosae is a thin layer of smooth muscle that separates mucosa from submucosa. Muscularis mucosae can be difficult to recognize in poorly ori-

ented sections (not shown in Fig. 4a). The submucosa is the small space between muscularis mucosae and muscularis interna; it contains blood vessels and lymphoid aggregates (Peyer's patches). Lymphoid aggregates can sometimes extend into the mucosa. In contrast to human bowel, submucosa is not well-visualized in mouse intestine. Muscularis interna and externa are thicker bundles of muscle and have ganglion cells and nerve bundles in between (Fig. 4a). These (muscularis interna and externa) correspond to the muscularis propria in humans. Serosa is the most outer layer, mostly lined by peritoneum.

24. In contrast to small intestine that has villi, the colonic (large intestine) mucosa is flat and does not have villi. In proximal mouse colon mucosal folds are most commonly seen as horizontal ridges. The colonic mucosa is composed of crypts lined by Goblet cells, enterocytes, and rare endocrine cells. Lamina propria has minimal to no inflammation (Fig. 4b). Scattered lymphoid aggregates can be seen. The remainder of the wall components in colon is identical to small intestine, but muscularis mucosae is very poorly formed and not well-visualized in most cases (Fig. 4b) [25].

25. The histologic lesions in the *Apc^min/+* mouse closely mimic the FAP lesions in humans [13, 26]. The morphologic spectrum of the intestinal lesions that can be seen in *Apc^min/+* mice varies from adenoma to invasive carcinoma. In *Apc^min/+* models of colorectal cancer, adenomas develop mostly in small intestine, and colonic lesions are rare. An adenoma is characterized by cigar-shaped nuclei with an increased Nucleus–Cytoplasmic ratio populating the crypts (Figs. 5a, b and 6a). Adenomas are low-grade dysplasia by definition. Adenoma and tubular adenoma terms are used interchangeably. Microadenomas, which are single crypt adenomas, can be seen in both FAP patients and *Apc^min/+* mice. Single crypt adenomas (Fig. 5a) will not be visible to naked eye with gross examination, whereas larger lesions will be grossly visible (Fig. 5b). Therefore, both gross and microscopic examinations of the intestines are important.

26. Adenomas can be divided into low- and high-grade dysplastic lesions. High-grade lesions are characterized by architectural complexity including cribriform glandular structures, dirty luminal necrosis, and loss of nuclear polarity (Fig. 6b).

27. Intramucosal carcinomas are characterized by single cells invading into lamina propria. In contrast to colon, intramucosal carcinomas in human small intestine have a potential to metastasize to lymph nodes. Submucosal invasion is characterized by invasion of the tumor through muscularis mucosae and the presence of desmoplasia (Fig. 6c).

28. Even though ideal for both H&E and immunohistochemistry applications, you must use charged slides for immunohistochemical staining.

29. Select proper positive and negative tissue controls for each antibody run. This information is available at antibody manufacturer's website.

Acknowledgment

Work cited in this review was supported by the following grants: HL38180, DK56260 and Digestive Disease Research Core Center P30DK52574 to N.O.D.

References

1. American Cancer Society (2014) Colorectal cancer facts & figures 2014–2016. American Cancer Society, Atlanta, GA

2. American Cancer Society (2015) Surveillance research, 2015. American Cancer Society, Atlanta, GA

3. Jasperson KW, Tuohy TM, Neklason DW et al (2010) Hereditary and familial colon cancer. Gastroenterology 138:2044–2058

4. Chan AT, Giovannucci EL (2010) Primary prevention of colorectal cancer. Gastroenterology 138:2029–2043

5. Ullman TA, Itzkowitz SH (2011) Intestinal inflammation and cancer. Gastroenterology 140:1807–1816

6. Walter V, Jansen L, Hoffmeister M et al (2014) Smoking and survival of colorectal cancer patients: systematic review and meta-analysis. Ann Oncol 25:1517–1525

7. Pino MS, Chung DC (2010) The chromosomal instability pathway in colon cancer. Gastroenterology 138:2059–2072

8. Liang J, Lin C, Hu F et al (2013) APC polymorphisms and the risk of colorectal neoplasia: a HuGE review and meta-analysis. Am J Epidemiol 177:1169–1179

9. Taketo MM, Edelmann W (2009) Mouse models of colon cancer. Gastroenterology 136:780–798

10. Moser AR, Pitot HC, Dove WF (1990) A dominant mutation that predisposes to multiple intestinal neoplasia in the mouse. Science 247:322–324

11. Bilger A, Shoemaker AR, Gould KA et al (1996) Manipulation of the mouse germline in the study of Min-induced neoplasia. Semin Cancer Biol 7:249–260

12. Khazaie K, Zadeh M, Khan MW et al (2012) Abating colon cancer polyposis by Lactobacillus acidophilus deficient in lipoteichoic acid. Proc Natl Acad Sci U S A 109:10462–10467

13. Shoemaker AR, Gould KA, Luongo C et al (1997) Studies of neoplasia in the Min mouse. Biochim Biophys Acta 18:F25–48

14. Cooper HS, Everley L, Chang WC et al (2001) The role of mutant Apc in the development of dysplasia and cancer in the mouse model of dextran sulfate sodium-induced colitis. Gastroenterology 121:1407–1416

15. Giammanco A, Blanc V, Montenegro G et al (2014) Intestinal epithelial HuR modulates distinct pathways of proliferation and apoptosis and attenuates small intestinal and colonic tumor development. Cancer Res 74:5322–5335

16. Tanaka T, Kohno H, Suzuki R et al (2006) Dextran sodium sulfate strongly promotes colorectal carcinogenesis in Apc(Min/+) mice: inflammatory stimuli by dextran sodium sulfate results in development of multiple colonic neoplasms. Int J Cancer 118:25–34

17. Xie Y, Matsumoto H, Nalbantoglu I et al (2013) Intestine-specific Mttp deletion increases the severity of experimental colitis and leads to greater tumor burden in a model of colitis associated cancer. PLoS One 8, e67819

18. Baltgalvis KA, Berger FG, Pena MM et al (2009) The interaction of a high-fat diet and regular moderate intensity exercise on intestinal polyp development in Apc Min/+ mice. Cancer Prev Res 2:641–649

19. Day SD, Enos RT, McClellan JL et al (2013) Linking inflammation to tumorigenesis in a mouse model of high-fat-diet-enhanced colon cancer. Cytokine 64:454–462

20. Newberry EP, Xie Y, Kennedy SM et al (2006) Protection against Western diet-induced obesity and hepatic steatosis in liver fatty acid-binding protein knockout mice. Hepatology 44:1191–1205

21. Cao H, Luo S, Xu M et al (2014) The secondary bile acid, deoxycholate accelerates intestinal adenoma-adenocarcinoma sequence in Apc (min/+) mice through enhancing Wnt signaling. Fam Cancer 13:563–571

22. Mahmoud NN, Dannenberg AJ, Bilinski RT et al (1999) Administration of an unconjugated bile acid increases duodenal tumors in a murine model of familial adenomatous polyposis. Carcinogenesis 20:299–303

23. Kerr TA, Ciorba MA, Matsumoto H et al (2012) Dextran sodium sulfate inhibition of real-time polymerase chain reaction amplification: a poly-A purification solution. Inflamm Bowel Dis 18:344–348

24. Dharmarajan S, Newberry EP, Montenegro G et al (2013) Liver fatty acid-binding protein (L-Fabp) modifies intestinal fatty acid composition and adenoma formation in ApcMin/+ mice. Cancer Prev Res 6:1026–1037

25. Atlas of Laboratory Mouse Histology (2004) Texas histopages. http://ctrgenpath.net/static/atlas/mousehistology

26. Preston SL, Leedham SJ, Oukrif D et al (2008) The development of duodenal microadenomas in FAP patients: the human correlate of the Min mouse. J Pathol 214:294–301

Chapter 28

Modeling Murine Gastric Metaplasia Through Tamoxifen-Induced Acute Parietal Cell Loss

Jose B. Saenz, Joseph Burclaff, and Jason C. Mills

Abstract

Parietal cell loss represents the initial step in the sequential progression toward gastric adenocarcinoma. In the setting of chronic inflammation, the expansion of the mucosal response to parietal cell loss characterizes a crucial transition en route to gastric dysplasia. Here, we detail methods for using the selective estrogen receptor modulator tamoxifen as a novel tool to rapidly and reversibly induce parietal cell loss in mice in order to study the mechanisms that underlie these pre-neoplastic events.

Key words Tamoxifen, Parietal cell loss, Metaplasia, Oxyntic atrophy, Spasmolytic polypeptide-expressing metaplasia (SPEM)

1 Introduction

Gastric adenocarcinoma remains one of the leading causes of cancer-related deaths worldwide [1]. The sequence of events leading to the development of gastric dysplasia and neoplasia begins with the loss of acid-secreting parietal cells, a process known as oxyntic atrophy, followed by the expansion of pre-neoplastic changes in the setting of chronic inflammation [2]. The early mucosal response to oxyntic atrophy includes reorganization of the gastric unit, characterized initially by an increased proliferation of gastric progenitor cells and the reprogramming of post-mitotic chief cells at the base of the gastric gland into a proliferating population of metaplastic cells [3]. Overall, the pattern of gastric unit reorganization that characterizes the response to oxyntic atrophy is known as spasmolytic polypeptide-expressing metaplasia (SPEM), as the metaplastic chief cells express spasmolytic polypeptide (also known as trefoil factor family 2; TFF2). SPEM can either be a transient alteration in the gastric landscape, followed by repair and restoration of normal architecture, or it can represent a crucial

Andrei I. Ivanov (ed.), *Gastrointestinal Physiology and Diseases: Methods and Protocols*, Methods in Molecular Biology, vol. 1422, DOI 10.1007/978-1-4939-3603-8_28, © Springer Science+Business Media New York 2016

pre-neoplastic event en route to gastric dysplasia in the setting of chronic inflammation. The study of the mechanisms underlying the development and evolution of SPEM has been accelerated by recently developed tools [4–6] which rapidly induce SPEM in animal models of gastric dysplasia. Here, we describe the discovery and use of the selective estrogen receptor modulator, tamoxifen, as a model for studying SPEM.

In addition to its widespread therapeutic use as hormonal therapy, tamoxifen has recently found a role in conditional gene targeting in the mouse [7]. Notably, the development of a ligand-dependent Cre-ER recombinase, in which the Cre enzyme is fused to a mutated hormone-binding domain of the estrogen receptor, has allowed for the use of tamoxifen to modulate gene expression in a spatiotemporal fashion [8]. As a result, tamoxifen now serves as a tool for regulating tissue-specific Cre activity.

However, the use of tamoxifen for induction of the Cre-ER recombinase led to a serendipitous discovery in the mouse stomach that has broadened its role beyond the Cre-ER system and implicated tamoxifen as a unique agent for studying the early events following oxyntic atrophy [9, 10]. Serial intra-peritoneal injections of various strains of wild-type mice with tamoxifen induced apoptosis in the vast majority of parietal cells, metaplastic changes in the chief cells at the bases of the gastric glands, and an increased proliferation of gastric progenitor cells, changes characteristic of and consistent with SPEM. This effect is reproducible [11], estrogen-independent, and reversible, with a normalization of gastric histology within weeks of tamoxifen discontinuation [10]. The tamoxifen administration protocol described below therefore offers a unique method for reproducing oxyntic atrophy and dissecting early pre-neoplastic events leading to gastric dysplasia.

2   Materials

2.1   Preparation of Tamoxifen Stock

1. Tamoxifen (*see* **Note 1**).
2. Sterile sunflower seed oil (*see* **Note 2**).
3. Ethanol (200 proof).
4. Sonic dismembrator with microtip (2 mm).
5. Eppendorf tubes (1.5 mL).
6. Benchtop vortex machine.
7. Pipettor.
8. Protective headphones.

2.2 Mouse Injection

1. Insulin syringe with needle (0.5 mL, 27 G×0.5 in.).

2. Balance.

3. Alcohol wipes.

3 Methods

Carry out all procedures at room temperature unless otherwise specified.

The following protocol corresponds to a tamoxifen solution dissolved in 10 % ethanol and 90 % sunflower seed oil (*see* **Note 3**).

3.1 Preparation of Tamoxifen Stock

1. Weigh mice (*see* **Note 4**).

2. Weigh out 25 mg of dry tamoxifen and place it in a 1.5 mL Eppendorf tube (*see* **Note 5**).

3. Slowly add 100 µL of 100 % ethanol, trying to keep the tamoxifen at the bottom of the tube. Do not shake, mix, or pipet.

4. Measure 900 µL of sterile sunflower seed oil in a separate 1.5 mL Eppendorf tube.

5. Sonicate the tamoxifen/ethanol mixture in the Eppendorf tube at 40 % amplitude in 20-s pulses until the tamoxifen is completely dissolved (*see* **Note 6**).

6. Immediately combine the tamoxifen/ethanol mixture with the sunflower seed oil. Cap and vortex the solution to ensure adequate mixing (*see* **Note 7**).

7. The tamoxifen mixture can be stored at 4 °C for up to 3 days or at −20 °C indefinitely (*see* **Note 8**). Allow the mixture to warm to room temperature prior to injection.

3.2 Tamoxifen Treatment

1. Using the insulin syringe needle, measure out the appropriate amount of the tamoxifen mixture so as to inject 5 mg tamoxifen for every 20 g mouse body weight (*see* **Note 9**).

2. Sanitize the injection site by wiping the mouse abdomen with an alcohol wipe. Intra-peritoneally inject the vehicle (10 % ethanol/90 % sunflower seed oil) or tamoxifen mixture (*see* **Note 10**).

3. Repeat the injection for 3 consecutive days using the same tamoxifen stock, stored at 4 °C.

4. Mouse stomachs can be harvested at any time following the first injection or thereafter, and tissue can be processed accordingly (Figs. 1 and 2, and *see* **Note 11**).

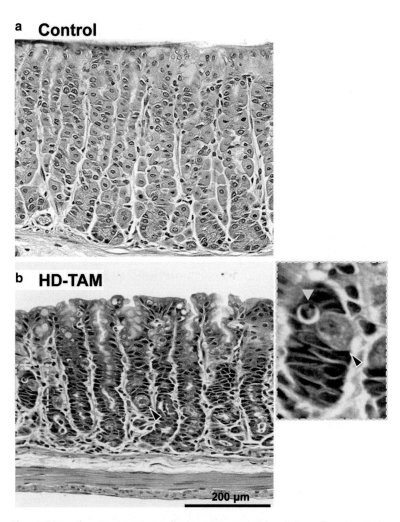

Fig. 1 Tamoxifen treatment results in acute parietal cell loss. Representative hematoxylin and eosin stain of gastric corpus from wild-type C57BL/6 mice after intra-peritoneal injection with 3 days of either vehicle (**a**; Control) or 5 mg/20 g body weight tamoxifen (**b**; HD-TAM). Note the relative decrease in parietal cells (*black arrowhead*) compared to the vehicle-treated mouse. An apoptotic body (*yellow arrowhead*) adjacent to a dying parietal cell is highlighted (*inset*)

Fig. 2 (continued) are highlighted by *dashed lines*. (**b**) A representative immunostain of the gastric corpus from a mouse intra-peritoneally injected with 5 mg/20 g body weight tamoxifen for 3 days (HD-TAM) shows an acute loss of parietal cells, as demonstrated by the relative paucity of VEGFb-staining cells (*green*). Fragments of parietal cells are highlighted by the *white arrowheads*. In addition, note the shift in GSII expression (*red*) toward the bases of glands in tamoxifen-treated mice compared to the vehicle-treated controls. Nuclei are stained with Hoescht (*blue*), and representative gastric units are highlighted by *dashed lines*

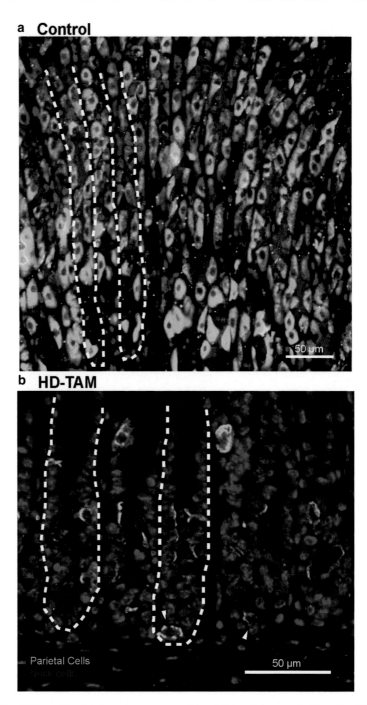

Fig. 2 Tamoxifen treatment causes acute parietal cell loss and alters the GSII expression pattern in gastric units. (**a**) Representative immunostain of the gastric corpus of a mouse intra-peritoneally injected with vehicle alone (Control) for 3 days demonstrates normal-appearing gastric units, highlighted by abundant parietal cells (stained with H⁺/K⁺ ATPase; *green*) and neck cells (stained with GSII; *red*). Nuclei are stained with Hoescht (*blue*), and representative gastric units

4   Notes

1. The source of tamoxifen has no appreciable effect on the ability to induce parietal cell loss. Tamoxifen stocks from three separate commercial suppliers, Sigma (St. Louis, MO), Cayman Chemical Company (Ann Arbor, MI), and Toronto Research Company (Toronto, Canada), have demonstrated similar efficacy [10]. In addition, parietal cell toxicity is specific to tamoxifen and not a general toxic effect of selective estrogen receptor modulators, as treatment with raloxifene, a member of the estrogen receptor modulator family with pro- and anti-estrogenic effects, had no appreciable toxicity at a comparable dose [10].

2. To sterilize the sunflower seed oil, heat an appropriate amount in an Erlenmeyer flask on a hot plate to 85–90 °C for 15–20 min. Do not boil. Allow the flask to cool and store 40-mL aliquots at room temperature. Alternatively, the sunflower seed oil can be autoclaved prior to use.

3. The free base form of tamoxifen and one of its commonly used active metabolites, 4-hydroxytamoxifen (*see* **Note 12**), are largely insoluble in water. The original formulation for intraperitoneal injection was found to be soluble in 60 % ethanol [12], and its solubility has since been optimized in a sunflower seed oil/ethanol mixture (*see* **Note 13**). Tamoxifen citrate, an oral formulation that has been developed for administering tamoxifen to mice via chow ([13]; *see* **Note 10**), is soluble in water at 0.3 mg/L at 20 °C. Tamoxifen-free base powder should be stored at −20 °C in the dark.

4. Our experience has shown that three different wild-type mouse strains (*C57BL/6*, *BALB/c*, and *FVB/N*; all purchased from the Jackson Laboratory) have similar gastric mucosal responses to tamoxifen treatment [10]. In our limited experience with the strain, *BALB/c* mice are particularly sensitive to tamoxifen treatment, with mice commonly dying of unknown causes during treatment. Mice are typically used at 6–8 weeks of age, but SPEM is effectively induced in mice as old as 6 months of age. The effects on older mice are less obvious, potentially due to increased body fat causing changes in tamoxifen metabolism and distribution.

5. An injection dose of 5 mg/20 g mouse weight over 3 days results in a dramatic phenotype, with >90 % loss of parietal cells, a significant increase in gastric progenitor cells, and morphologic changes in the chief cells at the bases of glands in the gastric corpus, histologic changes consistent with the induction of SPEM ([10], Figs. 1b and 2b). However, we have previously shown that tamoxifen injections at lower doses (≤1 mg/20 g body weight) can be used for efficient,

inducible Cre-mediated recombination in the context of the Cre-ERT/*loxP* system, without the development of SPEM [9]. It is thus possible to obtain specific recombination of floxed alleles in tamoxifen-inducible Cre lines in a dose-dependent manner, while avoiding the stomach-altering effects seen at higher tamoxifen doses. Interestingly, though this has not been formally tested, SPEM induction by tamoxifen seems to have an all-or-none response, where no detectable damage can be seen at ≤1 mg/20 g mouse body weight, but ≥3 mg/20 g mouse body weight causes near complete SPEM, without an intermediate phenotype.

6. Make sure to wear protective headphones when using the sonicator.

7. Vortex the solution for at least 20 s. Allow the solution to sit at room temperature for several minutes. Proper mixing is crucial, and the mixture should be homogeneous. If it looks cloudy or layered, discard the mixture and start over.

8. No appreciable decline in the ability of tamoxifen to induce SPEM has been seen for tamoxifen mixtures stored at 4 °C over the duration of injections. Similarly, the tamoxifen stock can be stored at –20 °C until further use. Our lab, however, makes a fresh tamoxifen stock prior to each treatment regimen and uses this stock for the duration of the treatment.

9. Given the viscosity of the tamoxifen mixture, aspiration into the syringe can take 10–15 s.

10. Various tamoxifen formulations and numerous modes of tamoxifen administration have been reported. We focus here on intra-peritoneal administration, which we use most commonly, though we also observe SPEM induction with oral gavage. It is worth noting that other methods, in addition to oral gavage [14], for inducing Cre recombinase activity via tamoxifen have been used, including via drinking water [15], chow [13, 16], and subcutaneous implantation [17]. In that respect, we can only attest that oral gavage and intra-peritoneal administration of tamoxifen cause SPEM and have not tested the effects of other modes of tamoxifen administration.

11. The effects of tamoxifen on the mouse stomach can be seen within 12–24 h of the first intra-peritoneal injection [10]. Our laboratory nomenclature designates the first day of tamoxifen injection as day 0 (D0), with the last day of injection corresponding to day 2 (D2). A recent report found that a single intra-peritoneal injection at 4 mg/25 g mouse body weight induced a 57 % loss of parietal cells in the gastric corpus [11]. In our experience, the peak effect (i.e., maximal parietal cell loss, *see* Figs. 1b and 2b) is seen at 1 day following the third tamoxifen injection (D3). We have also achieved ≥90 % loss of

parietal cells at D3 even after a single injection of tamoxifen at 5 mg/20 g mouse body weight. The single-injection protocol, however, shows more variability between mice than the 3-day injection protocol. A recovery of the gastric epithelium and a return to normal histology are seen within 14–21 days [10]. Like previously described pharmacologic induction of SPEM (*see* **Note 14**), the effects of tamoxifen on the mouse stomach are transient. Many studies using tamoxifen-inducible Cre lines wait at least 2 weeks prior to assessing recombination, by which point parietal cells have largely recovered. This may explain how tamoxifen-induced parietal cell loss is often missed by investigators using tamoxifen to induce gene recombination in the stomach.

12. Tamoxifen is a prodrug that is hepatically metabolized to produce two predominant active metabolites, 4-hydroxytamoxifen and *N*-desmethyl-4-hydroxytamoxifen [18]. Multiple studies have used 4-hydroxytamoxifen for induction of the Cre-ERT2 system. It is worth noting that 4-hydroxytamoxifen has shown higher affinity for the estrogen receptor than tamoxifen in vitro [19] and a greater inhibitory effect on proliferation of normal human breast cells as well as breast cancer cell lines in culture [20–22]. Differential effects between tamoxifen and 4-hydroxytamoxifen have also been reported in apoptosis of human mammary epithelial cells [23] and uterine gene expression in rats [24]. In our limited experience, intra-peritoneally administered 4-hydroxytamoxifen, at commonly used doses for Cre recombinase induction, induces less SPEM in mice compared to similar doses of tamoxifen.

13. It has been speculated that the observed effects of intra-peritoneal tamoxifen administration could be unrelated to the tamoxifen itself and rather an effect of the ethanol solvent on parietal cells. Though the effect of ethanol on parietal cell membranes and H^+/K^+ ATPase function has been reported [25], our experience has shown that intra-peritoneal injection of mice with ethanol does not induce substantial parietal cell loss. On the other hand, excluding ethanol as a solvent results in poor solubility of tamoxifen. Oral gavage or intra-peritoneal injection of the resulting suspension, rather than solution, might cause suboptimal absorption and less substantial and/or consistent SPEM induction. Differences in solubilization methods may also explain why some investigators using tamoxifen for Cre recombinase induction may not have observed SPEM in their control mice.

14. Previous methods for inducing oxyntic atrophy and SPEM have been described, varying in their mechanism of action, onset of effect, and degree of inflammation (Table 1). Chronic infection of mice with *Helicobacter felis* [26] or of Mongolian

Table 1
Characteristics of various inducers of SPEM

Agent	Route of administration	Time to onset of SPEM	Inflammatory response?	Reversibility
Helicobacter pylori[a]	Oral gavage	Months	Yes	No
Helicobacter felis[a]	Oral gavage	Months	Yes	No
DMP-777	Oral gavage	10-14 days	No	Yes
L-635	Oral gavage	3 days	Yes	Yes
Tamoxifen	Intra-peritoneal, oral gavage[b]	3 days	Scant	Yes

See text and associated references for more details
[a]Reported in C57BL/6 mice and Mongolian gerbils
[b]Other forms of tamoxifen administration have been described and are referenced in the text

gerbils with *Helicobacter pylori* [27] results in the emergence of SPEM within months of infection. In contrast to these chronic infectious models, pharmacologic induction of SPEM has provided a more rapid and reversible means for achieving the same result. The neutrophil elastase inhibitor DMP-777 has been shown to cause a rapid loss of parietal cells in rats and mice within 3–4 days of daily dosing [4, 5]. Treatment with this parietal cell-specific apical membrane protonophore leads to the emergence of SPEM within 7–10 days, in the absence of inflammation. Like tamoxifen [10], the effects on parietal cell loss can be mitigated by pretreatment with omeprazole. More recently, a related variant of DMP-777, known as L-635, was found to produce a more rapid onset of SPEM in mice within 3 days of treatment [6]. The mechanism of action of L-635 is similar to that of DMP-777, though, unlike DMP-777, the onset of SPEM is accompanied by an exuberant inflammatory response.

Acknowledgments

The protocol described here is based on previous studies from our lab, all performed according to protocols approved by the Washington University School of Medicine Animal Studies Committee. We acknowledge the Advanced Imaging and Tissue Analysis Core of the Washington University Digestive Disease Core Center (DDRCC) for histological preparation.

This work was supported by grants from the National Institutes of Health to Jason C. Mills (R01 DK094989, 2P30 DK052574) and to Jose B. Saenz (T32 DK007130-42).

References

1. Ferlay J, Soerjomataram I, Dikshit R et al (2015) Cancer incidence and mortality worldwide: sources, methods, and major patterns in GLOBOCAN 2012. Int J Cancer 136:E359–E386

2. Ernst PB, Peura DA, Crowe SE (2006) The translation of *Helicobacter pylori* basic research to patient care. Gastroenterology 130:188–206

3. Goldenring JR, Nam KT, Mills JC (2011) The origin of pre-neoplastic metaplasia in the stomach: chief cells emerge from the Mist. Exp Cell Res 317:2759–2764

4. Goldenring JR, Ray GS, Coffey RJ et al (2000) Reversible drug-induced oxyntic atrophy in rats. Gastroenterology 118:1080–1093

5. Nomura S, Yamaguchi H, Ogawa M et al (2005) Alterations in gastric mucosal lineages induced by acute oxyntic atrophy in wild-type and gastrin-deficient mice. Am J Gastrointest Liver Physiol 288:G362–G375

6. Nam KT, Lee HJ, Sousa JF et al (2010) Mature chief cells are cryptic progenitors for metaplasia in the stomach. Gastroenterology 139:2028–2037

7. Saunders TL (2011) Inducible transgenic mouse models. Methods Mol Biol 693:103–115

8. Hayashi S, McMahon AP (2002) Efficient recombination in diverse tissues by a tamoxifen-inducible form of Cre: a tool for temporally regulated gene activation/inactivation in the mouse. Dev Biol 244:305–318

9. Huh WJ, Mysorekar IU, Mills JC (2010) Inducible activation of Cre recombinase in adult mice causes gastric epithelial atrophy, metaplasia, and regenerative changes in the absence of "floxed" alleles. Am J Physiolo Gastrointest Liver Physiol 299:G368–G380

10. Huh WJ, Khurana SS, Geahlen JH et al (2012) Tamoxifen induces rapid, reversible atrophy, and metaplasia in the mouse stomach. Gastroenterology 142:21–24

11. Sigal M, Rothenberg ME, Logan CY et al (2015) *Helicobacter pylori* activates and expands Lgr5+ stem cells through direct colonization of the gastric glands. Gastroenterology 148(7):1392–404.e21

12. Sohal D, Nghiem M, Crackower MA et al (2001) Temporally regulated and tissue-specific gene manipulations in the adult and embryoinc heart using a tamoxifen-inducible Cre protein. Circ Res 89:20–25

13. Casanova E, Fehsenfeld S, Lemberger T et al (2002) ER-based double iCre fusion protein allows partial recombination in forebrain. Genesis 34:208–214

14. Park EJ, Sun X, Nichol P et al (2008) System for tamoxifen-inducible expression of Cre-recombinase from the Foxa2 locus in mice. Dev Dyn 237:447–453

15. Jones ME, Kondo M, Zhuang Y (2009) A tamoxifen inducible knock-in allele for investigation of E2A function. BMC Dev Biol 9:51

16. Kiermayer C, Conrad M, Schneider M et al (2007) Optimization of spatiotemporal gene inactivation in mouse heart by oral application of tamoxifen citrate. Genesis 45:11–16

17. Sheh A, Ge Z, Parry NM et al (2011) 17β-estradiol and tamoxifen prevent gastric cancer by modulating leukocyte recruitment and oncogenic pathways in *Helicobacter pylori*-infected INS-GAS male mice. Cancer Prev Res (Phila) 4:1426–1435

18. Poon GK, Walter B, Lonning PE et al (1995) Identification of tamoxifen metabolites in human HepG2 cell line, human liver homogenate, and patients on long-term therapy for breast cancer. Drug Metab Dispos 23:377–382

19. Robertson DW, Katzenellenbogen JA, Hayes JR, Katzenellenbogen BS (1982) Antiestrogen basicity – activity relationships: a comparison of the estrogen receptor binding and antiuterotrophic potencies of several analogues of (Z)-1,2-diphenyl-1-[4-[2-(dimethylamino)ethoxy]ephenyl]-1-butene (tamoxifen, Nolvadex) having altered basicity. J Med Chem 25:167–171

20. Malet C, Gompel A, Spritzer P et al (1988) Tamoxifen and hydroxytamoxifen isomers versus estradiol effects on normal human breast cells in culture. Cancer Res 48:7193–7199

21. Coezy E, Borgna JL, Rochefort H (1982) Tamoxifen and metabolites in MCF7 cells: correlation between binding to estrogen receptor and inhibition of cell growth. Cancer Res 42:317–323

22. Vignon F, Bouton MM, Rochefort H (1987) Antiestrogens inhibit the mitogenic effect of growth factors on breast cancer cells in the total absence of estrogens. Biochem Biophys Res Commun 146:1502–1508

23. Dietze EC, Caldwell LE, Grupin SL et al (2001) Tamoxifen but not 4-hydroxytamoxifen initiates apoptosis in p53(-) normal human mammary epithelial cells by inducing mitochondrial depolarization. J Biol Chem 276:5384–5394

24. Reed CA, Berndtson AK, Nephew KP (2005) Dose-dependent effects of 4-hydroxytamoxifen, the active metabolite of tamoxifen, on estrogen receptor-alpha expression in the rat uterus. Anticancer Drugs 16:559–567

25. Mazzeo AR, Nandi J, Levine RA (1988) Effects of ethanol on parietal cell membrane phospholipids and proton pump function. Am J Physiol 254:G57–G64

26. Wang TC, Goldenring JR, Dangler C et al (1998) Mice lacking secretory phospholipase A2 show altered apoptosis and differentiation with *Helicobacter felis* infection. Gastroenterology 114:675–689

27. Yoshizawa N, Takenaka Y, Yamaguchi H et al (2007) Emergence of spasmolytic polypeptide-expressing metaplasia in Mongolian gerbils infected with *Helicobacter pylori*. Lab Invest 87:1265–1276

Chapter 29

The Hamster Buccal Pouch Model of Oral Carcinogenesis

Siddavaram Nagini and Jaganathan Kowshik

Abstract

The hamster buccal pouch (HBP) carcinogenesis model is one of the most well-characterized animal tumor models used as a prelude to investigate multistage oral carcinogenesis and to assess the efficacy of chemointervention. Hamster buccal pouch carcinomas induced by 7,12-dimethylbenz[a]anthracene (DMBA) show extensive similarities to human oral squamous cell carcinomas. The HBP model offers a number of advantages including a simple and predictable tumor induction procedure, easy accessibility for examination and follow-up of lesions, and reproducibility. This model can be used to test both chemopreventive and chemotherapeutic agents.

Key words Carcinogenesis, Chemointervention, DMBA, Hamster, Oral cancer

1 Introduction

Oral squamous cell carcinoma (OSCC) is one of the major global health problems with 300,000 cases diagnosed every year [1]. Asians are at high risk for oral cancer because of various factors such as environmental, social, and behavioral effects [2]. Tobacco consumption has been reported as the single most important risk factor accounting for 90 % of oral cancer [3]. Despite several advances in molecular diagnostics and therapeutics, the 5-year survival rate of oral cancer is among the lowest of the major cancers [3, 4]. Biologically and clinically relevant animal tumor models that closely mimic events in the development and progression of human OSCC are therefore of paramount importance to identify early biomarkers and to evolve therapeutic strategies. The hamster buccal pouch (HBP) carcinogenesis model, one of the most well-characterized tumor efficacy models, functions as a paradigm for oral oncogenesis [5].

The hamster has one pouch under the cheek muscles on each side of the mouth that opens into the anterior part of the oral cavity. The pouches are extended backwards along the oral cavity and lined with keratinizing squamous epithelium similar to the human

Andrei I. Ivanov (ed.), *Gastrointestinal Physiology and Diseases: Methods and Protocols*, Methods in Molecular Biology, vol. 1422, DOI 10.1007/978-1-4939-3603-8_29, © Springer Science+Business Media New York 2016

palate or the gingiva. HBP carcinomas induced by 7,12-dimethylbenz[a]anthracene (DMBA) exhibit extensive similarities to human OSCC with respect to the morphology, histology, pre-neoplastic lesions, propensity to invade and metastasize, expression of biochemical and molecular markers, and genetic and epigenetic alterations [5–8].

The HBP model first developed by Salley in 1954 [9] was subsequently modified and standardized by Morris [10] and Shklar [11]. In the original protocol adopted by Salley [9], DMBA dissolved in acetone or benzene was painted on the pouch three times a week for 16 weeks. Morris [10] demonstrated that a 0.5 % solution of DMBA in acetone produced maximum tumor yield with minimum latency and morbidity. The HBP model has been extensively used in our laboratory for chemointervention studies [6, 12, 13]. Topical application of 0.5 % DMBA in liquid paraffin induces SCCs in 14 weeks. The HBP mucosa progresses sequentially through four histologically discernible stages: hyperplasia, dysplasia, carcinoma in situ, and squamous cell carcinoma [14]. Recently, Bampi et al. [15] reported that application of DMBA together with carbamide peroxide gel significantly reduced the latency period in inducing HBP carcinomas.

The HBP model offers a number of advantages as an animal model system for oral cancer. The tumor induction protocol is simple, reproducible, and the lesions are grossly visible at all stages. The hamster cheek pouch is easily accessible for tumor induction and application of test agent without the need for anesthesia. The pouch can be readily subjected to macroscopic examination and follow-up of lesions. This model is ideal for analyzing the stepwise evolution of oral cancer and the effect of chemointervention [6–8]. Sequential changes in the vascular architecture and the effect of anti-angiogenic therapies can be recorded using a variety of techniques [15–17]. Most importantly, HBP carcinomas show a number of similarities to human OSCC. DMBA, the carcinogen used to induce HBP tumors, is a prototype of the polycyclic aromatic hydrocarbons (PAH) implicated in the development of human oral cancer. DMBA is a procarcinogen that is metabolized to form electrophilic diol epoxides that can form adducts with DNA causing mutations and oncogenic transformation [6–8]. This chapter describes in detail the protocol utilized to induce oral cancer in the HBP model.

2   Materials

2.1   Animals

1. Five- to eight-week old male or female golden Syrian hamsters weighing 80–120 g (*see* **Note 1**).

2.2 Reagents

1. DMBA solution: Prepare 0.5 % DMBA by weighing 0.5 g in 100 ml of liquid paraffin (*see* **Note 2**). Mix the solution properly to dissolve DMBA and store in a brown bottle (*see* **Note 3**). DMBA is toxic and adequate precautions should be taken while handling (*see* **Note 4**).

2. Neutral buffered formalin, 10 %: This can be purchased as a prepared solution or made from stock solutions. A standard stock solution of unbuffered formalin is 37–40 % formaldehyde (*see* **Note 5**). Prepare a 10 % formalin solution by dissolving 1 part of stock formalin with 9 parts of distilled water. Neutralize the unbuffered formalin by adding 4 g of NaH_2PO_4 and 12 g of Na_2HPO_4 to 1 l of 10 % formalin solution (*see* **Note 6**).

3. Harris Hematoxylin.

4. Eosin Phloxine stain.

5. Acid Ethanol: Add 1 ml of concentrated HCl to 400 ml of 70 % ethanol.

6. Mounting medium.

2.3 Supplies

1. Paint brush, No. 4.

2. Cotton.

3. Forceps.

4. Vernier caliper.

5. Brown bottle.

6. Polylysine-coated glass slides.

3 Methods

3.1 Maintaining and Handling Hamsters

The Syrian or Golden hamsters are stocky, primarily granivorous animals have been used for experimental studies more recently compared to guinea pigs, rats, and mice. Maintenance and handling of hamsters differs slightly from other laboratory animals [18].

3.1.1 Housing

1. House the hamsters three or four to a cage and provide standard pellet diet and water ad libitum.

2. Ensure that hamsters have cages of appropriate size for free mobility, enough feed and water, and other essentials for stress-free housing (*see* **Note 7**).

3. Label the cages clearly with the following details: group name and number, control or experimental, date of DMBA painting, chemopreventive/chemotherapeutic agent administered. Do not interchange the cages.

4. Monitor the health of the hamsters regularly during the study. Any health problems should be addressed immediately to avoid mortality.

3.1.2 Handling

1. Give a week or so for the hamsters to get used to their new surroundings before they are handled.

2. Males are generally more docile than females.

3. Cannibalism is often encountered among hamsters. Isolate aggressive animals at the earliest.

4. To remove hamsters from the cage, make them enter into a small can or cup, or scoop out with cupped hands or grasp by the abundant loose skin over the dorsal cervical region.

5. You can restrain a hamster by placing the palm over it with the thumb near the head, and gathering the excess skin into the hand. This will immobilize the animal preventing it from turning its head and biting.

3.2 DMBA Painting

1. Restrain the hamster with the left hand and paint the buccal pouch with DMBA using the right hand (*see* **Note 8**).

2. Apply DMBA to either the right or left buccal pouch using the wiped brush method. After dipping the brush into the DMBA solution (0.5 %), wipe the brush once against the side of the container before painting the buccal pouch of the hamster. The side selected for painting should be constant throughout the experimental period [14].

3. Paint the DMBA solution in a slow and steady circular motion on the anterior wall of the buccal pouch. This will ensure uniform distribution of the carcinogen and also prevent damage to the buccal mucosa. Each application leaves approximately 0.4 mg DMBA.

3.3 Tumor Induction Protocol

1. Apply DMBA as described above three times a week for 14 weeks and monitor changes in the buccal pouch once in 2 weeks (*see* **Note 9**). Topical application of DMBA induces hyperplasia in 4–6 weeks, dysplasia in 6–8 weeks, and well-differentiated SCCs in 12–14 weeks [14] (*see* **Note 10** and Figs. 1 and 2).

2. Note the type of changes observed and the area of alteration in an animal's individual record together with photographs. This will enable correlation with subsequent changes. The data will be helpful for standardization of the experiment and result interpretation.

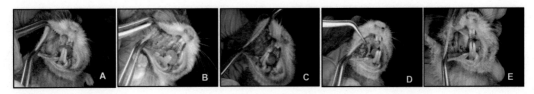

Fig. 1 Gross appearance of control (A) and hamsters painted with DMBA for 4 (B), 8 (C), 12 (D), and 16 (E) weeks

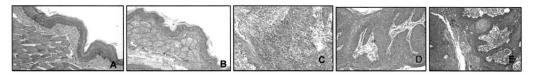

Fig. 2 Histopathological images of control and DMBA painted hamsters

3. At the end of the experiment, fast the hamsters overnight before euthanasia.

4. Calculate the tumor volume, multiplicity, and burden (*see* **Note 11**).

3.4 Chemo-intervention Protocol

1. The mode of administration of the chemopreventive/therapeutic agent may be decided based on the agent selected.

2. For chemoprevention studies, paint 0.5 % DMBA three times a week for 14 weeks for tumor induction. Administer the putative chemopreventive agent prior to or concurrently with the carcinogen or even at 8 weeks when dysplastic lesions appear as illustrated in Fig. 3 [6, 19, 20].

3. Administer the putative chemotherapeutic agent after 12–14 weeks when tumors are visible and continue up to 18 weeks [21, 22]. At the end of the 18th week, in addition to tumor multiplicity and tumor burden calculate the tumor growth delay (*see* **Note 12**). Table 1 shows a typical protocol for induction of HBP carcinomas and testing a putative chemotherapeutic agent.

3.5 Histopathology

1. After sacrifice, immediately fix the tissues in 10 % formalin, embed in paraffin, and mount on polylysine-coated glass slid.

2. Deparaffinize and rehydrate the sections by the following sequence of steps:

(a) Dry the sections in a 60 C oven for 30 min to 1 h (*see* **Note 13**).

(b) Place the slides in xylene for 5 min. Repeat the process three times. Remove the excess of xylene before going into ethanol.

(c) Dip the slides in 100 % ethanol three times for 3 min each.

(d) Place the slides three times for 1 min each in 95 % ethanol. Repeat the process in 80 % ethanol.

(e) Rinse the slides in deionized water for 5 min. Remove excess of water before going into hematoxylin.

3. Stain the slides with hematoxylin for 3 min (*see* **Note 14**). Rinse in deionized water followed by tap water for 5 min.

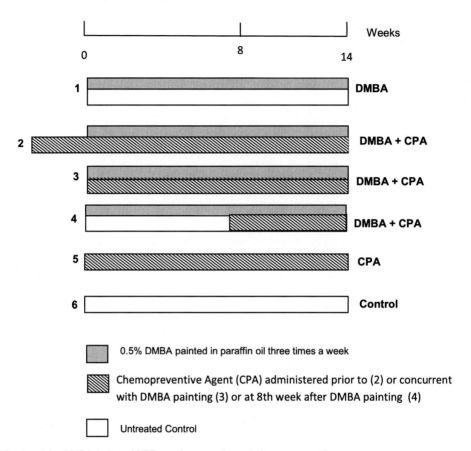

Fig. 3 Protocol for DMBA-induced HBP carcinogenesis and chemoprevention

4. Destain in acid ethanol. Rinse in tap water followed by deionized water. Remove excess of water before staining with eosin.

5. Stain the slides with eosin for 30 s (*see* **Note 15**).

6. Dehydrate with 95 % ethanol three times for 5 min each, followed by 100 % ethanol three times for 5 min each. Remove excess of ethanol.

7. Dip the slides in xylene for 15 min (three times).

8. Place coverslips on the slides and add a drop of Permount. Dry the slides overnight.

9. Observe the H&E-stained slides under a microscope for basal cell hyperplasia, dysplasia, and squamous cell carcinoma (*see* **Note 16**).

Table 1
The experimental protocol for tumor induction and chemotherapy in the HBP model

Group no.	Treatment	1 week	2 weeks	3 weeks	4 weeks	5 weeks	6 weeks	7 weeks	8 weeks	9 weeks	10 weeks	11 weeks	12 weeks	13 weeks	14 weeks	15 weeks	16 weeks	17 weeks	18 weeks
1.	Control	–	–	–	–	–	–	–	–	–	–	–	–	–	–	–	–	–	–
2.	DMBA	D	D	D	D	D	D	D	D	D	D	D	D	–	–	–	–	–	–
3.	DMBA+ Chemotherapeutic Agent (CTA)	D	D	D	D	D	D	D	D	D	D	D	D	CTA	CTA	CTA	CTA	CTA	CTA
4.														CTA	CTA	CTA	CTA	CTA	CTA

Tumor induction   *Therapy*

4 Notes

1. Five- to eight-week old hamsters are best suited for the induction of tumor. Older animals tend to be resistant to tumor induction by DMBA.

2. While preparing 0.5 % DMBA solution, proper mixing is required, otherwise the volume of DMBA will tend to differ with each painting.

3. DMBA must be stored in a brown bottle and exposure to light must be avoided.

4. DMBA is a potent mutagen and carcinogen after metabolic activation. It is absorbed through the skin and respiratory and intestinal tracts; and by intravenous and intraperitoneal injection, ingestion, and inhalation. Care should be taken to avoid contact with the skin or breathing of dusts during preparation. Wash hands thoroughly after handling DMBA, animals painted with DMBA, or soiled bedding.

5. Formalin is formaldehyde gas dissolved in water that reaches saturation at 37–40 % and therefore regarded as 100 % formalin. The exact amount of dissolved formaldehyde in the 10 % formalin is only 3.7–4.0 % because it represents 10 % of the 37–40 % stock.

6. Unbuffered formalin can convert hemoglobin in the tissues to produce dark brown acid formaldehyde hematin precipitates that can interfere with histological interpretation.

7. Hamsters navigate with their noses and strong smells tend to cause distress. The bedding should be changed thrice a week to avoid distress and infection.

8. Three or four hours prior to the painting, food should be removed from the cage and the buccal pouch. Painting DMBA in the presence of food in the pouch will lead to experimental errors.

9. The cage bedding, where hamsters painted with DMBA are housed, may be contaminated with the carcinogen. Because of this, it should be disposed carefully.

10. Care should be taken, while painting after the eighth week to avoid bleeding from the buccal pouch. In case of bleeding, stop the painting and return the animal back to the cage.

11. The mean tumor burden is determined by multiplying the number of tumors in each group by the mean tumor volume in millimeters. Tumor volume is calculated using $4/3\pi r^3$, where r represents ½ tumor diameter in mm. Tumor multiplicity refers to the number of tumors per hamster.

12. Tumor growth delay (TGD) used as treatment outcome can be calculated by the formula. %TGD = Q1/D1 × 100. Q1 is the difference between tumor volume measured at the end of the 18th week and at the end of the 12th week in DMBA painted hamsters administered the putative chemotherapeutic agent (Group 3). D1 is the difference between tumor volume measured at the end of the 18th week and at the end of the 12th week in hamsters painted with DMBA alone (Group 2).

13. Xylene is immiscible in water. Sections must therefore be dried before being placed in xylene. Inadequate drying will hamper deparaffinization with xylene as well as staining.

14. Hematoxylin should be filtered or the surface of the solution wiped with a paper towel to remove oxidized particles that could precipitate on the sections. Hematoxylin is a basic dye that stains nucleic acids blue.

15. Eosin is a red or pink acidic stain that binds to proteins.

16. Hyperplasia of the buccal pouch epithelium is indicated by increased number of basal cells. Irregular epithelial stratification, increased number of mitotic figures, increased nuclear-to-cytoplasmic ratio, and loss of polarity of basal cells characterize oral dysplastic lesions. SCC is diagnosed by the invasion of underlying tissues, nuclear pleomorphism, and increased mitoses.

References

1. Ferlay J, Soerjomataram I, Dikshit R, Eser S et al (2015) Cancer incidence and mortality worldwide: sources, methods and major patterns in GLOBOCAN 2012. Int J Cancer 136:E359–E386

2. Krishna Rao SV, Mejia G, Roberts-Thomson K, Logan R (2013) Epidemiology of oral cancer in Asia in the past decade- an update (2000–2012). Asian Pacific J Cancer Prev 14:5567–5577

3. Zhang Y, Wang R, Miao L (2015) Different levels in alcohol and tobacco consumption in head and neck cancer patients from 1957 to 2013. PLoS One 10:e0124045

4. Kao SY, Mao L, Jian XC, Rajan G, Yu GY (2015) Expert consensus on the detection and screening of oral cancer and precancer. Chin J Dent Res 18:79–83

5. Mognetti B, Di Carlo F, Berta GN (2006) Animal models in oral cancer research. Oral Oncol 42:448–60

6. Nagini S (2009) Of humans and hamsters: the hamster buccal pouch carcinogenesis model as a paradigm for oral oncogenesis and chemoprevention. Anticancer Agents Med Chem 9:843–852

7. Shklar G (1999) Development of experimental oral carcinogenesis and its impact on current oral cancer research. J Dent Res 78:1768–1772

8. Gimenez-Conti IB, Slaga TJ (1993) The hamster cheek pouch carcinogenesis model. J Cell Biochem 17(Suppl):83–90

9. Salley JJ (1954) Experimental carcinogenesis in the cheek pouch of the Syrian hamster. J Dent Res 33:253–262

10. Morris AL (1961) Factors influencing experimental carcinogenesis in the hamster cheek pouch. J Dent Res 40:3–15

11. Shklar G (1972) Experimental oral pathology in the Syrian hamster. Prog Exp Tumor Res 16:518–538

12. Kavitha K, Thiyagarajan P, Rathna-Nandhini J, Mishra RK, Nagini S (2013) Chemopreventive effects of diverse dietary phytochemicals against DMBA-induced hamster buccal pouch carcinogenesis via the induction of Nrf2-mediated cytoprotective antioxidant, detoxification and

DNA repair enzymes. Biochimie 95:1629–1639

13. Vidya Priyadarsini R, Kumar N, Khan I, Thiyagarajan P, Kondaiah P, Nagini S (2012) Gene expression signature of DMBA-induced hamster buccal carcinomas: modulation by chlorophyllin and ellagic acid. PLoS One 7:e34628

14. Vidya Priyadarsini R, Senthil Murugan R, Nagini S (2012) Aberrant activation of Wnt/β--catenin signalling pathway contributes to the sequential progression of DMBA-induced HBP carcinomas. Oral Oncol 48:33–39

15. Bampi FV, Ferreira Vilela W, Vilela Gonçalves R et al (2014) The promoting effect of carbamide peroxide teeth bleaching gel in a preclinical model of head and neck cancer in hamster buccal pouch. Clin Exp Otorhinolaryngol 7:210–215

16. Kreimann EL, Itoiz ME, Dagrosa A, Garavaglia R, Farías S, Batistoni D, Schwint AE (2001) The hamster cheek pouch as a model of oral cancer for boron neutron capture therapy studies: selective delivery of boron by boronophenylalanine. Cancer Res 61:8775–8781

17. Lurie AG, Tatematsu M, Nakatsuka T, Rippey RM, Ito N (1983) Anatomical and functional vascular changes in hamster cheek pouch during carcinogenesis induced by 7, 12-dimethylbenz(a)anthracene. Cancer Res 43:5986–5994

18. Hoosier V Jr, McPherson CW (1987) Laboratory hamsters. Academic Press, London

19. Kowshik J, Baba AB, Giri H, Deepak Reddy G, Dixit M, Nagini S (2014) Astaxanthin inhibits JAK/STAT-3 signaling to abrogate cell proliferation, invasion and angiogenesis in a hamster model of oral cancer. PLoS One 9:e109114

20. Subapriya R, Bhuvaneswari V, Ramesh V, Nagini S (2005) Ethanolic leaf extract of neem (*Azadirachta indica*) inhibits buccal pouch carcinogenesis in hamsters. Cell Biochem Funct 23:229–238

21. Vidya Priyadarsini R, Vinothini G, Senthil Murugan R, Manikandan P, Nagini S (2011) The flavonoid quercetin modulates the hallmark capabilities of hamster buccal pouch tumours. Nutr Cancer 63:218–226

22. Kavitha K, Kranthi Kiran Kishore T, Bhatnagar RS, Nagini S (2014) Cytomodulin-1, a synthetic peptide abrogates oncogenic signaling pathways to impede invasion and angiogenesis in the hamster cheek pouch carcinogenesis model. Biochimie 102:56–67

INDEX

Andrei I. Ivanov (ed.), *Gastrointestinal Physiology and Diseases: Methods and Protocols*, Methods in Molecular Biology,
vol. 1422, DOI 10.1007/978-1-4939-3603-8, © Springer Science+Business Media New York 2016